P9-EEP-916

About Island Press

Island Press is the only nonprofit organization in the United States whose principal purpose is the publication of books on environmental issues and natural resource management. We provide solutions-oriented information to professionals, public officials, business and community leaders, and concerned citizens who are shaping responses to environmental problems.

In 1998, Island Press celebrates its fourteenth anniversary as the leading provider of timely and practical books that take a multidisciplinary approach to critical environmental concerns. Our growing list of titles reflects our commitment to bringing the best of an expanding body of literature to the environmental community throughout North America and the world.

Support for Island Press is provided by The Jenifer Altman Foundation, The Bullitt Foundation, The Mary Flagler Cary Charitable Trust, The Nathan Cummings Foundation, The Geraldine R. Dodge Foundation, The Charles Engelhard Foundation, The Ford Foundation, The Vira I. Heinz Endowment, The W. Alton Jones Foundation, The John D. and Catherine T. MacArthur Foundation, The Andrew W. Mellon Foundation, The Curtis and Edith Munson Foundation, The National Fish and Wildlife Foundation, The National Science Foundation, The New-Land Foundation, The David and Lucile Packard Foundation, The Surdna Foundation, The Winslow Foundation, The Pew Charitable Trusts, and individual donors.

Green Versus Gold

To the people of California

Green Versus Gold

Sources in California's Environmental History

Edited by

Carolyn Merchant

ISLAND PRESS
Washington, D.C. • Covelo, California

Good-faith efforts have been made to secure permission to reprint the following material. If the copyright holder will contact the publisher, any corrections necessary can be made to future printings. "A Historian Chronicles the Return of the Sea Otter," from Augustin S. Macdonald, *Pacific Pelts: Sea Otters Choose California Coast* (Oakland, Calif., 1938); "Yosemite Indians Recount Their Origin Story, Recorded in 1927," from Elinor Shane Smith, *Po-ho-no and Other Yosemite Legends* (Monterey, Calif., 1927); "Samuel Wood and Alfred E. Heller on California Going, Going. . . , 1962" from *California Going, Going. . . .: Our State's Struggle to Remain Beautiful and Productive* (Sacramento, Calif.: California Tomorrow, 1962).

Library of Congress Cataloging-in-Publication Data
Green versus gold : sources in California's environmental history /
edited by Carolyn Merchant
p. cm.
Includes bibliographical references and index.
ISBN 1-55963-579-7. — ISBN 1-55963-580-0
1. California—Environmental conditions—History. 2. Human ecology—California—History. 3. Nature—Effect of human beings on—History. I. Merchant, Carolyn.
GE155.C2G74 1998 98-3346
333.7'09794—dc21 CIP

Printed on recycled, acid-free paper 

Manufactured in the United States of America
10 9 8 7 6 5 4 3 2 1

Contents

Chapter 9. Preserving Parks

Chapter 10. Battles Over Energy

Chapter 11. Second Nature: California Cities

Chapter 12. The Rise of Environmental Science

Chapter 13. Revisioning California: Contemporary Environmental Movements

List of Maps and Illustrations

Maps

Illustrations

Introduction

The Fate of Nature in the Golden State

Carolyn Merchant

Flying over California can be a startling experience. Densely populated areas contrast with open, less inhabited spaces. The Central Valley corridor and the population centers of southern California and the San Francisco Bay–Delta region stand out in sharp distinction to the deserts of the eastern portion of the state and the forests of the coast range and Sierra Nevada mountains. Yet all have been transformed by human presence, first by native Californians and more dramatically by nineteenth- and twentieth-century settlers. Only remnants of the once-vast coastal and mountain redwood and Ponderosa forests remain. Riparian forests of the Central Valley have almost vanished. Flocks of geese and ducks that once blackened inland skies, and gulls and cormorants amassed along craggy coastal outcrops, are few compared to their presettlement populations, as are grizzly bears, mountain lions, tule elk, and sea otters.

For some people this new California of the third millennium is a paradise lost; for others it is a paradise reclaimed from the jaws of nature. To relive the changes in forests, grasslands, mountains, and valleys over time, as presented in this book, is to participate in California's environmental history. The sense of history reclaimed in the words of travelers, settlers, gold seekers, novelists, and scientists brings alive the immediacy of contact with a sometimes benign, sometimes majestic, but often harsh nature, a nature that is itself an actor in the drama. At the same time, the perspectives of recent commentators are equally illuminating. Changes

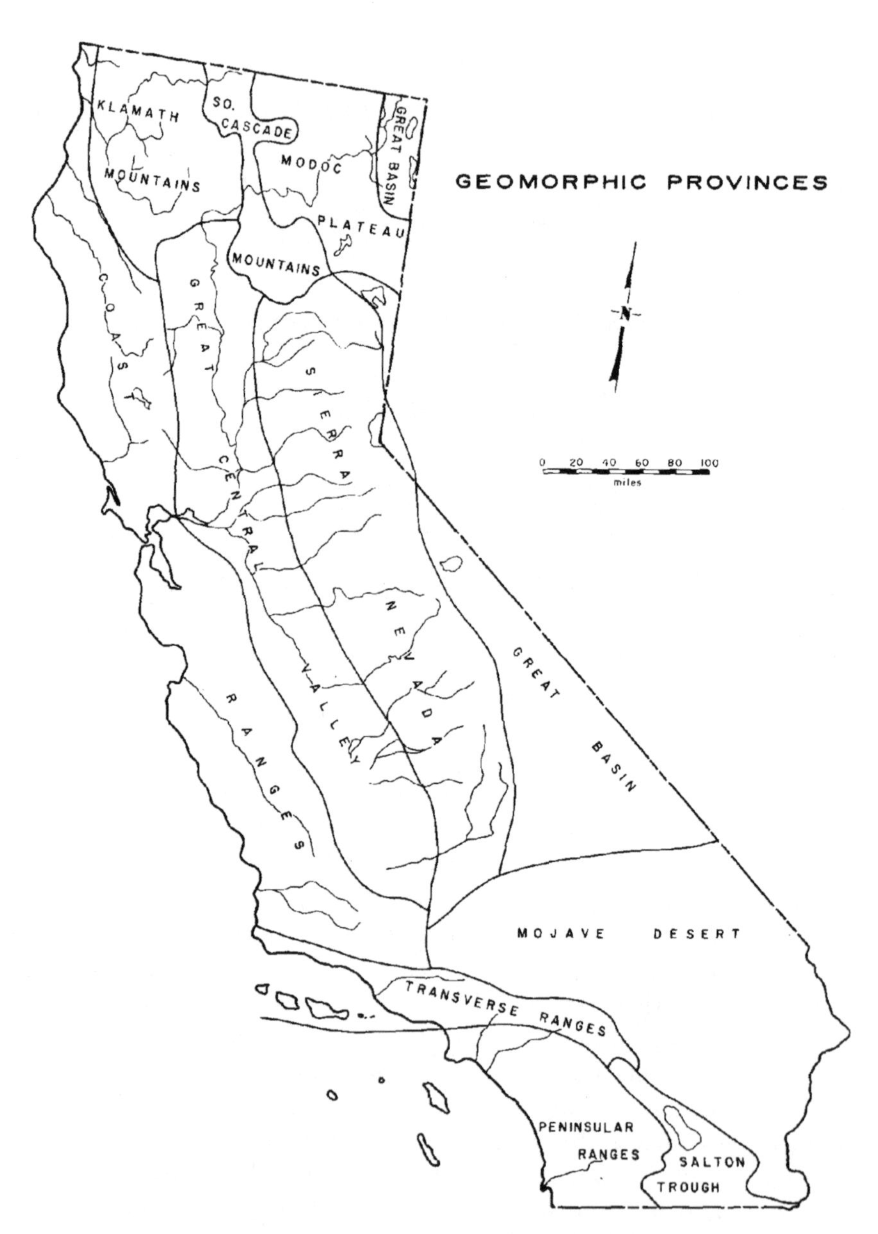

MAP 1. GEOMORPHIC PROVINCES OF CALIFORNIA, SHOWING MOUNTAIN RANGES, VALLEYS, AND NATURAL LAKES.
From Warren A. Beck and Ynez D. Haase, Historical Atlas of California. *Norman: University of Oklahoma Press, 1974, map 3. Reprinted by permission.*

wrought inexorably over time, the sense of past and present spatial contrasts, and the processes and reasons for those transformations as interpreted by historians, geographers, philosophers, and natural scientists can be startling and raise deeply troubling questions.

The book's title, *Green Versus Gold*, characterizes the many changes and tensions between environment and economy and between nature and humanity that took place in California's natural and human history. California evolved from an ecologically green dominion in the native American and Hispanic eras to a dominion of gold created by the 1848 discovery that gave the Golden State its reputation for wealth and opportunity. From the post–Gold Rush era to the present, green and gold often came into tension as settlers and entrepreneurs exploited the environment for its wealth of timber, grass, soil, and water and conservationists extolled and preserved its vanishing ecosystems. The colors of green and gold also reflect nature's seasonal changes between rainy, winter months and dry, summer months within the arid west's longer cycles of wet and dry weather patterns. Finally green and gold suggest renewal and synthesis as nature and culture come together in new visions and appreciation for a potentially green *and* golden state.

In exploring the interactions of green and gold over time, the voices of many participants from many areas of the state are reproduced in the documents that introduce each chapter. Included are Native Americans, whose words and ways were recorded by explorers, anthropologists, and historians; traders and settlers who wrote down their impressions of often formidable, yet lush and attractive, environments; prospectors, ranchers, and lumberers who settled and reaped nature's resources; nature writers, novelists, poets, and artists who put words and images on paper; and entrepreneurs, scientists, and visionaries who saw the larger pictures that emerged from local features.

The perspectives of environmental historians, geographers, and scientists presented in the essays show how people have both used and developed the environment and saved and appreciated it. For many, nature itself is a major participant in the stories enacted over time. Nature brings earthquakes, droughts, diseases, and debris flows, as well as rainfall, fertility, and bountiful harvests. Interpreters of history do not always agree on their approaches, and hence the essays often present conflicting points of view. Readers must approach the materials critically and formulate their own perspectives.

Environmental historians look at history from the ground up. On the most general level they are concerned with interactions between nature and humanity. How have people over time affected their surroundings and what has been the outcome? And how, in turn, have changes in the environment affected humans, their ideas, and institutions? In the realm of ideas, people use their intellect and imagination to conceptualize or personify nature as an actor. Over time, Native Americans have thought of nature as alive and various parts of it as powerful spirits or animals who bring about change in the world. Europeans at times conceptualized nature as an assistant who carried out God's will in the world, a mother who cared for her needy children, a set of forces that acted on matter, and an integration of various manifestations of energy. In California, ideas such as these have influenced the way people act in the world and their ethic toward it.

People sustain their lives by obtaining subsistence from nature and by trading commodities on local, national, and international markets. The place where the ax meets the tree and the plow upturns the soil is the site of most immediate transformation of the material world. The first Californians used weirs and baskets to catch and store fish and fires to encourage particular plants to root in special, fertile places, while Spanish missionaries introduced horses, sheep, cattle, and European vegetables and fruits to provide subsistence for padres and Indian neophytes. After the Gold Rush, international trade in gold and other commodities rapidly introduced a market economy along with laborers of many nationalities and races. These forms of production altered the environment sometimes in minimal, but often, drastic and rapid ways.

People also interact with the material world through domestic, economic, legal, and political institutions. They reproduce themselves and their societies over time in accordance with social norms of gender, race, and class. How people are socialized within their homes, churches, schools, and communities affects their attitudes and permissible actions toward nature. In California, the kinds of organizations people formed, such as nature and hiking groups, conservation organizations, and citizens' action groups, have been instrumental in using and setting aside land or halting perceived forms of pollution or degradation. Similarly the political process and various government bureaus and departments have produced laws and bond measures that transform the landscape in the form of freeways, bridges, dams, and water projects. Powerful bureaucracies in tension with an active citizenry can open up major environmental conflicts over issues such as redwoods, wild rivers, incinerators, and nuclear power plants.

All of these modes of interaction between human society and nature are illustrated in the thirteen chapters of documents and essays that comprise this book. Each chapter focuses on particular natural resources or environments as experienced, transformed, admired, and preserved by human users. Chapter 1 presents the natural environment as a resource to be encountered for its potential economic or aesthetic values, as seen through the eyes of early travelers. Euro-American men and women who experienced the sierras, deserts, central valley, coasts, and lakes recorded their impressions of the terrain, its dangers, and beauties. They engage us in their anxieties, joy, awe, and excitement over a new land and its impressive array of plants and animals. They involve us in their concerns over travel and settlement in the landscape that became famous as the state of California. The words of geologists, ecologists, and historians help us appreciate the massive natural forces that created the terrain, the animals that once inhabited it, and the kinds of changes that European settlement would bring to it.

Native Californians, whose voices are heard in Chapter 2, gathered subsistence and found spiritual meanings in the natural world, as cultivators, fishers, hunters, shamens, or star gazers. They were extraordinarily knowledgeable about local environments, astute managers of plants, animals, fish, and fire, and remarkable storytellers whose explanations about places and events created tribal cohesion and social identification. Many of these first peoples, however, found their lives irrevocably changed by the introduction of new ideas such as Christianity, of new animals such as horses and cattle released on to the land by European explorers, and of lethal diseases. During the eighteenth and nineteenth centuries, these new animals, plants, pathogens, and peoples began arriving together as an ecological complex, tied together in interdependent ways and having a synergistic effect on native Californian peoples and ecosystems, the results of which often were deplored, but sometimes appreciated by native inhabitants.

In the eighteenth and nineteenth centuries, as revealed in Chapter 3, Spanish, Russian, and American explorers, missionaries, and traders, from their own point of view, found the California waters filled with desirable seals, otters, and whales, while inland coastal valleys supplied water and nutrients for mission gardens, horses, and cattle. Mercantile trade and economic and spiritual impacts on Indians brought about ecological changes on the land, while the depletion of fur-bearing animals initiated a long decline in animal populations and their associated coastal waters. Spanish missionaries brought with them accustomed ways of meting out land and water, and Russian traders set up California outposts and entered into contracts with American sea captains and Mexican ranchers. With the end of the United States' war with Mexico, the 1848 signing of the Treaty of Guadalupe Hidalgo and discovery of gold, and California's 1850 admission into the United States, the state's environment embarked on a more rapid phase of development.

Beginning with the 1849 Gold Rush, as detailed in Chapter 4, prospectors and entrepreneurs from around the world transformed California to a degree unprecedented in the early trading and mission eras. Not only gold, but also timber, grasslands, and soils brought settlers to the state. Chapter 4 shows how the Sierra foothills succumbed to the onslaught of hydraulic mining as great mountains of debris were washed down the rivers and whole hillsides vanished under the nozzle and hose. Simultaneously, as Chapters 5 and 6 reveal, giant redwoods and fir trees yielded to the ax and saw, while on rangelands perennial grasses gave way to annuals and then weeds as cattle boomed "on a thousand hills." The voices in these chapters, however, do not portray individuals in black-and-white terms or ecological change as wholly adverse. Lumberers often admired the magnificence of the redwoods and deplored their transient future under the crosscut saw, and ranchers appreciated and saved the vanishing tule elk. In addition, ecological change could sometimes be reversed as rivers and hillsides eventually recovered from mining debris and vegetation loss, while introduced grasses often proved nutritious to reduced numbers of cattle.

The lessons learned from water transport in hydraulic mining were not lost, however, when "hydraulicking," was banned in 1884. Farmers demanded water for arid soils, and engineers had become well versed in the construction of dams, reservoirs, and pipelines. As documented in Chapter 7, California's San Joaquin Valley, irrigated by farmers during the 1880s, saw the advent of a vast hydraulic system through the federal Bureau of Reclamation's Central Valley Project in the 1930s through the 1950s and the State Water Project in the 1950s through the 1970s. With the construction of enormous dams and water distribution systems, the potential for family farms gradually yielded to the complex operations of agribusiness, intensifying issues of farm labor and soil alkalization. Chapter 8, on the transition to agribusiness, reveals the increasing diversity of workers of both genders and many nationalities, as well as problems of poverty and environmental health faced by workers during the Central Valley's reinvention as the green–gold food capital of the world.

With California's intensive environmental transformation, many people recognized that its natural wonders would soon vanish. A movement to set aside land as parks began as early as the 1860s with the granting of Yosemite Valley to the state and subsequent movements to preserve the remaining giant redwoods and pines for their beauty and majesty. Chapter 9 portrays state and nationwide efforts at park preservation, as well as conflicts within the conservation movement itself over the best uses of public land and water.

Water was needed not only as a resource for the green and gold agricultural empire and as a drinking supply for cities, but also as an important energy source for the state's continuing population growth. As animal muscle was supplemented on a vast scale by hydropower and oil in the late nineteenth century and by nuclear power in the twentieth century, environmental repercussions arose. Oil drilling in southern California and nuclear power in central and northern California became major environmental issues. Chapter 10 explores environmental opposition to energy development and the rise of citizen protest movements.

With the growth and development of urban centers in southern California and the San Francisco Bay Area, a new set of environmental concerns emerged during the twentieth century. Water and air pollution were of primary importance, but the rise of automobile transportation and the concomitant explosion of freeway systems compounded such problems. The freeway system was itself vulnerable to yet another of nature's unpredictable actions, earthquakes, and the denudation of hillsides brought devastating debris flows to ever-expanding suburban communities. Chapter 11 depicts these concerns, along with the new urban environmental movements to save suburbs from growth, to save bays from fill, and to prevent air and water pollution.

Throughout the history of California, scientists and engineers have played a major role in creating the state's environmental knowledge base. From geographical surveys, to engineering feats, to natural history collections, from efforts to save crops from pest depredations and soils from exhaustion to plant and animal restoration, scientists have contributed to and shaped environmental outcomes. Chapter 12 reveals these scientific advances and responses to ecological problems and suggests that much work remains to be done.

As California entered the 1980s and 1990s, ecological visionaries began to propose ways to retain and recover California's diverse natural heritage and to make the state responsive to the needs of its multicultural population. Ideas of ecotopians, bioregionalists, deep ecologists, ecofeminists, and ecojustice advocates that evolved in response to the state's environmental needs are presented in Chapter 13.

No collection can cover every region, resource, and subculture in a place as vast and diverse as California. Much is by necessity omitted to present broad themes and interpretations. The bibliographies that close each chapter suggest further sources for both enjoyment and research. To make the collection accessible and cost-effective, footnotes have been omitted from the documents and essays. Citations with page numbers from the original publications are included for the benefit of those wishing to pursue further reading and research.

It is hoped that *Green Versus Gold* will appeal to a wide audience. Historians, scientists, planners, policymakers, writers, educators, citizen environmentalists, and community leaders in California and the nation will be able to draw on its documents and historical analysis. High school, college, and university students will find a different picture of California's history than that portrayed in standard textbooks. The traveling public and visitors to parks and natural areas will find a compelling historical backdrop for the places they are seeing. The documents also will be useful as practical tools for policymakers who wish to learn the historical background of a current issue, restoration ecologists who need an overview of the state's ecological history, and citizen activists who need to know how the land has been altered by past generations.

Many people have assisted in the preparation of this book and I wish to acknowledge their efforts and thank them for their contributions. David Igler, John Keilch, Richard Orsi, Charles Sellers, Thomas Wellock, Donald Worster, and two anonymous referees reviewed the proposed selections, offered suggestions and interpretations, and helped obtain source materials. Jessica Teisch prepared the bibliographies with major contributions from Tamara Whited. Elisa Cooper assisted in research and copying, and Hilary Goldstein, Andrea Clark, and Jessica Teisch spent many long hours preparing the documents and essays for editing and suggested sources and illustrations. The research for this book was supported by the Agricultural Experiment Station at the University of California, Berkeley, project CA-B*-ESH-5647-H. I am grateful to the librarians at the University of California for assistance in obtaining materials on interlibrary loan, to Barbara Dean, Barbara Youngblood, and Island Press for editorial and copyediting assistance, and to Celeste Newbrough for preparing the index. Responsibility for the final outcome is, of course, my own.

Chapter 1
California's Natural Environment

Documents

Sarah Royce Encounters the Sierra Nevada, 1849

The great Sierra Nevada Mountains were still all before us, and we had many miles to make, up this [Carson] River, before the ascent was fairly begun. . . . All the clothing and personal conveniences we had in the world were in our wagon, and we had neither a sufficient number of sound animals nor those of the right kind, to pack them across the mountains. So the only way was to try to keep on. But it looked like rather a hopeless case when, for this whole day, we advanced but a few miles. The next morning, Friday the 12th of October, we set out once more. . . .

We were now so near the foot of the hills that we could distinctly see a stretch of road leading down a very steep incline to where we were moving so laboriously along. Presently at the head of this steep incline appeared two horsemen, clad in loose, flying garments that flapped, like wings on each side of them, while their broad-brimmed hats blown up from their foreheads, revealed hair and faces that belonged to no Indians.

From Sarah Royce, A *Frontier Lady* (New Haven: Yale University Press, 1932), pp. 61, 62–67, 68–69, 72–73. Reprinted by permission.

Their rapidity of motion and the steepness of the descent gave a strong impression of coming down from above, and the thought flashed into my mind, "They look heaven-sent." As they came nearer we saw that each of them led by a halter a fine mule, and the perfect ease with which all the animals cantered down that steep, was a marvel in our eyes. My husband and myself were at the heads of the lead cattle, and our little Mary was up in the front of the wagon, looking with wonder at the approaching forms.

As they came near they smiled and then said "Well sir, you are the man we are after!" "How can that be?" said my husband, with surprise. "Yes, sir," continued the stranger, "you and your wife, and that little girl, are what brought us as far as this. You see we belong to the Relief Company sent out by order of the United States Government to help the late emigrants over the mountains. We were ordered only as far as Truckee Pass. When we got there we met a little company that had just got in. They'd been in a snow storm at the summit; most got froze to death themselves, lost some of their cattle, and just managed to get to where some of our men had fixed a relief camp. There was a woman and some children with them. . . . And she kept at me so, I couldn't get rid of her. You see I've got a wife and little girl of my own; so I felt just how it was; and I got this man to come with me and here we are, to give you more to eat, if you want it, let you have these two mules, and tell you how to get right over the mountains the best and quickest way. . . .

Having made their hasty explanation, our new friends advised us to keep on some little distance farther, to a point where there was a spring in the hills, and excellent camping, to which they would guide us. There we were to rest the remainder of the day, while they would help to select, put into proper shape and pack, everything in the wagon that could be packed. The rest we must be content to leave. As we moved leisurely on to our camping place, they explained more fully the details of our situation—which they understood so much better than we could—and told us what we were to do. There had been two nights of snow storm at the summit: had there come much more they could not have got through. But the weather had cleared, the snow was fast going off the roads as they came over; and, if no other storm occurred, the pass would be in good order when we reached it. But we must hasten with all possible dispatch, for, when the storms once again set in, they were not likely at that season to give any more chance for crossing the mountains. As to keeping on with the wagon, even supposing the cattle to grow no weaker than now, it would take us two weeks at the least to ascend the Carson Valley to the cañon. That cañon could not in several places be traversed by wheels. Wagons had been taken through; but only by taking them apart and packing, at the most difficult points; which of course could only be done by strong companies with plenty of time. Our only hope, therefore, was to pack. They then went farther into details about packing. The oxen, they said, could easily be made to carry, each, two moderate sized bundles, if snugly packed and well fastened on. Then the old horse could carry something though not very much. And the mule the young men had brought along, they said must carry most of the provisions. . . .

The programme for the afternoon was successfully carried out. Every thing was arranged for an early morning start; and, at night I lay down to sleep for the last time in the wagon that had proved such a shelter for months past. I remembered well, how dreary it had seemed, on the first night of our journey (which now looked so long ago) to have only a wagon for shelter. Now we were not going to have even that. But, never mind, if we might only reach in safety the other foot of the mountains, all these privations would in their turn look small; and the same rich Providence that had led, and was still so kindly leading us, would, in that new land, perhaps, show us better things than we had seen yet.

So, when morning came, I hailed it with cheerful hope, though with some misgivings because I had not ridden horseback for several years, and, whenever I had it had been with side-saddle, and all the usual equipments for lady's riding, and, certainly, with no baby to carry. Now, I was to have only a common Spanish saddle, I must have Mary in front of me, and, it turned out, that several things needed for frequent use would have to be suspended from the pommel of my saddle, in a satchel on one side and a little pail on the other. At first, I was rather awkward, and so afraid Mary would get hurt, that at uneven places in the road I would ask my husband to get up and take her, while I walked. But in a few hours this awkwardness wore off; and the second day of our new style of traveling I rode twenty-five miles, only alighting once or twice for a brief time. Our friends, the government men, had left us the morning we left our wagon; taking the road to the Truckee, where they felt themselves emphatically "due," considering their orders. I have more than once since wished I could see and thank them again; for . . . only ten days after we crossed the summit, the mountains were all blocked with snow, and the stormiest winter California had known for years was fully set in. . . .

On the 17th of October we reached the head of Carson Valley, and, just after noon, entered the great cañon. Here the road soon became so rough and steep as to make it very difficult for me to hold Mary and keep my seat. The men had hard work to drive the cattle and mules over the boulders at the frequent crossing of the stream, and in between the great masses of rock where the trail sometimes almost disappeared. As the cañon narrowed, the rocky walls towered nearly perpendicular, hundreds of feet; and seemed in some places almost to meet above our heads. At some of the crossings it was well nigh impossible to keep the trail, so innumerable were the boulders; and the scraggy bushes so hid the coming-out place. The days were shortening fast, and, in this deep gulch, darkness began to come on early. . . .

That night we slept within a few yards of snow, which lay in a ravine; and water froze in our pans not very far from the fire, which, however, was rather low the last part of the night. But the morning was bright and sunny. "Hope sprang exultant"; for, that day, that blessed 19th of October, we were to cross the highest ridge, view the "promised land," and begin our descent into warmth and safety. So, without flinching I faced steeps still steeper than yesterday: I even laughed in my little one's upturned face, as she lay back against my arm, while I leaned forward almost to the neck of the mule, tugging up the hardest places. I had purposely hastened, that morning, to start ahead of the rest; and not far from noon, I was rewarded by coming out, in advance of all the others, on a rocky height whence I looked down, far over constantly descending hills, to where a soft haze sent up a warm, rosy glow that seemed to me a smile of welcome; while beyond, occasional faint outlines of other mountains appeared; and I knew I was looking across the Sacramento Valley.

California, land of sunny skies—that was my first look into your smiling face. I loved you from that moment, for you seemed to welcome me with loving look into rest and safety. However brave a face I might have put on most of the time, I knew my coward heart was yearning all the while for a home-nest and a welcome into it, and you seemed to promise me both. A short time I had on those rocks, sacred to thanksgiving and prayer; then the others came, and boisterous shouts, and snatches of song made rocks and welkin ring.

William Brewer Explores the Central Valley, 1861

Salinas Valley and Monterey

Nacimiento River, May 4, 1861. It is a lovely afternoon, intensely hot in the sun, but a wind cools the air. A belt of trees skirts the river. I have retreated to a shady nook by the water, alike out of the sun and wind; a fine, clear, swift stream passes within a few rods of camp, a belt of timber a fourth of a mile wide skirts it—huge cotton woods and sycamores, with an undergrowth of willow and other shrubs. We have been here three days. . . .

The grizzly bear is much more dreaded than I had any idea of. . . . They will kill and eat sheep, oxen, and horses, are as swift as a horse, of immense strength, quick though clumsy, and very tenacious of life. A man stands a slight chance if he wounds a bear, but not mortally, and a shot must be well directed to kill. The universal advice by everybody is to let them alone if we see them, unless we are well prepared for battle and have experienced hunters along. They will generally let men alone, unless attacked, so I have no serious fears of them.

Less common than bear are the California lions, a sort of panther, about the color of a lion, and size of a small tiger, but with longer body. They are very savage, and I have heard of a number of cases of their killing men. But don't be alarmed on my account—I don't court adventures with any such strangers. Deer are quite common. Formerly there were many antelope, but they are very rapidly disappearing. We have seen none yet. Rabbits and hares abound; a dozen to fifty we often see in a single day, and during winter ate many of them.

There are many birds of great beauty. One finds the representatives of various lands and climes. Not only the crow, but also the raven is found, precisely like the European bird; there are turkey-buzzards, also a large vulture something like the condor—an immense bird. Owls are very plenty, and the cries of several kinds are often heard the same night.

From William H. Brewer, *Up and Down California, 1860–1864* (Berkeley: University of California Press, 1974), pp. 91, 95–96, 380–82, 387, 510, 513–15.

Hawks, of various sizes and kinds and very tame, live on the numerous squirrels and gophers. I see a great variety of birds with beautiful plumage, from humming birds up.

But it is in reptile and insect life that this country stands preeminent. There are snakes of many species and some of large size, generally harmless, but a few venomous. Several species of large lizards are very abundant. Salamanders and chameleons are dodging around every log and basking on every stone. Hundreds or thousands may be seen in a day, from three inches to a foot long. Some strange species are covered with horns like the horned frogs.

But insects are the most numerous. They swarm everywhere. House flies were as abundant in our tent in winter as at home in summer. Ticks and bugs get on us whenever we go in the woods. Just where we are now camped there are myriads of bugs in the ground, not poisonous, but annoying by their running over one. Last night I could scarcely sleep, and shook perhaps a hundred or two hundred out of my blankets this morning. . . .

Tejon-Tehachapi-Walker's Pass

Visalia, April 12 [1863]. About six miles from Kings River we struck a belt of scattered oaks—fine trees—and what a relief! For, except a few cragged willows, shrubs rather than trees, in places along the sloughs, we had seen no trees for the last 130 miles of the trip! We crossed Kings River, a swift deep stream, by ferry, and stopped at a house on the bank, the most like a home of anything we had seen for two hundred miles. The owner was a Massachusetts Yankee, and his wife a very intelligent woman —I noticed an atlas of the heavens hanging up in the sitting room. . . .

[On] Friday, April 10, we [arrived] here, [after] twenty-five miles, crossing an open plain of nearly twenty miles. The morning was clear, and the view of the snowy Sierra most magnificent. Tomorrow we push on, and anticipate a rough time for the next four or six weeks.

Visalia is a little, growing place, most beautifully situated on the plain in an extensive grove of majestic oaks. These trees are the charm of the place. Ample streams from the mountains, led in ditches wherever wanted, furnish water for irrigating. We have stopped here two days to allow our animals to rest and get inspiration for our trip ahead. . . .

May 5 [1863]. Tuesday we came on thirty miles and stopped at Coyote Springs, about six or seven miles from White River. The road this day was through a desolate waste—I should call it a desert—a house at Deer Creek and another at White River were the only habitations. The soil was barren and, this dry year, almost destitute of vegetation. A part of the way was through low barren hills, all rising to about the same height—in fact, a tableland washed down into hills. We stopped at a miserable hut, where there is a spring and a man keeps a few cattle. He was not at home, but his wife was, and she gave us something to eat, and we slept out upon the ground. . . .

Wednesday we came on thirty-five miles to Kern River, the most barren and desolate day's ride since leaving Fresno, and for thirty miles we saw no house. We continued among the low barren hills until we came near Kern River—here we had to leave the road and go down the river nine or ten miles to find a ford. We followed a few wagon tracks, left the hills, and struck down the plain. The soil became worse—a sandy plain, without grass, in places very alkaline—a few desert or saline shrubs growing in spots, elsewhere the soil bare—no water, no feed. We saw some coyotes (wolves) and antelope. Night came on, and still we found neither grass nor river ford. Long after dark, when we began to get discouraged and to fear we would have to stop without water or feed for ourselves or animals, we heard some dogs bark. Soon we saw a light and soon afterward struck a cabin. Here we found some grass, went into the house, made some tea, and then slept on the river bank. Here in a cabin lived a man, wife, and several children, all ragged, dirty, ignorant—not one could read or write—and Secessionists, of course. . . .

May 6 [1863], we left Fort Tejon and crossed through the mountains south, to the Liebre Ranch. The pass is a very picturesque one, 4,256 feet high, with peaks on each side rising several thousand feet higher. The valleys are green, the region beautiful, but all changes on crossing the chain. We passed down a valley, dry and alkaline. Two little salt lakes were dry—the salt and alkali produced by the evaporation covering the ground like a crust of ice. For several miles we followed a line of earthquake cracks which were formed in 1856. The ground had opened for several feet wide, no one knows how deep, and partially closed again. We hear that these cracks extend nearly one hundred miles. In the valley we passed down a woman was killed by her house falling in the earthquake.

San Joaquin Valley—Giant Sequoias

June 1 [1864], we came on to Firebaugh's Ferry, on the San Joaquin, twenty-five miles. Portions of this day's ride, for miles together, not a vestige of herbage of any kind covered the ground; in other places there was a limited growth of wire grass or alkali grass, but not enough to make it green. Yet cattle live here—we passed numbers during the day, and countless carcasses of dead animals. We camped at Firebaugh's, where we got hay for our animals and took a grateful bath in the cold San Joaquin. The bad water, dust, alkali, and our change of diet begin to tell on the boys, but all are cheerful.

June 2, to Fresno City. For the first ten miles the ground was entirely bare, but then we came on green plains, green with fine rushes, called wire grass, and some alkali grass. The ground is wetter and cattle can live on the rushes and grass. We now came on thousands of them that have retreated to this feed and have gnawed it almost into the earth.

The air is very clear this day; on the one side the Coast Range loomed up, barren and desolate, its scorched sides furrowed into canyons, every one of which was marvelously distinct; on the other side the distant Sierra, its cool snows glistening in the sun and mocking us on our scorching trail. We camped by a slough of stinking, alkaline water, which had the color of weak coffee. It smelt bad and tasted worse, and our poor animals drank it protesting. We drank well water, which looked better and tasted better, but I think it smelt worse. But in this dry, hot, and dusty air we must drink, and drink much and often. . . .

Before reaching Visalia we again struck timber. The region about Visalia is irrigated from the Kaweah River, and is covered with a growth of scattered oaks—fine, noble, old trees.

We had now got up some three thousand feet, had passed lower, dry foothills, and had just struck the region of pines. Grand old trees grew in the valley where we camped and over the neighboring ridges, large, but scattered, hardly forming forests. And how delicious the cool, pure mountain water tasted—our first real good water for many a long day! In the afternoon I climbed a high point above camp, commanding a fine view of the surrounding region.

Friday, June 10 [1864], we came on but four miles to this camp. Up, up, up, over a high ridge, and at last into a dense forest of spruces, pines, firs, and cedars. We then sank into a little depression where there is a beautiful grassy meadow of perhaps two hundred acres, surrounded by dense, dark forests. Here there is a steam sawmill, where two or three families live.

And here let me describe this delightful camp, so refreshing after the monotony, heat, dust, alkali, discomfort, and tedium of the great plain. The level, grassy meadow lies in front, with a rill of pure, cold water. Ridges are all around, clothed with dark pines and firs, with here and there the majestic form of some scattered Big Trees, the giant sequoias that abound here, although so rare elsewhere. We are at an altitude of over five thousand feet, or just about one mile above the sea. We are far above the heat and dust of the plain. It has been cold every night—from 23° to 32°—the days cloudless, the sky of the clearest blue, the air balmy and so cool that it is just comfortable without our coats. You cannot imagine the relief we feel both by day and night after the discomfort of the previous two weeks.

As I have said, the Big Trees are abundant here, scattered all along between the Kings and Kaweah rivers. We are on the south branches of Kings River. Saturday we all went up on the ridges about a thousand feet above to see the largest trees. . . . The largest one standing is 106 feet in circumference at the ground and 276 feet high. But it swells out at the base, so that at twelve feet from the ground it is only seventy-five feet in circumference. It is finely formed, and you can but imperfectly imagine its majesty. It has been burned on one side, and were it entire its circumference at the base would be 116 to 120 feet! . . .

About six miles east of this is a high bald mountain about eight thousand feet, which we ascended, and a description of the view will answer for any of the higher points near. It commands a view of the whole western slope of the Sierra, the snowy peaks on one side, the great plain on the other.

A Traveller on Settling in California, 1873

In making his selection [of land], [the stranger] should bear in mind these things, among others:

> 1. California is subject to droughts. Experience shows, so far, that there are about seven good years out of ten; that is to say, in ten years the farmer may, in almost any part of the State fit for agriculture, expect to get seven good field crops without irrigation. This is the general testimony of careful and experienced farmers to whom I put the question. There are bottoms, as in the Pajaro Valley, and there are

From Charles Nordhoff, *California for Travellers and Settlers*, (Berkeley, CA: Ten Speed Press, 1973), pp. 132–33.

tracts of land in the northern part of the State and elsewhere, which are never affected by drought. But of the great bulk of the arable land in California what I have said above is true.

2. Moreover, the farmer in Southern California, as in the San Joaquin Valley, who should plant the orange, lemon, almond, and other sub-tropical fruits, needs water to irrigate these.

3. Water is also needed, except in seasons when the rain-fall is above the average, to get two good crops from the same land in a year. With water this is easy and certain, and you may follow your crop of wheat or barley, sown in December and reaped in May, with a crop of corn planted in May or June on the same land.

4. For all these reasons, it is a very great advantage to have a water supply on your place, or at least within reach. "Be more careful to buy water than land," said an experienced and successful California farmer to me—a man who, beginning with but a small capital fifteen years ago, has now an income of fifteen thousand dollars a year from his farm and orchards. Water is not scarce in California; but there are tracts of land which have it not, and these it is best to avoid. . . . To an Eastern man few things are more surprising than the ease, skill, and cheapness with which a small stream is tapped by half a dozen Californian farmers according to a plan matured at a ditch-meeting," led into a reservoir, and made available for irrigation.

5. If there is a proper irrigating canal or ditch available to the land you prefer, that is sufficient. You have only to ascertain the price of the water. The company which has now built forty miles of canal in the San Joaquin Valley, and whose extensive plans I spoke of above, charges one dollar and a quarter per acre per crop, which is a very light burden; far cheaper than manure on an Eastern farm.

6. On the eastern aside of the San Joaquin Valley, in the San Bernardino Valley, and in other parts also, artesian wells are easily and cheaply made. A flowing well, wherever it can be got at moderate cost, answers admirably for irrigating purposes; and a well of seven inch bore will water a considerable piece of land. Gardens and pleasure-grounds are commonly irrigated in this State by means of windmills, which pump water into small tanks. The windmill is universal in California; the constant breezes make it useful; and as there is no frost to break pipes, water is led from the tank into the house and stable, which is a very great convenience, at a small cost.

7. The level or plain land is probably the richest; it is certainly the

most easily cultivated. . . . But the foot-hills have a peculiar value of their own, which has been overlooked by the eager California farmers. The vine, and, I believe, most of the sub-tropical fruits, grow best in the foothills. The soil is somewhat lighter; it will probably not bear such heavy crops of grain; but a homestead on the hills has a fine look-out; water is probably more easily obtainable; the air is fresher than on the plains; and, for my own part, I have seen, in the more settled parts of the State, that the cheapest lands—the foot-hill lands, namely—were, on many accounts, preferable. Vine-growers begin to perceive that the best wine comes from these higher lands; and ten or fifteen years hence it is believed that the principal and most profitable vineyards in the State will be in the foot-hills.

8. California is a breezy State; the winds from the sea draw with considerable force through the cañons or gorges in the mountains and sweep over the plains. This is no doubt one of the chief causes of its remarkable healthfulness; and it gives to the workman, in the summer, the great boon of cool nights. No matter how warm the day has been in any part of the State, the night is always cool, and a heavy blanket is needed for comfort. Now there are places where the wind is too severe, where a constant gale sweeps through some cañon and is an injury to the farmer. Such places should be avoided and are easily avoidable. In many parts of the State farms would be benefited by trees, planted as wind-breaks; and, fortunately, the willow or sycamore forms, in two years, in this climate, a sufficient shelter, besides furnishing fire-wood to the farmer.

9. Where one man has selected land for himself and several friends, he can easily and quickly prepare the way for them. Fences and houses can be built by contract in every part of the State. Men make it their business to do this; and at the nearest town the intending settler can always have all his necessary "improvements" done by contract, even to ploughing his land and putting in his first crop. In this respect labor is admirably organized in California. You will see, then, that your pioneer may make ready for those who are to come after, so as to save them much delay and inconvenience.

10. In some parts of the State Indians hire themselves out as farm laborers. They usually live on the place where they work, and they are a harmless and often a skillful laboring population, though somewhat slow. They understand the management of horses, are ploughmen, and know how to irrigate land. The Chinese also make useful farm laborers, and are every year more used for this purpose. They learn

very quickly, are accurate, painstaking, and trustworthy, and especially as gardeners and for all hand-labor they are excellent. White laborers are—as in every thinly-settled country—unsteady, and hard to keep.

Mark Twain on Tahoe and Mono Lakes, 1903

If there is any life that is happier than the life we led on our timber ranch for the next two or three weeks, it must be a sort of life which I have not read of in books or experienced in person. We did not see a human being but ourselves during the time, or hear any sounds but those that were made by the wind and the waves, the sighing of the pines, and now and then the far-off thunder of an avalanche. The forest about us was dense and cool, the sky above us was cloudless and brilliant with sunshine, the broad lake [Tahoe] before us was glassy and clear, or rippled and breezy, or black and storm-tossed, according to Nature's mood; and its circling border of mountain domes, clothed with forests, scarred with landslides, cloven by cañons and valleys, and helmeted with glittering snow, fitly framed and finished the noble picture. The view was always fascinating, bewitching, entrancing. The eye was never tired of gazing, night or day, in calm or storm; it suffered but one grief, and that was that it could not look always, but must close sometimes in sleep.

We slept in the sand close to the water's edge, between two protecting boulders, which took care of the stormy night winds for us. We never took any paregoric to make us sleep. At the first break of dawn we were always up and running foot-races to tone down excess of physical vigor and exuberance of spirits. . . . We watched the tinted pictures grow and brighten upon the water till every little detail of forest, precipice, and pinnacle was wrought in and finished, and the miracle of the enchanter complete. Then to "business."

That is, drifting around in the boat. We were on the north shore. There, the rocks on the bottom are sometimes gray, sometimes white. This gives the marvelous transparency of the water a fuller advantage

From Mark Twain, *Roughing It* (New York: Harper & Row, 1903), pp. 161–63, 259–62.

than it has elsewhere on the lake. We usually pushed out a hundred yards or so from the shore, and then lay down on the thwarts in the sun, and let the boat drift by the hour whither it would. We seldom talked. It interrupted the Sabbath stillness, and marred the dreams the luxurious rest and indolence brought. The shore all along was indented with deep, curved bays and coves, bordered by narrow sand-beaches; and where the sand ended, the steep mountainsides rose right up aloft into space—rose up like a vast wall a little out of the perpendicular, and thickly wooded with tall pines.

So singularly clear was the water, that where it was only twenty or thirty feet deep the bottom was so perfectly distinct that the boat seemed floating in the air! Yes, where it was even eighty feet deep every little pebble was distinct, every speckled trout, every hand's-breadth of sand. Often, as we lay on our faces, a granite boulder, as large as a village church, would start out of the bottom apparently, and seem climbing up rapidly to the surface, till presently it threatened to touch our faces, and we could not resist the impulse to seize an oar and avert the danger. But the boat would float on, and the boulder descend again, and then we could see that when we had been exactly above it, it must still have been twenty or thirty feet below the surface. Down through the transparency of these great depths, the water was not merely transparent, but dazzlingly, brilliantly so. All objects seen through it had a bright, strong vividness, not only of outline, but of every minute detail, which they would not have had when seen simply through the same depth of atmosphere. So empty and airy did all spaces seem below us, and so strong was the sense of floating high aloft in mid-nothingness, that we called these boat excursions "balloon voyages."

Mono Lake lies in a lifeless, treeless, hideous desert, eight thousand feet above the level of the sea, and is guarded by mountains two thousand feet higher, whose summits are always clothed in clouds. This solemn, silent, sailless sea—this lonely tenant of the loneliest spot on earth—is little graced with the picturesque. It is an unpretending expanse of grayish water, about a hundred miles in circumference, with two islands in its center, mere upheavals of rent and scorched and blistered lava, snowed over with gray banks and drifts of pumicestone and ashes, the winding-sheet of the dead volcano, whose vast crater the lake has seized upon and occupied.

The lake is two hundred feet deep, and its sluggish waters are so strong with alkali that if you only dip the most hopelessly soiled garment into them once or twice, and wring it out, it will be found as clean as if it had been through the ablest of washerwomen's hands. While we camped there our laundry work was easy. We tied the week's washing astern of our boat, and sailed a quarter of a mile, and the job was complete, all to the wringing out. If we threw the water on our heads and gave them a rub or so, the white lather would pile up three inches high. This water is not good for bruised places and abrasions of the skin. We had a valuable dog. He had raw places on him. He had more raw places on him than sound ones. He was the rawest dog I almost ever saw. He jumped overboard one day to get away from the flies. But it was bad judgment. In his condition, it would have been just as comfortable to jump into the fire. The alkali water nipped him in all the raw places simultaneously, and he struck out for the shore with considerable interest. He yelped and barked and howled as he went—and by the time he got to the shore there was no bark to him—for he had barked the bark all out of his inside, and the alkali water had cleaned the bark all off his outside, and he probably wished he had never embarked in any such enterprise. . . . A white man cannot drink the water of Mono Lake, for it is nearly pure lye. It is said that the Indians in the vicinity drink it sometimes, though. It is not improbable, for they are among the purest liars I ever saw. [There will be no additional charge for this joke, except to parties requiring an explanation of it. This joke has received high commendation from some of the ablest minds of the age.]

There are no fish in Mono Lake—no frogs, no snakes, no polliwogs—nothing, in fact, that goes to make life desirable. Millions of wild ducks and seagulls swim about the surface, but no living thing exists under the surface, except a white feathery sort of worm, one-half an inch long, which looks like a bit of white thread frayed out at the sides. If you dig up a gallon of water, you will get about fifteen thousand of these. They give to the water a sort of grayish-white appearance. Then there is a fly, which looks something like our house-fly. These settle on the beach to eat the worms that wash ashore—and any time, you can see there a belt of flies an inch deep and six feet wide, and this belt extends clear around the lake—a belt of flies one hundred miles long. If you throw a stone among them, they swarm up so thick that they look dense, like a cloud. You can hold them under water as long as you please—they do not mind it—they are only proud of it. When you let them go, they pop up to the surface as dry as a patent-office report, and walk off as unconcernedly as if they had been educated especially with a view to affording instructive entertainment to man in that particular way. Providence leaves nothing to go by chance. All things have their uses and their part and proper place in Nature's economy: the ducks eat the flies—the flies eat the worms—the Indians eat all three—the wildcats eat the Indians—the white folks eat the wildcats—and thus all things are lovely.

Mary Austin Appreciates the California Desert, 1903

East away from the Sierra, south from Panamint and Amargosa, east and south many an uncounted mile is the Country of Lost Borders.

Ute, Paiute, Mojave, and Shoshone inhabit its frontiers, and as far into the heart of it as man dare go. Not the law, but the land sets the limit. Desert is the name it wears upon the maps, but the Indian's is the better word. Desert is a loose term to indicate land that supports no man; whether the land can be bitted and broken to that purpose is not proven. Void of life it never is, however dry the air and villainous the soil.

This is the nature of that country. There are hills, rounded, blunt, burned, squeezed up out of chaos, chrome and vermilion painted, aspiring to the snow-line. Between the hills lie high level-looking plains full of intolerable sun glare, or narrow valleys drowned in blue haze. . . .

The sculpture of the hills here is more wind than water work, though the quick storms do sometimes scar them past many a year's redeeming. . . . Since this is a hill country one expects to find springs, but not to depend upon them; for when found they are often brackish and unwholesome, or maddening, slow dribbles in a thirsty soil. Here you find the hot sink of Death Valley, or high rolling districts where the air has always a tang of frost. Here are the long heavy winds and breathless calms on the tilted mesas where dust devils dance, whirling up into a wide, pale sky. Here you have no rain when all the earth cries for it, or quick downpours called cloudbursts for violence. A land of lost rivers, with little in it to love; yet a land that once visited must be come back to inevitably. If it were not so there would be little told of it.

This is the country of three seasons. From June on to November it lies hot, still, and unbearable, sick with violent unrelieving storms; then on

From Mary Austin, *Land of Little Rain* (Boston: Houghton Mifflin, 1950 [1903]), pp. 1–8.

until April, chill, quiescent, drinking its scant rain and scantier snows; from April to the hot season again, blossoming, radiant, and seductive. These months are only approximate; later or earlier the rain-laden wind may drift up the water gate of the Colorado from the Gulf, and the land sets its seasons by the rain. . . .

If you have any doubt about it, know that the desert begins with the creosote. This immortal shrub spreads down into Death Valley and up to the lower timber-line, odorous and medicinal as you might guess from the name, wandlike, with shining fretted foliage. . . .

Nothing the desert produces expresses it better than the unhappy growth of the tree yuccas. Tormented, thin forests of it stalk drearily in the high mesas, particularly in that triangular-slip that fans out eastward from the meeting of the Sierras and coastwise hills where the first swings across the southern end of the San Joaquin Valley. The yucca bristles with bayonet-pointed leaves, dull green, growing shaggy with age, tipped with panicles of fetid, greenish bloom. . . .

Above the lower tree-line, which is also the snow-line, mapped out abruptly by the sun, one finds spreading growth of piñon, juniper, branched nearly to the ground, lilac and sage, and scattering white pines. . . .

Go as far as you dare in the heart of a lonely land, you cannot go so far that life and death are not before you. Painted lizards slip in and out of rock crevices and pant on the white hot sands. Birds, hummingbirds even, nest in the cactus scrub; woodpeckers befriend the demoniac yuccas; out of the stark, treeless waste rings the music of the night-singing mockingbird. If it be summer and the sun well down, there will be a burrowing owl to call. . . .

None other than this long brown land lays such a hold on the affections. The rainbow hills, the tender bluish mists, the luminous radiance of the spring have the lotus charm. They trick the sense of time, so that once inhabiting there you always mean to go away without quite realizing that you have not done it. Men who have lived there, miners and cattle-men, will tell you this, not so fluently, but emphatically, cursing the land and going back to it. For one thing there is the divinest, cleanest air to be breathed anywhere in God's world. Some day the world will understand that, and the little oases on the windy tops of hills will harbor for healing its ailing, house-weary brood. . . .

For all the toll the desert takes of a man it gives compensations, deep breaths, deep sleep, and the communion of the stars. It comes upon one with new force in the pauses of the night that the Chaldeans were a desert-bred people. It is hard to escape the sense of mastery as the stars move in the wide clear heavens to risings and settings unobscured. They look large and near and palpitant; as if they moved on some stately service not needful to declare. Wheeling to their stations in the sky, they make the poor world fret of no account. Of no account you who lie out there watching, nor the lean coyote that stands off in the scrub from you and howls and howls.

Essays

California's Geological Past

David Alt and Donald Hyndman

California is the product of a prolonged head-on collision between the leading western edge of North America and the floor of the Pacific Ocean. It consists of rocks from the deep ocean bottom and mud scraped off them as the continent overrode the ocean basin. The story of rocks from deep in the earth, the muddy sediment that covered them, the crumpling collision that destroyed them both, and the birth of new rocks that resulted, is the story of California's rocks. The details are complicated but the broad picture is not. It is a picture that anyone can see in the rocks of the state. . . .

Continents are really rafts of light rocks, granite, for the most part, floating embedded in the heavy black rocks of the earth's mantle—the rocks that make the bedrock sea floor. . . .

The entire ocean floor is . . . a sort of giant conveyor belt rising from its source in the mid-ocean ridge and sinking as a great slab to its destiny in the deep ocean trench. This great planetary conveyor belt seems to move about two inches per year, ponderously slow in human terms but unseemly haste for a geologic process. Movement at that rate can carry the sea floor from its source in mid-ocean to the edge of the widest ocean in less than 200 million years and nowhere is sea floor known to be older than that.

The sea floor does not move in one piece but in a series of segments that behave almost as though they were rigid. These slide past each other

From David Alt and Donald Hyndman, "The Great Collision: California's Geological Past," in *Roadside Geology of Northern California* (Missoula, MT: Mountain Press Publishing, 1975), pp. 1, 3, 5–6, 7–9. Reprinted by permission.

along great faults and where they carry the continent along with them they tear it into large slabs. One of these large sea floor faults, called the Mendocino fracture, projects across northern California along the line of the southern boundaries of the Klamath Mountains and Modoc Plateau. The San Andreas fault is another such boundary between slabs of continent riding on two moving segments of the mantle. . . .

When the moving floor of the Pacific Ocean began to slide beneath the edge of the continent, thick deposits of sediment near shore were crushed together and jammed onto the old continent to make the crumpled metamorphic rocks of the Sierra Nevada and Klamath Mountains, the wreck of what was once a quiet coastal plain. Generous slices of the black sea floor itself were incorporated within them to become the broad belts of dark rocks that lace through both ranges.

As the action continued, the growing welt of new coastal mountains built steadily westward as the moving sea floor stuffed slice after slice of muddy sediment under its seaward edge. And each new slice of sediment was brought in from farther out at sea. So the rocks in California tend to become younger westward and to include sediments deposited farther from shore.

While the younger slices of deep sea sediment were being stuffed into a marginal trench to make the rocks that later became the Coast Range, the coastal mountains already formed were torn into two segments that moved about 60 miles apart to become the present Sierra Nevada and Klamath Mountains. Although no one can be sure, it seems likely that the Klamath Mountain block moved west opening a seaway gap behind it in the northeastern part of California, that part destined to later become the Modoc Plateau.

Also while the younger sediments were accumulating to become the Coast Range, molten magma rose from the descending slab of sea floor into the Sierra Nevada and Klamath Mountain rocks that had already formed. Long chains of volcanoes rose above them and part of the older sedimentary rocks also melted. Some of the new magma erupted at the surface but most of it solidified within the crust to become the enormous masses of granite that we see in those mountains today now that they are deeply exposed by erosion. The tremendous quantity of heat in the crumpled sedimentary rocks cooked them and welded them into solid metamorphic rocks. . . .

Slow processes of erosion removed the volcanoes and reduced the Sierra Nevada to a province of low, rolling hills barely above sea level. The Great Valley and the Modoc seaway in the northeast both filled

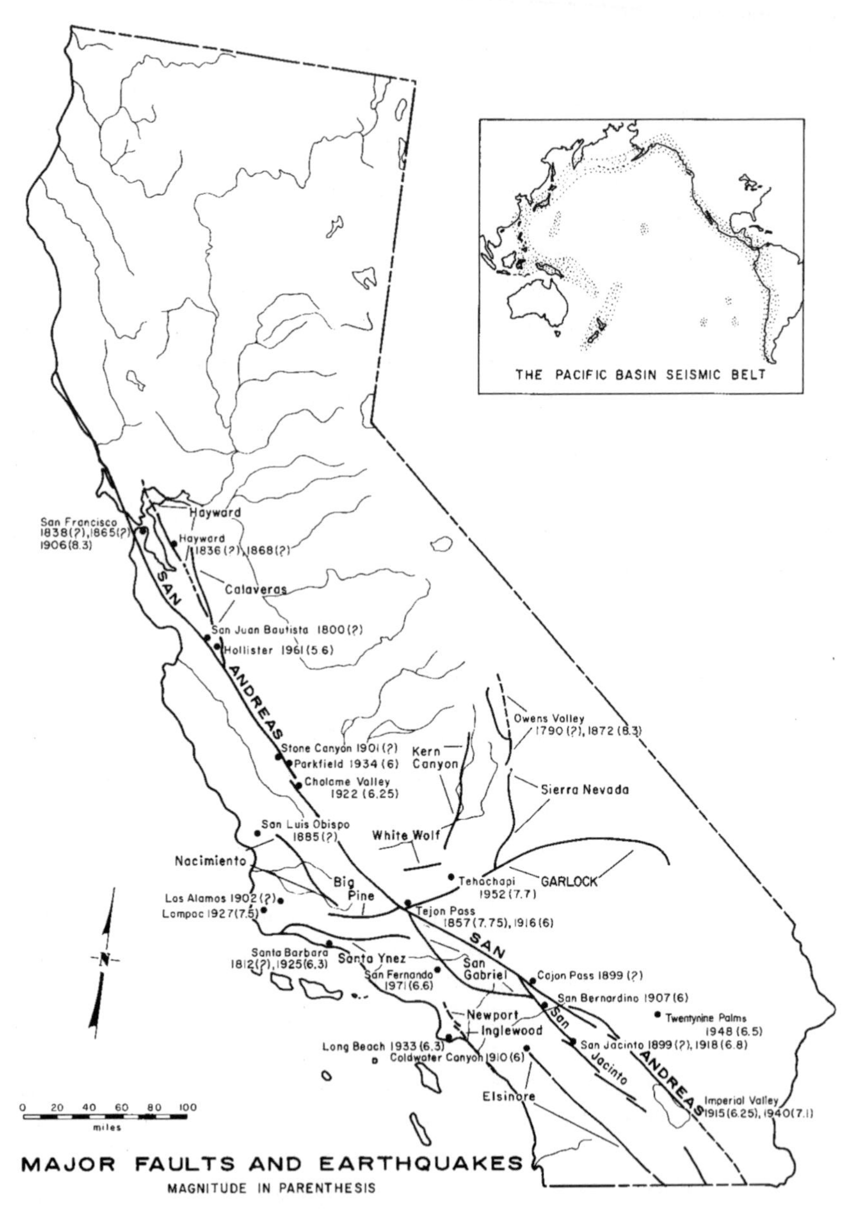

MAP 2. MAJOR FAULTS AND EARTHQUAKES IN CALIFORNIA.
From Warren A. Beck and Ynez D. Haase, Historical Atlas of California. *Norman: University of Oklahoma Press, 1974, map 4. Reprinted by permission.*

completely with sediment eroded from the surrounding hills and became dry land.

Instability resumed about 30 million years ago with uplift and stretching of a large portion of the western part of our continent. Action began in California with eruption of a series of enormous basalt lava flows. Much like those normally associated with the mid-ocean rise, they spread onto the area of the former Modoc seaway. These converted what had been a level plain similar to the present Great Valley into a high volcanic plateau several thousand feet above sea level.

Uplift and stretching of the continental crust broke it into a series of great blocks that moved like sections of a concrete sidewalk set on unstable ground. This involved all of the eastern Sierra Nevada and the Modoc Plateau. Movement of the blocks was accompanied by renewed volcanic activity, eruption of smaller quantities of black basalt and extensive blankets of white volcanic ash. The blocks that moved up are now mountain ranges, the largest of them is the Sierra Nevada, and those that moved down became large basins like the Sierra Valley east of the northern Sierras.

While the continental crust east of the Great Valley has been stretching and moving vertically to make mountains and basins, other crustal movements have been rearranging the geography of the Coast Range. A large piece of continent, essentially the southern end of the Sierra Nevada, once in southern California, is moving northward west of the San Andreas fault. Evidently it is attached to a northward-moving segment of the earth's mantle.

During the last several million years, very recently as geologic events go, the big Cascade volcanoes have built a curving chain of peaks extending from the broken northern end of the Sierra Nevada to the Canadian border. Evidently another descending current of sea floor off the coast of the Pacific Northwest is now carrying the sea floor beneath the continent where it melts to form magmas. Almost surely, the Cascades are today as the Sierra Nevada was over 100 million years ago. So in California we see all stages of the processes that convert muddy sediments of the sea floor into continental rocks which rise as mountains, erode and return to the sea as sediments, and then are swept back onto the continents again by the constant motion of the sea floor conveyor belt.

Wildlife in Transition

Raymond F. Dasmann

Of all the environmental changes that have taken place in California, perhaps none had such an immediately obvious and dramatic effect as the decline in wildlife from abundance to scarcity. This resulted from the impact of European colonists and was to touch off one of the first conservation movements in California, a movement that is stronger today than heretofore. . . .

Of California's six biotic provinces, two are almost exclusively in California (they extend into Baja California), and four are generally outside the state. This division means that California had not only its own unique fauna, but also representatives of the Great Basin, Sonoran, Oregonian, and Sierra-Cascade fauna. Furthermore, because of its climate and the originally great extent of marshes and inland waters, California is a major wintering ground of the waterfowl and other birds that come down the Pacific Flyway from breeding areas in Alaska and Canada. This diversity of fauna is not to be equaled elsewhere in America outside of the tropics. . . .

To say that the wildlife of California has declined seriously in abundance is true, if we mean that California now supports a lesser biomass of wild animal life than it did at the time of European settlement. But in this general statement we lose sight of the fact that some species—like the raccoon, the mule deer, the ground squirrel, the mockingbird, Brewer's blackbird—are probably more abundant today than in the past, whereas others—like the grizzly, wolf, wolverine, and California condor—are extinct or near extinct. To illustrate this, a few species or groups of species will be considered in some detail.

Undoubtedly the most spectacular land animal to survive the post-Pleistocene extinctions in California was the grizzly bear (*Ursus arctos*). This bear was the same species, but a different race of the grizzly, that once occupied a large area of North America and Eurasia. The grizzly was

From Raymond F. Dasmann, "The Displacement of Wildlife," in *California's Changing Environment* (San Francisco: Boyd & Frazer, 1981), pp. 9–19. By permission of Materials for Today's Learning, Reno, Nevada.

sufficiently distinct in appearance to be considered a separate species when it was described by taxonomists. In fact, C. Hart Merriam, who first classified the grizzly, named seven different species in California on the basis of their skulls and teeth, of which the southern California type was considered the largest, with males weighing over 1400 pounds. Like top predators anywhere the grizzly was fearless, for it had nothing to fear. The Indians could kill a grizzly, but more often in an encounter the grizzly killed the Indian. Being large and bold, the grizzlies impressed early Europeans with their abundance.

Bears of great size, many bear tracks, and bear trails were reported all along the south coastal region that was visited by Spanish explorers, and particularly in the San Francisco Bay area. . . . In their classic *Fur-Bearing Mammals of California*, Joseph Grinnell, Joseph Dixon, and Jean Linsdale commented: "Great numbers of people, pioneers or those living of later years in the less settled parts of the State, have been alert to seize any and every opportunity to *kill* bears. Their attitude has been adventurous, as a rule, or else founded on determination to destroy a marauder or a supposedly dangerous enemy."

Nevertheless, we have sufficient accounts of bears being in abundance. In Humboldt County Calvin Kinman recorded that he and his father counted forty bears in sight at once from a high point in the Mattole country. J. S. Newberry in 1857 also noted that bears "are rather unpleasantly abundant in many parts of the Coast Range and Sierra Nevada in California, where large numbers are annually killed by the hunters, and where not a few of the hunters are annually killed by the bears." Jedediah Smith in 1838 encountered grizzlies frequently as he traveled down the Sacramento River while trapping beaver, and he had little doubt that they were common throughout the Central Valley. . . .

James Capen Adams, in the early 1850s, made his living capturing and selling bears and traveled around the state with two tame grizzlies he had captured and raised as cubs. By the 1860s, when William Brewer traveled through California, bears were still relatively common, but the decimation had long been under way and was having its effects.

The last grizzly in Humboldt County was killed in 1868, in Mendocino in 1875, in Santa Cruz in 1886, in Monterey in 1886, in Los Angeles County in 1916. The last bear in captivity died in 1911. The last grizzly reported killed in California was in the Sierra Nevada in Tulare County in 1922. The last grizzly reported to be seen in the wild in California was in Sequoia National Park in 1925. They were big, they threatened, they ate livestock, and they killed people, and in consequence they were poisoned, shot, and exterminated.

By contrast the black bear has managed to survive. Perhaps because the species was not a top predator and lived with experience of grizzly bears, it was not particularly aggressive toward people. . . . The black bear belongs to the widespread American species *Euarctos americanus*, of which two separate races occur, one in the northern coast ranges and the other in the Sierra Nevada and Tehachapis. They average less than half the size of the grizzly, since the largest grizzly could weigh a ton, and black bears rarely exceed 500 pounds. They occupied the higher elevations of the mountains, above the chaparral and woodland belt where grizzlies were most numerous. . . .

The black bear has not escaped the harassment to which all large predators are subjected, since it kills sheep and occasionally other livestock. Sheep owners still detest bears and kill them whenever possible, even though the species is given some protection by law and may be taken by sport hunters only in prescribed seasons and by specified methods. They have survived because they were wary and willing to hide and because protection came soon enough.

The large wild grazing animals of California, unlike the carnivores, in no way threatened man directly, but they occupied space he wanted to use. They were good to eat, their hides were useful, and their heads in the opinion of some people seemed suitable for adorning the walls of homes. Their abundance at the time of European arrival was noted by all who recorded the California wildlife scene.

The Spanish, traveling northward from San Diego, first encountered the tule elk in the vicinity of Monterey Bay and then found them in great abundance on the plains and hills surrounding San Francisco Bay. To the extent that they visited the Central Valley, they noted the presence of elk there also. Later visitors were more ecstatic about the great abundance of game in the Central Valley, where the tule elk was the most conspicuous species. This was the stronghold of the tule elk, which roamed throughout, feeding in the grasslands and the tule marshes. They roamed as far south as the Tehachapis and across the south Coast Range to the Salinas Valley and Monterey Bay. They were numerous around San Francisco Bay and northward to the Russian River. There they were replaced by the Roosevelt elk which occupied the forested mountains and river valleys northward into Oregon and Washington.

The onslaught against elk began with the Gold Rush and the market for meat in the mining camps. In some areas they were rapidly hunted into extinction. From tens of thousands they dwindled to a few survivors which were protected on the Miller and Lux ranch in the Buttonwillow area north of Bakersfield. These few elk and their descendants were to help write an interesting chapter in the history of wildlife restoration, since they are now being brought back to their original native range. The Roosevelt elk were equally hard hit and wiped out throughout most of their California range with a few surviving in the redwood-covered north coastal ranges of Humboldt and Del Norte counties. Some of their descendants were to reappear much later as the Prairie Creek herd in the Redwood National Park.

The pronghorn antelope was probably more numerous than the elk. This species occurred in suitable grassy areas throughout the Mojave region and along the coast of southern California. It extended up the Great Basin side of the Sierra and in the north followed the Great Basin sagebrush vegetation halfway across Siskiyou County. The pronghorn were perhaps most numerous in the Central Valley, but they ranged into the lowlands around San Francisco Bay and across the south coast ranges into the Salinas Valley, Monterey Bay area, and the Santa Barbara–San Luis Obispo coastal plain. In 1924 a census taken by the California

FIGURE 1. TULE ELK. *Numbering in the tens of thousands in the Central and Owens Valleys in aboriginal times, tule elk neared extinction in the early twentieth century. They have now been reestablished in much of their former range. Reprinted by permission of the California Department of Fish and Game.*

Academy of Sciences indicated that there were only a thousand left, living in the northeastern sagebrush region and in Fresno and Los Angeles counties. By the 1920s they were further reduced and the last of the valley and southern California antelope were dead. The remnants in Modoc and Lassen counties were saved and later contributed to the recent story of wildlife recovery and restoration.

The bighorn sheep story is equally depressing. Once they were reasonably abundant through the mountains along the Great Basin side of the state and into the Cascades. Another population occupied the high Sierra. A third was distributed across the deserts and into the Tehachapi Mountains. A fourth occupied the higher mountains of Riverside and San Diego counties. The northern lava beds populations were exterminated; the Sierra sheep were drastically reduced. The desert sheep and southern peninsula bighorn survived in small numbers. The causes of the decline are believed to be severe depletion of essential forage resulting from heavy and uncontrolled use by livestock, the transference of diseases and parasites from domestic livestock, the increase of human activity, including appropriation of water of desert springs, and, finally, poaching. The bighorn, like the antelope, have been brought back, in part. . . .

By contrast, the last of the species of large wild grazing animals is a story of survival and success. The small deer of California, belonging to the mule deer (*Odocoileus hemionus*) species, are adaptable. In pre-European times they were not especially abundant, being clearly overshadowed by the elk and antelope. They were fairly numerous around the more settled areas, where the Indians kept the brush burned and the forests open. They are successional species not inclined to favor either dense forests, tall brush, or the open bunch-grass prairies. Cutting of the forests, burning of the brush, and overgrazing of the prairies and the Great Basin plains favored the deer. Despite heavy hunting, which continued well into the 1930s despite the existence of game laws, deer not only survived, but increased. In a 1950 survey, there were an estimated million deer in California. It was a conservative estimate. . . .

Just as great herds of elk and antelope moved across the plains of the Central Valley in pre-European days, so also did great herds of sea mammals travel above the plains of the continental shelf, moving up the slopes of the islands and occasionally down into the depths of the submarine canyons. The abundance and variety of these sea mammals were greater than those of their terrestrial counterparts. However, they included no grazing herbivores equivalent to deer or mountain sheep. Some fed on the floating plankton, some on the fish who fed on the plankton, and some on the mammals that fed on the fish that fed on the plankton. The largest of them all, the baleen whales, fed on the smallest planktonic prey.

Few people realize even today, with the current interest in whales, how many kinds of sea mammals occur in California waters. There are twenty-six species of cetaceans, the whales and dolphins, of which seven are the baleen whales, three are toothed sperm whales, three are beaked whales, also with teeth, and thirteen are in the dolphin–porpoise group which includes the killer whale. There are seven species of seals and sea lions, and one sea-going otter. It is impossible to know the relative abundance of these species in Indian times, since the existence of many was not known. The sea-going Chumash were most familiar with them—dolphins played a role in their religion, and the killer whale, represented in remarkable soapstone carvings, was apparently regarded as a friendly being—he is shown smiling. Sea otters provided skins for clothing and trade. Seals and sea lions were food sources. Sea mammals were apparently abundant and were easy prey later to those who had dollar signs in their eyes and slaughtered them for profit.

Even around Monterey Bay, the Ohlone Indians, who were not seafarers, received a high percentage of their animal protein from sea mammals. . . . The most abundant species among the larger whales originally included the right whale, which migrated to and from Arctic waters in the spring and fall. It is now severely endangered. The humpback whale may have used Monterey Bay as a breeding area, similar to what is now found in Hawaiian waters. Summers were spent in the Arctic. Grey whales did not remain in California but passed through its waters from December to February on their way to breed in Scammon's Lagoon (*Ojo de Liebre*) in Baja, California, or in other bays along the coast of Mexico. In March and April they moved north again along the coast of California. These are still the whale-watcher's whale, easily visible from shore as they swim through the Santa Barbara and San Pedro channels and along the San Diego coast. They too spend the summer in Arctic waters feeding on the abundant plankton. The fin whale, one of the largest whales, was abundant off the coast in the summertime and is still the most likely whale to be seen in summer. The sperm whale was reasonably numerous, but like the fin, less likely to be close to the shore. Three other species, the blue whale, largest of them all; the sei whale; and the small minke whale (up to 33 feet long) were present, probably in lesser numbers.

News of the abundance of whales in California had reached New England by the end of the eighteenth century. In the early 1800s, New Bedford whaling ships were in California waters pursuing the right and sperm whales—the former for whalebone and the latter for its oil. In 1851, shore-based whaling began in California, using small boats which could operate near shore and haul the whales to shore stations. These whalers concentrated on the humpback and later the grey whale. By 1875, the *Handbook to Monterey and Vicinity* stated that "The whale fishery, which for the last twenty-five years has constituted one of the most important of our local industries, is likely soon to become a thing of the past. The whales are gradually becoming scarcer." The greys and humpbacks were soon to join the right and the blue whales on the endangered species list. Still, whaling continued intermittently in California. Shore stations in San Francisco Bay opened as late as the middle 1950s, with killer boats operating off the coast. During their years of activity, until all American whaling was stopped by federal legislation in 1970, they killed: 1054 fins, 841 humpbacks, 379 sets, 783 sperms, 317 greys, 48 blues, and 21 smaller whales, along the continental shelf and slope from Point Arena to the Golden Gate.

The grey whale had been given international protection in 1938 (though shore stations continued to kill them). This protection proved effective and their numbers increased to 12,000. It was estimated that there were once 30,000, but this is a guess. The blue, humpback, fin, and sei are considered threatened species. The sperm is still present but in reduced numbers. Maybe they will recover, though it will be a long time before they crowd the boats in Monterey Bay and the pollution from their bad breath takes the place of petroleum fumes in the Monterey air. . . .

Other examples could be cited documenting the decline of species from abundance to a low level, and then the subsequent recovery (for many species) to a safe, if not abundant, population level. . . . It seems unlikely that the original levels of wildlife abundance will again be reached. It is also unlikely that the close relationship that once existed among people and animals can be restored except in places set aside from usual human use.

The Changing Face of the San Joaquin Valley

Felix E. Smith and Anne Sands

The San Joaquin Valley discovered by early settlers was a vast grassland dotted with oaks, with riparian woodlands along the perennial rivers and extensive wetlands. In less than one hundred years conversion of the San Joaquin Valley to agricultural uses was nearly complete.

Today almost the entire valley floor is in some kind of agricultural, urban, or industrial use. About 5,350,000 acres are under irrigation and another 150,000 are dry-farmed. Most remaining irrigable land is used for grazing. A variety of crops is grown.

Overgrazing and introduction of plants from foreign countries have altered the species and character of what is left of the once vast native grasslands, the remnants of which are most easily recognized as the almost treeless ring bordering the cultivated valley floor. The once extensive riparian forests are now narrow bands or pockets of trees along the rivers. The managed wetlands of the Grasslands Water District and adjacent areas in Merced County, the Department of Fish and Game's Wildlife Management Areas, and the Fish and Wildlife Service's National Wildlife Refuges contain most of what is left of the once vast valley wetlands.

The pronghorns are gone, incompatible with fences and intensive agriculture. Tiny remnant herds of tule elk are confined to reserves. The flights of geese and ducks are greatly diminished. The once common valley quail are much reduced. The kit fox now seeks to exist on the valley's edges. The golden beaver is condemned as a threat to the levee systems. Chinook salmon, their migration to historical spawning grounds made difficult or impossible by dams, water diversion structures, pollution, and loss of water to irrigation, are gone from streams or reduced to only remnant populations. The blunt-nosed Leopard lizard, Fresno and giant kangaroo rats, San Joaquin antelope squirrel, and other plants and animals

From Felix E. Smith, "The Changing Face of the San Joaquin Valley," and Anne Sands, "The Value of Riparian Habitat," *Fremontia*, 10, no. 1 (1982): 24–25, 5–7. By permission of the California Native Plant Society, Sacramento, California.

are now rare or endangered. Few vernal pool ecosystems remain unleveled by the plow.

Prior to 1814, the riparian woodlands of California were seldom visited by Europeans. Although settlements had been established in southern and northern California and along the coast, the central valleys with their extensive marshes and nearly impenetrable riverine forests were left to native Americans and the abundant wildlife. Among the first outsiders to explore the Sacramento Valley were fur trappers from Hudson's Bay Company. They were followed by a rapid succession of explorers, including Luis Arguello, Jedediah Smith, Sir Edward Belcher, and Lieutenant George H. Derby. Diaries, field notes, maps, and topographic surveys from their expeditions are the only records we have of the once vast riparian jungles that flanked the bottomland rivers of California.

In 1840, Captain Belcher described the riparian forests of the Sacramento River below Red Bluff as follows:

> Its banks are well-wooded with oak, planes, ash, willow, walnut, poplar, and brushwood. Wild grapes in great abundance overhang the lower trees. . . . Our course lay between banks for the most part belted with willow, ash, oak, or plane, which latter of immense size, overhung the stream, without apparently a sufficient hold in the soil to support them, so much had the force of the stream denuded their roots. . . . Within, and at the very verge of the banks, oaks of immense size were plentiful. . . .

Most of the historical accounts do not indicate the depth of the forests, but some references suggest belts of trees averaging from two to four or more miles in width on both sides of the rivers; even tributaries had forests two or three miles wide.

The Gold Rush of the 1850s caused many significant changes in riverine ecosystems of the Central Valley. Hydraulic mining destroyed miles of streamside vegetation but the secondary effects of the Gold Rush—increased population and agricultural growth—were just as destructive. California's population soared. Rivers continued to be major transportation corridors and floodplain camps became cities, as people turned from gold mining to farming. Riparian trees were used for building materials and fuel, especially on the steam-powered paddle-wheelers that cruised the Sacramento-San Joaquin Delta carrying food, supplies, and passengers. Marshes were drained, levees cleared, and the rich alluvial soils planted to orchards and crops.

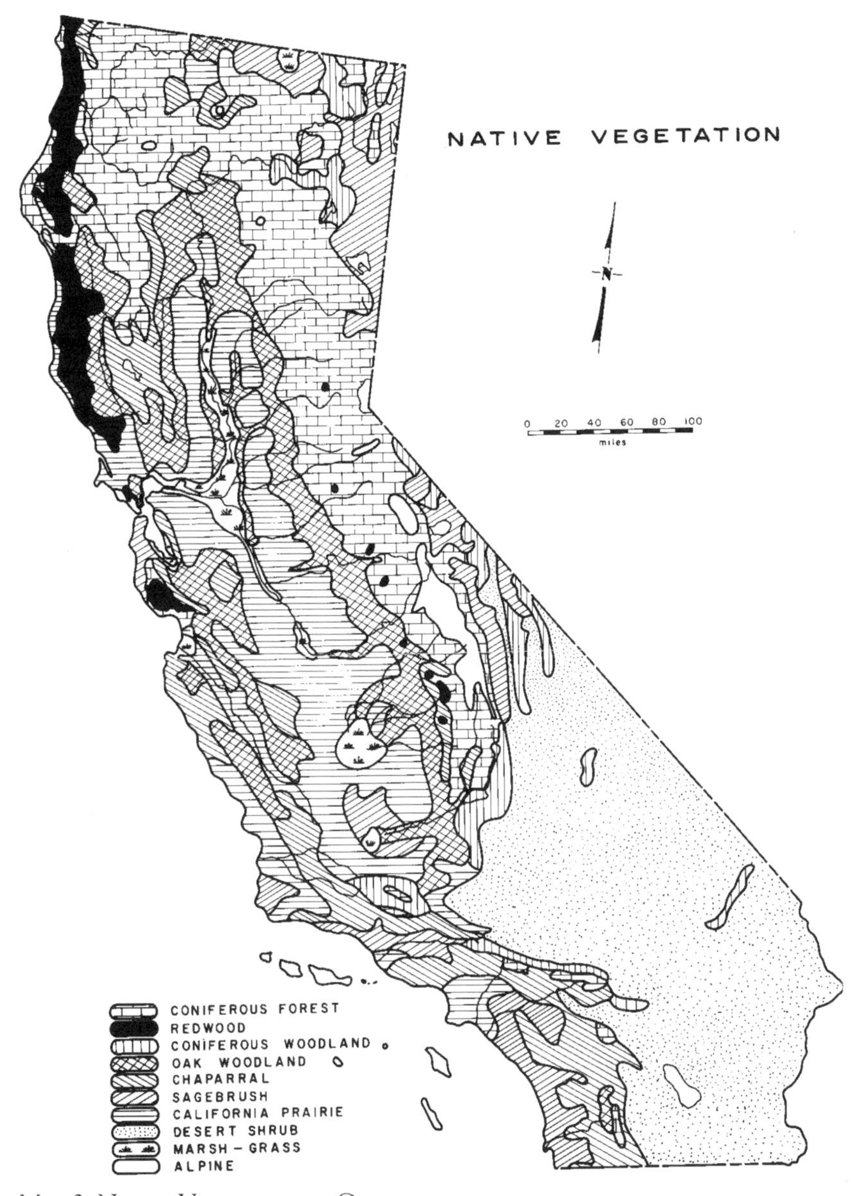

MAP 3. NATIVE VEGETATION OF CALIFORNIA.
These biomes, or biogeographical units, are classified by the dominant vegetation found within them. From Warren A. Beck and Ynez D. Haase, Historical Atlas of California. *Norman: University of Oklahoma Press, 1974, map 59. Reprinted by permission.*

Conversion of riparian lands to agriculture was facilitated by the construction of dams, levees, reservoirs, by-passes, and canals, which controlled the flow of water and promised to reduce flooding while providing irrigation water when needed. Confident of this protection, farmers removed the trees and shrubs that had helped stabilize the banks and planted more orchards.

It is estimated that there were still about 775,000 acres of riparian forests along the Sacramento River in the 1850s. By 1952, however, this figure had dropped to 27,000 and in 1972 there were less than 18,000 acres of riparian habitat remaining on the Sacramento River. Between 1972 and 1977 a further 20% reduction of riparian vegetation occurred between Redding and Colusa. Other California rivers have suffered similar impact. The California Department of Fish and Game estimates that less than 5% of the original riparian forests remain in the Central Valley. But no estimates have been attempted for the thousands of miles of foothill and coastal creeks that have been stripped of vegetation. Most Californians have forgotten, or have never known, that our rivers used to be flanked by magnificent jungles of vegetation, alive with song.

The rivers, however, continued to meander and undercut their banks, as they had always done. Seepage through levees became serious in some areas and drainage, too, became a problem. When rivers meandered and levees failed, landowners and county supervisors appealed to the state and federal governments for a solution. Construction of federal flood-control and bank-protection projects in California began in the 1920s and continues today.

But there are many . . . reasons . . . for preserving what remains of riparian forests. In many cases, they provide the only refuge for wildlife in an otherwise paved and developed landscape. Trees along a watercourse are essential for the health of the aquatic system itself. Fish use submerged roots and snags for cover. Insects dropping from overhanging vegetation are a source of food for fish. Trees provide shade that moderates water temperature, a critical factor for aquatic life.

Riparian forests shelter predators such as red-tailed hawks, coyotes, and gray foxes which feed on agricultural pests such as rodents and insects. Riparian bird species feed exclusively on insects and thus provide free pest control to those who allow their riparian forest to remain. . . .

When riparian lands are unwisely developed, the functions performed by nature must be replaced with expensive, often fuel-consuming, technological substitutes. If a floodplain is subdivided for homes, then flood-control dams, channels, and storage basins must be constructed. When natural areas are destroyed, man-made parks must substitute for natural recreation areas. The unwise destruction of natural drainage and vegetation, and the additional facilities necessitated as a result, represent a huge and unnecessary expense to the taxpayers.

Further Readings

Austin, Mary. *Land of Little Rain*. Boston: Houghton Mifflin, 1950 [1903].

Bakker, Elna. *An Island Called California: An Ecological Introduction to Its Natural Communities*. Berkeley: University of California Press, 1985.

Barbour, Michael, et al. *California's Changing Landscapes: Diversity and Conservation of California Vegetation*. Sacramento: California Native Plant Society, 1993.

Brewer, William H. *Up and Down California in 1860–64, the Journal of William H. Brewer*. Berkeley: University of California Press, 1966.

Chaney, Ralph W. *Redwoods of the Past*. Berkeley: Save-the-Redwoods League, 1964.

Crosby, Alfred W. *Ecological Imperialism: The Biological Expansion of Europe, 900–1900*. New York: Cambridge University Press, 1986.

Crosby, Alfred W. *The Columbian Exchange: Biological and Cultural Consequences of 1492*. Westport, CT: Greenwood, 1972.

Dasmann, Raymond F. *California's Changing Environment*. San Francisco: Boyd and Fraser, 1981.

Dasmann, Raymond F. *The Destruction of California*. New York: Macmillan, 1965.

Fradkin, Philip. *The Seven States of California: A Natural and Human History*. Berkeley: University of California Press, 1997.

Gordon, Burton L. *Monterey Bay Area: Natural History and Cultural Imprints*. Pacific Grove, CA: Boxwood Press, 1974.

Hornbeck, David. *California Patterns: A Geographical and Historical Atlas*. Palo Alto, CA: Mayfield Publishing, 1983.

Hundley, Norris, Jr. *The Great Thirst: Californians and Water, 1770s–1990s*. Berkeley: University of California Press, 1992.

Hyndman, David, and Donald Alt. *Roadside Geology of Northern California*. Missoula, MT: Mountain Press Publishing, 1975.

Johnston, Hank. *Thunder in the Mountains: The Life and Times of Madera Sugar Pine*. Los Angeles: Trans-Anglo Books, 1968.

King, Clarence. *Mountaineering in the Sierra Nevada, California*. Washington, DC: Department of the Interior, U.S. Geological Survey, Government Printing Office, 1902.

McCullough, Dale R. *The Tule Elk: Its History, Behavior, and Ecology*. Berkeley: University of California Press, 1971.

Muir, John. *My First Summer in the Sierra*. Boston: Houghton Mifflin, 1911.

Muir, John. *The Yosemite*. New York: Century Company, 1912.

Nordhoff, Charles. *California: For Health, Pleasure, and Residence*. Berkeley: Ten Speed Press, 1973.

Norris, Robert M., and Robert W. Webb. *Geology of California*. New York: John Wiley & Sons, 1990.

Palmer, Tim, ed. *California's Threatened Environment: Restoring the Dream*. Washington, DC: Island Press, 1993.

Powell, John Wesley. *Report on the Lands of the Arid Region of the United States*. Cambridge, MA: Harvard University Press, 1963.

Preston, William L. *Vanishing Landscapes: Land and Life in the Tulare Lake Basin*. Berkeley: University of California Press, 1981.

Pyne, Stephen J. *Fire in America: A Cultural History of Wildland and Rural Fire*. Seattle: University of Washington Press, 1997.

Skinner, J. E. *An Historical Review of the Fish and Wildlife Resources of the San Francisco Bay*. Sacramento: Water Project Branch Report, California Department of Fish and Game, 1962.

Storer, Tracy I., and Lloyd P. Tevis, Jr. *California Grizzly*. Berkeley: University of California Press, 1996.

Storer, Tracy I., and Robert L. Usinger. *Sierra Nevada Natural History*. Berkeley: University of California Press, 1963.

Strong, Douglas H. *Tahoe: An Environmental History*. Lincoln: University of Nebraska Press, 1984.

Twain, Mark. *Roughing It*. New York: Harper and Brothers, 1903.

Walton, John. *Western Times and Water Wars: State, Culture, and Rebellion in California*. Berkeley: University of California, 1992.

Warner, Richard E. *Riparian Resources of the Central Valley and California Desert*. Sacramento: Department of Fish and Game, 1985.

Chapter 2
INDIAN LANDS

DOCUMENTS

Kate Luckie (Wintu) Deplores the Soreness of the Land, Recorded in 1925

People talk a lot about the world ending. Maybe this child [pointing to her eldest child] will see something, but this world will stay as long as Indians live. When the Indians all die, then God will let the water come down from the north. Everyone will drown. That is because the white people never cared for land or deer or bear. When we Indians kill meat, we eat it all up. When we dig roots, we make little holes. When we build houses, we make little holes. When we burn grass for grasshoppers, we don't ruin things. We shake down acorns and pine nuts. We don't chop down the trees. We only use dead wood. But the white people plow up the ground, pull up the trees, kill everything. The trees say, "Don't. I am sore. Don't hurt me." But they chop it down and cut it up. The spirit of the land hates them.

They blast out trees and stir it up to its depths. They saw up the trees. That hurts them. The Indians never hurt anything, but the white people destroy all. They blast rocks and scatter them on the earth. The rock says, "Don't! You are hurting me." But the white people pay no attention.

From Cora DuBois, *Wintu Ethnography* (Berkeley: University of California Press, 1935), pp. 75–76. Reprinted by permission of the Regents of the University of California and the University of California Press.

When the Indians use rocks, they take little round ones for their cooking. The white people dig deep long tunnels. They make roads. They dig as much as they wish. They don't care how much the ground cries out. How can the spirit of the earth like the white man? That is why God will upset the world—because it is sore all over. Everywhere the white man has touched it, it is sore. It looks sick. So it gets even by killing him when he blasts. But eventually the water will come.

This water, it can't be hurt. The white people go to the river and turn it into dry land. The water says: "I don't care. I am water. You can use me all you wish. I am always the same. I can't be used up. Use me. I am water. You can't hurt me." The white people use the water of sacred springs in their houses. The water says: "That is all right. You can use me, but you can't overcome me." All that is water says this. "Wherever you put me, I'll be in my home. I am awfully smart. Lead me out of my springs, lead me from my rivers, but I came from the ocean and I shall go back into the ocean. You can dig a ditch and put me in it, but I go only so far and I am out of sight. I am awfully smart. When I am out of sight I am on my way home."

Karok Indians on Coyote and Salmon, Recorded in 1930

How Coyote Brought Salmon to the Klamath River (As Told by the Karok Indians of Humboldt County, California)

Although Kareya, in creating the world, had put many fishes into the ocean, none came up the Klamath river. The reason for this was that Kareya had built a great fish dam at the mouth of the Klamath river. He had closed it and given the keys to two old hags to keep, so that the salmon could not go up the river. The hags guarded the key night and day, and never slept for fear that someone might come and steal it.

The people did not know what to do to bring the salmon from the ocean up the river, and many were dying from lack of food. Coyote decided to help the people get the salmon up the river.

He went to an alder tree and broke off a piece of bark. Bark of an alder

From Edward Gifford, *Californian Indian Nights Entertainment* (Glendale, CA: The Arthur H. Clarke Co., 1930), pp. 176–77. Reprinted by permission of the publishers.

tree turns red and looks like salmon when broken off of the tree. So Coyote took this piece of red, salmon-looking bark and traveled down the river until he reached its mouth. Here he saw a cabin. It was the abode of the two hags.

He rapped at the door of the cabin, and when the hags opened it, he very politely said, "How do you do?"

The hags did not suspect that Coyote had come to steal the key, and so they invited him to sit by their fire. Coyote entered and seated himself in front of the fire to get warm. Very soon he took out his piece of alder tree bark and started nibbling on it.

"See, he has some salmon!" said one of the hags.

She arose and took down the key, which was hanging very high on the wall, and went out to get some salmon. Coyote saw her take the key, but this did not help him very much because it was too high for him to reach. When the hag returned with some salmon, she cooked it and then both of them ate it all without offering Coyote any.

Coyote stayed all night in the cabin. He pretended to sleep, but all the time he was thinking how he would get hold of the key. Morning came and he had thought of no plan.

One of the hags took down the key and started out again to get some salmon. Then quick as a flash a plan came to Coyote. He jumped up and darted under the hag, knocking her down, and causing her to fling the key a long way off. Coyote bounded over to it, seized it in his teeth, and ran out of the cabin. He raced quickly to the fish dam and opened it.

Thus all the salmon from the ocean were allowed to pass up the Klamath river, and the people had plenty of food after that.

Modoc People Recall Their Arrival in a Basket, Recorded in 1953

Kumush, Old Man of the Ancients, went down with his daughter to the underground world of the spirits. It was a beautiful world, reached by one long, steep road. In it were many spirits—as many as all the stars in the sky and all the hairs on all the animals in the world.

From "People Brought in a Basket," reported by Ella Clark (1953), in Richard Erdoes and Alfonso Ortiz, eds., *American Indian Myths and Legends* (New York: Pantheon, 1984), pp. 109–111. Reprinted by permission.

When night came, the spirits gathered in a great plain to sing and dance. When daylight came, they returned to their places in the house, lay down, and became dry bones.

After six days and six nights in the land of the spirits, Kumush longed for the sun. He decided to return to the upper world and to take some of the spirits with him to people in his world.

With a big basket in hand, he went through the house of the spirits and chose the bones he wished to take. Some bones he thought would be good for one tribe of people, others for another.

When he had filled his basket, Kumush strapped it to his back and together with his daughter started up the steep road to the upper world. Near the top he slipped and stumbled, and the basket fell to the ground. At once the bones became spirits again. Shouting and singing, they ran back to their house in the spirit world, lay down, and became dry bones.

A second time Kumush filled his basket with bones and started toward the upper world. A second time he slipped, and the spirits, shouting and singing, returned to the underground world. A third time he filled his basket with bones. This time he spoke to them angrily. "You just think you want to stay here. When you see my land, a land where the sun shines, you'll never want to come back to this place. There are no people up there, and I know I'll get lonesome again."

A third time Kumush and his daughter started up the steep and slippery road with the basket. When he came near the edge of the upper world, he threw the basket ahead of him, onto level ground. "Indian bones!" he called out.

Then he uncovered the basket and selected the bones for the kinds of Indians he wanted in certain places. As he threw them, he named them.

"You shall be the Shastas," he said to the bones he threw westward. "You shall be brave warriors."

"You also shall be brave warriors," he said to the Pit River Indians and the Warm Springs Indians.

To the bones he threw a short distance northward, he said, "You shall be the Klamath Indians. You'll be as easy to frighten as women are. You won't be good warriors."

Last of all he threw the bones which became the Modoc Indians. To them he said, "You will be the bravest of all. You will be my chosen people. Though you'll be a small tribe and though your enemies are many, you will kill all who come against you. You will keep my place when I have gone. I, Kumush, have spoken."

To all the people created from the bones of the spirits, Kumush said, "You must send certain men to the mountains. There they must ask to be made brave or to be made wise. There, if they ask for it, they will be given the power to help themselves and to help all of you."

Then Kumush named the different kinds of fish and beasts that the people should eat. As he spoke their names, they appeared in the rivers and lakes, on the plains and in the forests. He named the roots and the berries and the plants that the people should eat. He thought, and they appeared.

He divided the work of the people by making this law: "Men shall fish and hunt and fight. Women shall get wood and water, gather berries and dig roots, and cook for their families. This is my law."

So Kumush finished the upper world and his work in it. Then with his daughter, he went to the place where the sun rises, at the eastern edge of the world. He traveled along the sun's road until he reached the middle of the sky. There he built a house for himself and his daughter. There they live even today.

California Indians on "The Way We Lived," Recorded in 1981

The Creation of Turtle Island, Maidu

In the beginning there was no sun, no moon, no stars. All was dark, and everywhere there was only water. A raft came floating on the water. It came from the north, and in it were two persons—Turtle and Pehe-ipe. The stream flowed very rapidly. Then from the sky a rope of feathers, called Pokelma, was let down, and down it came Earth-Initiate. When he reached the end of the rope, he tied it to the bow of the raft, and stepped in. His face was covered and was never seen, but his body shone like the sun. He sat down, and for a long time said nothing.

At last Turtle said, "Where do you come from?" and Earth-Initiate answered, "I come from above." Then Turtle said, "Brother, can you not make for me some good dry land, so that I may sometimes come up out of the water?" Then he asked another time, "Are there going to be any

From Malcolm Margolin, *The Way We Lived: California Indian Reminiscences, Stories, and Songs* (Berkeley, CA: Heyday Books, 1981), pp. 122–23, 45. Reprinted by permission.

people in the world?" Earth-Initiate thought awhile, and then said, "Yes." Turtle asked, "How long before you are going to make people?" Earth-Initiate replied, "I don't know. You want to have some dry land: well, how am I going to get any earth to make it of?" Turtle answered, "If you will tie a rock about my left arm, I'll dive for some." Earth-Initiate did as Turtle asked, and then, reaching around, took the end of a rope from somewhere, and tied it to Turtle. When Earth-Initiate came to the raft, there was no rope there: he just reached out and found one. Turtle said, "If the rope is not long enough, I'll jerk it once, and you must haul me up; if it is long enough, I'll give two jerks, and then you must pull me up quickly, as I shall have all the earth that I can carry." Just as Turtle went over the side of the boat, Pehe-ipe began to shout loudly.

Turtle was gone a long time. He was gone six years; and when he came up, he was covered with green slime, he had been down so long. When he reached the top of the water, the only earth he had was a very little under his nails: the rest had all washed away.

Earth-Initiate took with his right hand a stone knife from under his left armpit, and carefully scraped the earth out from under Turtle's nails. He put the earth in the palm of his hand, and rolled it about till it was round; it was as large as a small pebble. He laid it on the stern of the raft. By and by he went to look at it: it had not grown at all. The third time that he went to look at it, it had grown so that it could not be spanned by the arms. The fourth time he looked, it was as big as the world, the raft was aground, and all around were mountains as far as he could see. The raft came ashore at Tadoiko, and the place can be seen today.

Acorn Trees, Nomlaki (Sacramento River)

The land does not belong to individuals. Dominic's grandfather [a chief], by being such a big and good man, was favored. He was left a big valley. He owned one big oak tree of a special kind. It was a singular tree called *nuis*. There was a village nearby, but old Dominic's grandfather owned that tree and got all the acorns from it. He also owned a valley of about 2,000 acres of open land. It was two or three miles away from his home. This valley was staked off—each different division [kin group] got a different part of the valley for themselves. They had poles to mark the different persons' territories. . . .

Where there is a tree of small acorns, some family owns that tree. He [the family head] will lean a stick against the tree on the side toward which he lives. Thus the people know what family owns it. He may set up too many and will give away the others to his relatives. This person kind of owns the tree—like you would a fruit tree. In those days the families owned them. They own trees in the mountains, too. They maintain border lines, but if you are friendly with them they may give you a tree in time of need.

Jeff Jones, Nomlaki

ESSAYS

Native Californian Cultivators

Kat Anderson

California has been sculpted by prehistoric human hands, as well as by earthquakes, lava flows, floods, lightning fires, and windstorms. Just as channel overflows revitalized the sandbar willow, California sycamore, and valley oak habitats, Native Americans—through the pattern and timing of harvests, as well as through the burning, pruning, weeding, and planting of places—favored certain mixtures and frequencies of plant and animal species.

Yet almost everywhere that Westerners have gone, they have underestimated the skill, ingenuity, and capability for shaping the landscape of the peoples they have displaced. Many of California's landscapes, which were culturally affected to a considerable extent, refute the idea that Native Americans "lived lightly on the land." However, the realization that parts of California bear the imprint of former human cultures does not necessarily imply that the land was therefore "corrupted" or "soiled." That self-defeating attitude keeps us from exploring a Native American partnership with nature that left the resource base intact. . . .

The long centuries (lasting at least 12,000 years) of successful adaptation to life in California suggest that certain tribal approaches to land use were far more sophisticated than Westerners suspect. When one talks with elders in various tribes today, it becomes clear that there was a realm, pattern, and scale to human use that was suited to wild places, accessible through the ancient knowledge of those elders' ancestors. . . .

Native Americans harvested resources in such a way that the plants continually thrived in the same locations. Many tribes understood that

From Kat Anderson and Thomas C. Blackburn, eds. *Before the Wilderness: Environmental Management by Native Californians* (Menlo Park, CA: Ballena Press, 1993), pp. 151–53, 154–55, 155–57, 158, 159–60, 162–64, 165, 168–70, 174. Reprinted by permission.

MAP 4. NATIVE GROUPS, 1770.
Geographical diversity and isolation precipitated the rise of over 135 dialects in over 20 basic linguistic families. From Warren A. Beck and Ynez D. Haase, Historical Atlas of California. *Norman: University of Oklahoma Press, 1974, map 77. Reprinted by permission.*

there were plant-use cycles. An important rule for most cultural groups was "do not take everything." Therefore, harvesting rates were not much in excess of tribal yearly needs, and whole plants were left to ensure next year's harvest. These rules are still remembered and practiced today:

> We used to go up to the mountains and my dad would spot the brodiaea. We would only take half of the plants in a cluster. We never would pick them all in one area. You only take what you need. [Hector Franco, Wukchumni Yokuts, 1991]

Black oak acorns were the best of all. It was a sin to cut a black oak tree down. "When gathering acorns, mother would leave some for seedlings" [Katie Appling, Southern Miwok-Paiute, 1989].

In addition, vegetative reproductive structures were left behind. The implication seems to be that prehistoric native peoples harvested the most important plants on a sustained-yield basis. Bulbs, corms, and tubers of wild onions (*Allium* spp.), yampah (*Perideridia* spp.), lilies (*Lilium* spp.), mariposa lilies (*Calochortus* spp.), and wild hyacinth (*Brodiaea* spp.) were harvested carefully, leaving bulblets, cormlets, or tuber fragments behind to ensure a population for the following year. . . .

There are still handfuls of native people today who have an intimate and respectful relationship with nature precisely because it is near at hand and because they continually tap its creative ferment and hidden potential through the use of such materials as leaves, branches, berries, flowers, seeds, roots, horns, fur, claws, meat, organs, and bones. When gathering or hunting plants or animals, they unite the concept of "home" and the notion of "use." The distinction between nature without humans (wilderness) and nature with humans (home) is not made: "I've always wondered why people call plants 'wild.' We don't think of them that way. They just come up wherever they are, and like us, they are at home in that place" [Clara Jones, Chukchansi Yokuts, 1990]. The fact that plants and animals are honored *through* human use reinforces the feeling that plants, animals, and humans all belong to a place:

> When a Karok woman went out to collect pine roots, hazel stems, and bear-lily roots for her baskets, she moved in an animate and indeed passionate world. . . . By plucking roots and stems she was not harming these people but rather honoring them, transforming them into beautiful baskets that would be displayed during ceremonies, sitting in glory before the rich people. The woman was thus helping the roots and stems fulfill their destiny. [Malcolm Margolin, 1981, p. 79]

Elders remember a time when mushrooms abounded, quail coveys were thick, and brodiaeas were dug "in a sea of blue." There is a common feeling among elders today that plants want to be used. This idea is similar to the concept that the fish and deer want to be caught and eaten. If not gathered they become scarce or disappear altogether. . . .

These examples suggest a respectful, attentive concern for the source of plant and animal abundance, a complete cooperation with nature's processes, and a yielding to its limits. They point to an intimacy and familiarity with the habits and requirements of valuable animals and plants that native peoples still continue to express daily.

Areas where desirable plants grew were transformed into favorite gathering sites through indigenous management. Native peoples increased the potential for plant resources in these areas. Wild plant populations at favored gathering sites persisted and flourished as a result of human manipulation, technology, labor requirements, and indigenous conservation rules, rather than purely as a result of natural processes. Abundance came with thrift and restraint—but also as a consequence of human disturbance.

For example, many of the plant species that were widely used in California are "weedy" perennial species that exhibit strong vegetative reproduction in the form of rhizomes, bulblets, cormlets, tuberous root fragments, and adventitious shoots. . . . Human groups, through tillage, pruning, coppicing, weeding, and burning, subjected these plants to no more disturbance than they were already accustomed to receiving.

Plant populations on important gathering sites could be destroyed by human mismanagement. Therefore, nondestructive gathering strategies involved technological factors as well as harvesting and management principles. For example, seed gathering was conducted in much the same manner by many tribes. Whether it involved Cahuilla women collecting chia seeds (*Salvia columbariae*), Pomo women beating the stalks of seedbearing grasses (*Elymus* spp.), or Miwok women harvesting the seeds of tarweed (*Madia* spp.), a procedure in which plants were struck with a stick or special seedbeater basket was used throughout California. . . . Seedbeating ensured that at least some seed fell at the source; in addition, perennial plants remained intact, thus ensuring next year's harvest. A strong wind will shake the ripening grain of seed culms onto the ground, thereby guaranteeing that any seeds ripening somewhat early (before harvesting) or late (after harvesting) will remain at the site.

Written descriptions of Indian women gathering seed are few. In early November, 1846, journalist Edwin Bryant observed California Indian women gathering grass seed for bread: "This process is performed with two baskets, one shaped like a round shield, and the other having a basin and handle. With the shield the top of the grass is brushed, and the seed by the motion is thrown into the deep basket held in the other hand." . . .

These gathering strategies were truly efficient. While harvesting grains with a seedbeater is labor-intensive and time-consuming, it was still efficient on a long-term basis, because it preserved the source of production—the plant population being harvested. A much more destructive method of harvest would be to break off the seed heads or uproot the plants, removing most of the seed from the site. According to the ethnohistorical record, this was rarely done. . . .

> After a trip to Torrey Pines, [Delfina Cuero, Diegueño] commented that there were more weeds, underbrush shrubs, carpets of dried pine needles, and broken branches than she had ever seen when she was young and with her family had gathered food and pine nuts there. She was afraid there was so much fuel that any chance fire would destroy the trees. . . . She said that burning every year never let enough fuel accumulate to damage the trees. She also said that she had been told that some pine nuts had been planted to increase the size of the grove.

Many California native shrubs have benefited from being tended for centuries. Weavers today remember their elders burning bushes, often "on the sly," to augment the production of long, straight shoots with no lateral branches. Sourberry thickets were burned by the North Fork Mono for baskets; oak stands were fired by the Western Mono for looped stirring sticks; Chukchansi Yokuts fired buttonwillow patches for arrows; and North Fork Mono burned white oak and buck brush for cradleboards:

> A burn brings the sourberry and the redbud up real nice for baskets or when we prune, it comes up nice every year. If we don't there's nothing there to use for baskets. [Anonymous North Fork Mono elder, 1991]

> They used to burn for buttonwillow to help it and make it nice for arrow shafts. It will be one of the first to be gone if we don't start preserving it. Horticulture will help. [Clara Charlie, Choinumni-Chukchansi, 1990]

> My uncle and grandfather burned all around the bluebrush [*Ceanothus cuneatus*] and white oaks [*Quercus douglasii*]. They'd pile brush on it and light the oaks and bluebrush on fire with a match. Mother, grandmother, and my great aunt would harvest the little ones the following August to January for the tops of the baby baskets. They preferred the shoots after a burn because these are the ones that grow right from the ground and they're straight and slender. [Norma Turner, Mono, 1991]

Hundreds of straight rhizomes and thousands of straight branches were needed to make the baskets produced by a single village, yet a search in the wilds for long, straight, slender switches with no lateral branching is largely in vain. To gather sufficient suitable branches for making the many kinds of baskets produced by adolescent and adult females in various villages, Native Americans had to manage and maintain abundant populations of certain plants at what was virtually an industrial level.

Because of the constant need for young growth, the burning of areas for the improvement of basketry materials was frequently conducted by Native Americans. . . . Lily Baker (Maidu) stated that fires were set around individual maple trees to promote the growth of new shoots for basketry. . . . The Karok and Wiyot burned to make hazel and willows grow better for manufacturing baskets. "Basketry material grows a lot better after a burn. The redbud and the sourberry they use for baskets. They know what they're looking for. They'd burn in the fall when the sap starts to go down. They'd light the whole hillside on fire" [Dan McSwain, Mono, 1991]. . . .

A survey of the ethnohistoric literature regarding burning by California Indian tribes shows that the major purposes for the burning that were recorded included (1) keeping the country open; (2) managing game; (3) stimulating the production of food crops; (4) decreasing insect pests and diseases; and (5) facilitating food gathering. The use of fire to induce the sprouting of many types of shrubs needed for making different cultural items was rarely documented. When it was recorded, it was often apparently considered unimportant, because the citation occurs as one obscure line buried in the text of a book, diary, or report. . . .

Is it possible that native groups practiced a form of tillage? The development of underground storage organs is an important means of propagation among perennial plants. The digging of bulbs may have "thinned" the resource, separating small bulbs or corms and activating their growth and, therefore, increasing the size of the tract; it would also have resulted in aerating the soil, lowering weed competition, and preparing the seedbed to increase seed germination rates.

Some of the various types of Indian potatoes are still highly valued or remembered by gatherers from various cultural groups, and recent interviews substantiate the fact that plant parts were and still are purposely left behind by Indian harvesters to ensure a future abundance:

> In digging wild potatoes we never take the mother plant. We just select the babies that have no flowers, just leaves. We are thinning the area out so that more will grow there next year. Also when harvesting wild onions with a little stick we would leave some of the young ones behind. They don't have a taste yet and will be ready the following year. [Clara Jones, Chukchansi Yokuts, 1989]

> We gathered Indian potatoes in May or June when the leaves are green and when in flower with digging sticks. . . . You boil them just like a potato and they're eaten plain. We'd go back to the same area and gather them. My mother and grandmother would only take the best and the biggest. They wouldn't harvest the smaller ones. They also gathered two or three kinds of wild onions in the foothills along streams. The plants were harvested in spring before flowering. They never cleaned everything out. They would always leave some behind. [Virgil Bishop, North Fork Mono, 1991]

The gardening or human management of bulbs and corms has been documented to a limited extent in the literature. Many of these species thrive on disturbed sites. . . . The harvesting of corms, bulbs, and tubers by Pomoans in the California Coast Range aerated the soil and resulted in the severing of bulblets from the parent bulb, thus increasing the size of the plant bed. The Cahuilla Indians in southern California harvested the large corms of *Brodiaea* spp. and replanted the smaller corms to ensure a crop the following year. The Yurok, Hupa, and Tolowa still harvest the bulbs of *Lilium* spp. and selectively harvest the biggest bulbs, replanting the small bulbs for later harvest. . . . The Maidu people steadfastly limited their gathering of the roots of wild carrot and camas, and always left some plants behind for seed. . . .

The various examples cited above may reflect an approach to plant harvesting that protected the desired ends. Expressed in Western terms, these examples illustrate a number of tenets that appear to have been operating within indigenous cultures to protect and conserve useful resources:

1. The quantity taken does not exceed the biological capacity of the plant population to regenerate or recover.

2. Gathering techniques sometimes mimic a parallel natural disturbance with which the plant has coevolved, thus maintaining and sometimes enhancing plant production.

3. The tool used is appropriate to the resource. It does not deplete the plant population of interest.

4. Horticultural techniques are used to give plants a competitive edge and put resources back into the system.

5. Often plants are chosen that exhibit remarkable vegetative reproduction.

6. Management is frequently at a scale that maintains the integrity of the plant community.

7. Taboos, codes, or other social constraints are in place to discourage depletion or overexploitation and avoid waste—thus reinforcing conservation-minded behaviors. . . .

The indigenous systems of resource management that effectively controlled our public lands five centuries ago are now as rare and priceless a resource as the plants, animals, and rocks that these systems of knowledge were designed to protect and enhance. A future challenge for us all will be to develop viable land-management strategies for California which sustain both the resource base and the cultural integrity of indigenous peoples. We can begin by integrating fully the figure of the Native American into our ecological vision of the American wilderness.

Aboriginal Fishers

Arthur McEvoy

Before the European and Asian immigrations of the nineteenth century, some 310,000 American Indians lived in what is now the State of California. So fertile were the habitats from which they hunted, gathered, and fished their subsistence that human beings populated the earth more densely here than anywhere on the continent north of central Mexico. Urban populations in aboriginal Mesoamerica grew as dense as they did by developing intensive agriculture and by taxing the resources of outlying, subjugated peoples, but in California each Indian community depended primarily on the particular combination of wild plants and animals available within its own locality. California's extreme ecological diversity thus encouraged the development of one of the most culturally variegated human populations the world has ever seen. Some of the most populous, materially wealthy, and culturally elaborate California Indian communities were those that harvested fisheries that supported commercial enterprise at a later time.

The history of the California fisheries thus began some 4,000 years before the Spanish friars arrived. Aboriginal groups faced the problem of environment no less squarely than their technologically more advanced successors did. Despite their relatively small numbers and simple techniques, Indian hunting and gathering economies apparently could strain their resources enough to damage them. There is archaeological evidence that aboriginal hunters drove the sea otter out of the Aleutian Archipelago, for example, drastically altering the ecology of those waters as they did so. Similar evidence exists that some California Indian communities died out because they overharvested their supplies of shellfish, while others tailored their exploitation to conserve their resources and their livelihoods. Indian weirs completely blocked the path of migrating salmon in some northwestern California rivers and would have prevented the fish from spawning had the natives not torn them down or let the rivers carry them away when they were through fishing. Being an Indian gave no one special advantage in confronting the problem, and different communities met it with varying degrees of success. In this, Indian fishers differed little from those who followed them.

Aboriginal fishers also shared with their successors their vulnerability to environmentally induced fluctuations in their harvests and the threat to economic stability that they posed. Many salmon-fishing groups worked at the southernmost limit of their prey's habitat, where slight changes in climate can drastically alter the magnitude of salmon runs from year to year. In southern coastal California, moreover, environmental change forced some Indian villages to change their fishing strategies and others to abandon fishing altogether, just as the sardine industry abandoned Cannery Row during the 1950s. Prehistoric fishing economies had to adapt in their own ways to problems that have plagued modern industries exploiting the same resources at the same localities.

They did so by deliberately managing their use of the fisheries on which they depended. . . . Indian communities eventually learned to balance their harvest of fish with their environment's capacity to yield them. Over time, fishing Indians carefully adjusted their use of resources so as to ensure the stability and longevity both of their stocks and of their economies. In some cases, they limited their production so that fish contributed a small but reliable share of a highly varied seasonal diet. In others, where fish were more critical to their economy, the Indians carefully circumscribed their harvests with complex systems of legal rights and religious observances. In general, the apparent antiquity and stability of California Indian societies at the time of contact with Westerners powerfully suggests the effectiveness of their strategies. . . .

In aboriginal California, fish generally formed part of a broadly based hunting and gathering economy. Indians consumed them wherever they could capture them easily with the help of simple tools. Where abundant fisheries coincided with scarcities of other wild foods, as they did along the Southern California Bight and in the lower reaches of the Klamath-Trinity river system in northwestern California, the original inhabitants depended on them more critically and developed a more advanced fishing technology accordingly.

Anadromous varieties, especially Chinook salmon, were by far the most important of the finfish. . . . Chinook salmon spawned in nearly every coastal stream from Monterey Bay to the Arctic. In California they were especially prolific in the Sacramento-San Joaquin and Klamath-Trinity systems, respectively the second and third largest streams on the coast below Puget Sound. The Smith, the Eel, and the Russian Rivers, all of which drain the coastal mountains north of the Golden Gate, also supported significant populations of salmon and of fishing Indians.

Two species of Pacific salmon frequent California rivers—the Chinook and the coho—although the Chinook variety is larger and more numerous. During their annual or semiannual runs to their spawning beds upstream, these species offered themselves for harvest in tremendous quantities at small expense of time, skill, and energy. They were therefore an ideal staple for the rainy winter season, when growing foods were in short supply. Indians usually fished them with harpoons, small hand nets, or simple weirs, although the specialized salmon-fishing tribes of the lower Klamath River Basin used distinctive A-frame scoop nets and more substantial weirs, some of which required many days and many hands to build.

There is no reliable way to estimate how much salmon Indians harvested from California rivers in the years before 1850. For groups inhabiting the lower Klamath-Trinity Basin, for example, the Bureau of Indian Affairs conservatively placed aboriginal consumption at about 36,000 Chinooks, or some half-million pounds, each winter; other estimates for the same groups range as high as 2 million pounds annually. One anthropologist, relying on estimates of Central Valley populations that were nearly doubled by later researchers, placed annual consumption in that region at nearly 9 million pounds. . . . On the Klamath, "the greatest run of salmon known to white men" produced only 1.4 million pounds of salmon, in 1912, while yields from the Sacramento peaked at roughly 10 million pounds per annum between 1880 and 1883. There is evidence, then, that Indian harvests were at least comparable to those of the commercial fishers who followed them, and thus that Indian fishing probably did exert significant pressure on the resource. What is most important is that, unlike modern fishers, the Indians sustained whatever yields they did take for centuries.

In addition to salmon, most coastal groups took smelt and other surf fish by hook and line and with handheld scoop nets. Few groups had the technical ability to fish extensively on the open ocean; those on the relatively sheltered southern coast ventured out short distances in flimsy reed boats to gather mackerel, sardine, and other nearshore species. Costanoan people fished San Francisco and Monterey Bays in similar craft as well. Specialized exceptions occurred, again, where other foodstuffs were in short supply. To compensate for an arid hinterland, coastal Chumash and Gabrielino people in southern California developed seaworthy plank canoes and carved shell fishhooks unique in aboriginal North America to take bonito, yellowtail, albacore, and other deep-sea fishes. Yurok, Tolowa, and Wiyot fishers from the northwestern redwood belt built dugout canoes to troll for salmon on the open sea and ventured as far as several miles out to hunt shellfish and marine mammals on the rocks off their coasts. No California group seems to have affected the abundance of the pelagic fishes or mammals it harvested, although fluctuations in the supply due to climatic conditions could be of great economic importance to them.

Shellfish were easy to take and were an important source of protein for those groups with access to them. They were especially abundant in sheltered waters such as Humboldt, San Francisco, and Monterey Bays and along the Southern California Bight. From Monterey south to Point Conception, high cliffs and rough seas denied access to any fishery resources whatever to the generally poor groups that lived along that section of the coast. Large sea mussels of the genus *Mytilus* were the most common shellfish species all along the coast. Next in importance were abalone, which thrived in the warmer waters of the southern coast but which thinned out north of Pomo territory in what is now Mendocino County, just north of the Golden Gate. The Indians also gathered other species of clam, oyster, and scallop where they could. . . .

Fisheries were far more important to the economies and cultures of Indians living in the far northwestern corner of California than they were to those elsewhere in the state. Here, from the Smith River south to Cape Mendocino, abundant salmon runs and the relative scarcity of other food resources forced the Indians to base their economies more squarely on fisheries. Consequently, while central and southern California tribes husbanded their fisheries primarily by keeping them to a limited share of a carefully adjusted seasonal round, Lower Klamath culture area groups took more direct steps to regulate their harvests. The social and religious systems controlling their fishery use wove through their entire culture. . . .

Woven through this peculiar social order, underwriting its legitimacy and sustaining its immediacy to its members, was a world-renewal religion that had its center among the three "climax" tribes and was the region's most distinctive cultural characteristic. Key to the cult was the life cycle of the Klamath River salmon. The Yurok and their neighbors lived in a universe dominated by the Klamath. Their cardinal directions were upriver and downriver, high country and low. Their world's center lay at Weitchpec, where the steep canyons of the Klamath and the Trinity joined a third, which led to the ocean. Through this universe, each year at the same time, rushed hordes of salmon appeared from somewhere out in the ocean and madly forced their way into the mountains, dissipating their great strength and destroying their bodies as they made their determined way upstream. Upon these heroic animals the Indians' lives depended. Their religion, therefore, served to propitiate the salmon and to guarantee their return from season to season.

Although Chinook salmon entered the Klamath system in two runs, one in spring and one in fall, the latter was most important because low water in the river made it easier to harvest the fish and because fall-run fish made up most of the Indians' winter food supply. The fall harvest was thus intensely ritualized. Just after the run got under way, a formulist performed a first salmon rite at Welkwau, near the mouth of the Klamath. He prayed "on behalf of the whole world, asking that everywhere there be money, fish and berries and food, and no sickness." When the first fish appeared, he acknowledged the salmon's paramount right to complete its life cycle by feigning a harpoon thrust at it and letting it pass. He speared the next fish he saw, killed and ate it according to prescription, and thanked it for its sacrifice "I am glad I caught you. You will bring many salmon into the river. Rich people and poor people will be happy. And you will bring it about that on the land there will be everything growing that there is to eat." Until this was done, salmon at the mouth of the river were taboo.

The key observance of the cult system took place at the peak of the fall run, some time after word of the Welkwau rite's completion made its way upriver. This was the construction of a huge weir near the Yurok village of Kepel, about 35 miles up from Welkwau and a few miles below the Klamath-Trinity confluence. Kepel was probably the best place from which to harvest large quantities of salmon. It had a good shallow stretch of river with a gravel bottom on which to build the weir, and there were only a few riffles, or rapids, below it. At this point the fish, having already had a week or so in which to burn off excess water in their flesh but not yet having expended much of their muscle and fat in struggling upstream, were in good condition for drying and eating. . . .

The entire round of ceremonies and dances that attended the construction took as much as fifty or sixty days to complete. The observance offered the region's people a chance to arbitrate conflicts with each other, to display wealth and status, and to share in the harvest. The construction itself required ten days and the cooperative effort of three communities to build. During the ten days in which the weir stood, local residents gathered a good share of the winter's supply of food. Thereafter, they dismantled it so that upstream tribes could take their shares of the run. The entire procedure took place under the strict supervision of a formulist, whose mumbled recitation was his secret personal property.

Neighboring communities were dutifully informed when the construction was about to begin, and in its final stages the ceremony drew participants from many widely scattered Yurok, Hupa, Karok, and Tolowa villages. . . .

World renewal permeated every aspect of the economic and social lives of the Lower Klamath culture area, integrating the production and distribution of their key resource with the ecological needs of the resource itself. It reaffirmed to all the ideological foundation on which their society rested. All the life forms with which the Indians shared the world had active and watchful spirits upon whose goodwill the continued stability and prosperity of the world depended. That goodwill, in turn, hinged on the Indians' reverent treatment of the salmon from which they drew their livelihood. This meant allowing enough of them to pass to sustain upriver peoples and, whether the Indians understood the mechanics of the process or not, allowing enough to reach their spawning grounds so that the stocks perpetuated themselves. Such protection extended to other of the Indians' prey species, as well: A hunter prospered only by maintaining the favor of his game "by respectful treatment." Although Indian society in the Klamath basin bore some remarkable similarities to the one that supplanted it, the natives' radically different view of the world and their place in it compelled them to behave in such a way as to save them from the environmental problems that so sorely beset their heirs.

The Impact of Disease

Sherburne F. Cook

From the first entrance of the Europeans into the new world the aboriginal inhabitants suffered one sweeping epidemic after another, each segment of the population undergoing in turn a cycle of devastating pestilence followed by gradual immunization and recovery. Among the

From Sherburne F. Cook, *The Conflict between the California Indian and White Civilization* (Berkeley: University of California Press, 1943), pp. 13–14, 17–20, 22–24, 31, 32, 34–35. Reprinted by permission of the Regents of the University of California and the University of California Press.

Indians of Lower California it has been estimated that disease was responsible for nearly one-half the observed reduction in population. One might therefore assume that in the neighboring province of Upper California a similar decline occurred.

Mortality in the missions. Since disease is the proximate cause of many deaths under all human conditions, we may begin with a consideration of the mortality figures in the missions. . . . From the earliest birth records and also from the results of careful extrapolation . . . it appears highly probable that the . . . Indian birth rate was approximately 45–50 per thousand per year. Now there is no reason to suppose, on any grounds whatever, that the premission or aboriginal population was suffering a material decline in numbers. Therefore the premission death rate must have equaled the birth rate. At least, it can have been no higher. This would imply a death rate of, let us say, 50 per thousand. But from the data it appears that the earliest mission death rate was definitely greater: approximately 70 per thousand as a mean for the decade 1774–1784, with the rate for the lowest year 54 per thousand. Granting some statistical inaccuracy due to relatively small numbers and possible errors of count and record, it is evident that the mortality jumped during the process of missionization. Part of this increment, but not all, was due to disease. . . .

Epidemics. Of true epidemics, carrying off hundreds or thousands in a few weeks or months, there were remarkably few in Upper California. In fact, there was only one of really great extent, and perhaps two of moderate intensity. This situation contrasts forcibly with that in Lower California where at least five serious epidemics occurred within a comparable period of time. . . . One must remember that commerce with the outside world was very small, that the West Coast was for many years almost completely closed to immigration, and, finally, that a watchful military government together with a competent and vigilant clergy closed the door to every obvious source of infection.

The first-notice of epidemic diseases in Upper California was contained in the works of Father Palou who mentions one in the vicinity of Santa Clara in 1777: "By the month of May of the same year (1777) the first baptisms took place, for as there had come upon the people a great epidemic, the Fathers were able to perform a great many baptisms by simply going through the villages. In this way they succeeded in sending a great many children (which died almost as soon as they were baptized) to Heaven." However, no details are given of the territorial extent, the numbers affected, or the type of disease.

No other record of a real epidemic occurs until 1802, when the inhabitants of the missions from San Carlos to San Luis Obispo were affected by some respiratory ailment. The children were the victims, to the almost entire exclusion of adults. . . . Clearly pneumonia and apparently diphtheria (*cerramiento de garganta*) are indicated. The greatest mortality was at Soledad with "many Christians and gentiles" and "more than seventy dead in that mission." "Great havoc" was also caused at Monterey and San Luis Obispo. At the peak of the epidemic at Soledad five or six died each day. Perhaps two to three hundred was the mortality at all the missions involved.

In 1806 occurred the first measles epidemic, by far the most serious witnessed in mission days. This was a clear example of a newly introduced malady attacking a fresh, unprotected population. Its mode of introduction is unknown (probably from Mexico by an incoming ship), but its spread was very rapid, and the damage very great among both children and adults. It was reported from San Francisco that from April 24 to June 27 the deaths had reached 234 in number, 163 adults and 71 children. In Santa Barbara during December, 44 neophytes died in 15 days. The total mortality may be reckoned from the general censuses. In 1806 the deaths in excess of the mean of the years 1805 and 1807 were 1,800. If we allow 200 as being due to other causes, we may still ascribe 1,600 to measles. The total population for 1806 was given at 18,665, a decrease of 1,693 from the previous year. But in 1806 there were 1,572 baptisms. Hence the affective reduction was 3,265. Granting half this as being due to measles, we again have a mortality of about 1,600. Although the adults were hard hit, the children suffered most. The mean child death rate in 1806 for all the missions was 335 per thousand. In San Francisco alone it was 880, the population under ten years of age being almost completely wiped out.

For the next twenty years no outstanding epidemic occurred, although the diseases already present flared up occasionally. About 1827, however, there was a recrudescence of measles, which, although of moderate intensity, did not approach the severity of the first outbreak. From the census figures it may be estimated that the mission mortality amounted to several hundred, perhaps a thousand. The incidence was spotty, the child death rate, which reached 577 per thousand at San Juan Bautista and 524 at Santa Clara, being quite low elsewhere. At San Diego it was noted that the measles had caused "some" deaths among the white population and "more damage" among the Indians. At San Buenaventura measles appeared at the end of 1827 and lasted till March, 1828. Many adults and children died. . . .

By the end of the decade 1820–1830 California was coming into much closer contact with the outside world; probably associated with this increased external intercourse, several new diseases appeared, which at times became epidemic in their scope. . . . In 1833 for the first time there was an alarming amount of smallpox. Scarlet fever may have been introduced along with other "contagious fevers," and it is certain that cholera reached menacing proportions in 1834. . . .

Syphilis. Without doubt the most important single component of this entire disease complex was syphilis. Indeed, so widespread and so devastating in its effects was venereal disease that it merits extended consideration. Among the natives of Lower California, and in direct contrast with the inhabitants of the west coast of the mainland, syphilis was universal in its occurrence and extremely severe in its effects. Upper California seems to have resembled Lower California rather than the mainland in this respect, although in the Franciscan missions of the north the disease may not have been so fatal, or its external manifestations so striking as in the older Jesuit missions of the southern peninsula. In both regions one might be inclined to discount the severity of the disease on grounds of exaggeration by those on the scene, were it not—and this is particularly true of Upper California—that there is absolute unanimity of opinion and emphasis on the part of priest and layman, soldier and civilian, contemporary reporter and later raconteur. After reviewing the evidence, one is impelled to the conclusion that venereal disease constituted one of the prime factors not only in the actual decline, but also in the moral and social disintegration of the population. These effects cannot be strictly assessed in numerical terms, but their weight can be appreciated if some of the evidence is reviewed.

Syphilis appeared in Upper California certainly within the first decade of settlement. The conventional story, which may or may not have been true, attributed its introduction to the Anza expedition to Los Angeles in 1777. Thus Miguel and Zalvidea state that this "putrid and contagious disease had its beginning with the time Don Juan Bautista de Anza stopped at the mission San Gabriel with his expedition." However, it may not be fair to lay the blame entirely on Anza's troops since there were numerous other means of introduction. The expeditionary force of Portola in 1769 and other troops entering the country were without doubt heavily infected, not to speak of the early civilian settlers. Indeed, irrespective of the social status of the immigrants, it would have been a miracle had the country escaped the pest. Once introduced, the spread was an easy matter. The relations of the soldiers with the Indian women were notorious, despite the most energetic efforts of both officers and clergy to prevent immorality. In fact, the entire problem of sexual relations between the whites and the natives, although one which was regarded as very serious by the founders of the province, has apparently escaped detailed consideration by later historians, both Californian and American. . . .

Environmental factors in disease. We may regard disease as a basic factor in the contact relation between any two races or species, such as those with which we are here dealing. This is particularly true of new maladies introduced by the new, invading race with which the established race has had no experience. . . .

Although at the outset most mission inmates were natives of the adjacent territory, as the establishments expanded, many were brought in from considerable distances. As a rule, with the exception of those in the Salinas Valley, the missions were located on or near the coast, and were hence subject to the cool, damp, foggy seashore climate. Indians brought here from the hills and valleys of the drier and warmer interior would doubtless have to become acclimated in the purely literal sense. During the process they would be somewhat more prone to infection than normally. Even the Spanish were impressed by this fact. Again and again in their reports they ascribed the unhealthy state of the neophytes, particularly with reference to tuberculosis, to the climate and pointed out that the maximum illness appeared to coincide with the rainy season of late winter and spring. . . .

In the independent family huts adjoining the mission the Indians were probably no worse off than when living in their own villages. However, the large rooms or compounds, where the unmarried men or women slept, must have been breeding-places for disease. With little ventilation, no heat, no protection from dampness in the rainy season, with the occupants packed in as closely as was humanly possible, it would have been amazing if respiratory infections had not been rampant. Thus a survivor of the period wrote: "The Indians in their wandering life as savages enjoyed good health. . . . Afterward very harmful to them was enclosure within infected walls, according to the system adopted by the missionaries."

To summarize, the mission Indians' lack of immunity to introduced infection . . . predisposed them to whatever new diseases happened to attack them. The effect of these diseases, in terms of population decline, was undoubtedly greater under the mission system than it would have been in the wild or natural state.

Native World Views

Robert F. Heizer

The Pit River Council proclamation to the president and people of the United States in June 1970 declared: "No amount of money can buy Mother Earth; therefore, the California Land Claims Case has no meaning. The Earth is our Mother, and we cannot sell her." What did the Pit River Indians mean by this?

The perception that individuals have of who they are in a world of other men and nature, what their existential and normative values comprise and the reasons why these values are important as a basis for behavior are all aspects of what is called "world view." . . .

California . . . was a region holding a large number of societies that had limited knowledge, understanding, experience, and tolerance of neighboring peoples. California Indians, while perhaps knowing individuals in neighboring tribelets, for the most part lived out their lives mainly within their own limited and familiar territory. . . . Living out the span of existence from birth to death within an area bounded by a horizon lying not more than 10 or 15 miles from one's village and not having talked to more than 100 different persons in a whole life must have made one's world small, familiar, safe, and secure. As if to emphasize or reinforce this sense of localization, many (perhaps all) tribes put their children through a drill of learning the group boundaries:

> The boundaries of all tribes . . . are marked with the greatest precision, being defined by certain creeks, cañon boulders conspicuous trees, springs, etc., each of which objects has its own individual name. It is perilous for an Indian to be found outside of his tribal boundaries, wherefore it stands him well in hand to make himself acquainted with the same early in life. Accordingly, the squaws teach these things to their chil-

From Robert F. Heizer, "Natural Forces and Native World View," in R. Heizer, ed., *Handbook of North American Indians: California* (Washington, DC: Smithsonian Institution, 1978), vol. 8, pp. 649–53. Reprinted by permission.

> dren in a kind of sing-song. . . . Over and over, time and again, they rehearse all these boulders, etc., describing each minutely and by name, with its surroundings. Then when the children are old enough, they take them around . . . and so faithful has been their instruction, that [they] generally recognize the objects from the descriptions of them previously given by their mothers. . . .

Indians not only lived close to nature but also felt intimately an integral part of it. The attribution to animals of higher intelligence and human qualities and emotions was common. In most mythologies it was animals who occupied the earth before man, and these myth-time beings lived, felt, and talked like men, created the earth for man, and provided man with many of the necessities of life (fire, food) as well as some of the undesirable features (death). . . . [But] Natural phenomena require some explanation, and in native California these usually were provided in the form of myths. Thus, the phases of the moon, eclipses, earthquakes, shooting stars, thunder, and the like are all accounted for. If these explanations do not agree with ones advanced by astronomers, meteorologists, and geologists that does not matter; as explanations they accounted satisfactorily for what would otherwise have been unknown and dangerous.

The Indians' relationship to the environment was guided by certain basic beliefs. . . . In each of the numerous forms of plants and animals there existed a soul or spirit, much like that of man's, so that all three were thought of as part of the whole of nature. Many tribes believed that men after death became transformed into animals. This idea may be linked with that of the immortality of game animals. When man hunted, he thanked the spirit of the deer for its assistance. By respecting other forms of life man did not abuse his relationship to them. Conservation of resources helped ensure the continued supply of all the things man needed, so that wasteful killing was rarely, if ever, practiced. One took what he needed and expressed appreciation, rather than acting as though what was available in the way of food and materials was simply there for the taking. . . .

All nature was capable of willful acts, usually potentially evil ones. The world of mountains, lakes, springs, caves, and forests was viewed as animate and therefore ever ready to intercede in human affairs. . . . To mitigate the potential of interference with human plans, ritual was practiced, even for the most practical and mundane human activities. The

Nomlaki said, "Everything in this world talks, just as we are [talking] now—the trees, rocks, everything. But we cannot understand them, just as the White people do not understand Indians."

The Wintu hunter must possess both skill and luck to kill deer. A man who had lost his hunting luck did not say "I cannot kill deer any more," but rather "Deer don't want to die for me." The Yurok believed that "a hunter's success is brought not by his own cunning, but by the favor he can win from his game by respectful treatment." A Pomo hunter before going out rubbed his body with aromatic angelica and pepper tree leaves, believing that by so doing he would not be molested by the spirit of the hills. . . .

The Indian concept of the mythical world [was] blended with the actual, physical earth. . . . The small bounded earth known to the Yurok [was a] land with the Klamath River flowing across it, [at] the edge of the great saltwater ocean. The rest is mythical, but in the Indian mind as real as the palpable earth. The Pomo [held] the . . . idea of the earth as floating in an ocean with the sky dome arching over it all like a gigantic bowl, and something very similar is reported for the Sierra Miwok. And far to the south among the Luiseno and Ipai-Tipai there are actual cosmological maps drawn with the colored earths that portray visible physical features of mountains the sea, islands, the moon, sun, and stars with supernatural and ordinarily invisible spirits. . . .

Eclipses of the sun or moon were generally thought to be due to a monster (often a frog or bear) devouring the luminary. In order to frighten away the monster people beat sticks on the houses, hit dogs to make them howl, and shouted their loudest. And, of course, the noise-making activity was always successful. The Wintu gathered up all the food and water in the village and threw it out after an eclipse, giving as the reason that some of the blood from the chewed sun or moon might have spattered on it and thus contaminated it. While nobody could see the blood, it might nevertheless be there, and the food-water disposal ritual was merely to play it safe. Throughout native California taboo and ritual served the purpose of controlling or stabilizing an unstable earth, which was populated both by men and animals as well as supernatural spirits. The earth-firming or world-renewal ritual was practiced as a formal cult in northwestern California basic to the Kuksu cult of north-central California. Native ritual in pre-White California was a continual check on and affirmation of the mechanism of natural forces.

Earthquakes were usually thought to be caused by a giant, at times the Creator who, while sleeping underground, rolled over. Shooting stars were thought by the Wintu to be the spirits of dead shamans who were doing a little traveling in the afterlife. The Northern Lights were thought to be a portent of epidemic illness that would later visit the village. A dust spiral near the house was interpreted as the presence of the spirit of a dead relative. . . . Such omens were part of the beliefs of all California tribes. . . .

Morality can be construed as an element of world view. This aspect of behavior rises to the fully conscious level in some groups where moralistic precepts were openly spoken of and formed the basis for interpersonal behavior. For the Luiseno, ethical principles were formulated in a ritualistic context in the form of addresses given to boys and girls who were being initiated into adulthood. The ethical or moral principles must be observed lest, as the people were warned, supernatural avengers would punish them with sickness or death. The impending punishment was, in short, concretely physical rather than merely one of incurring the displeasure of the supernatural spirit overseers of men. The "sermon" addressed to boys at their ritual induction into manhood when they are gathered around the sacred sandpainting went:

> The earth hears you, the sky and wood mountain see you. If you will believe this you will grow old. And you will see your sons and daughters, and you will counsel them in this manner, when you reach your old age. And if when hunting you should kill a hare or rabbit or deer, and an old man should ask you for it, you will hand it to him at once. Do not be angry when you give it, and do not throw it to him. And when he goes home he will praise you, and you will kill many, and you will be able to shoot straight with the bow. . . .

These are the kinds of concepts Native Californians developed to understand their worlds and to serve as guides for their behavior in them. Europeans destroyed or radically altered much of the environment and introduced by force or precept very different ways of conceiving the relations of man and nature. Ancient and efficient ecologies were disrupted before adequate and sympathetic records could be made that would promote understanding of what must have been a whole series of different integrated native philosophical systems.

Further Readings

Almaguer, Tomas. *Racial Fault Lines: The Historical Origins of White Supremacy in California.* Berkeley: University of California Press, 1994.

Anderson, Kat, and Thomas C. Blackburn, eds. *Before the Wilderness: Environmental Management by Native Californians.* Menlo Park, CA: Ballena Press, 1993.

Bean, Lowell J. *Mukat's People: The Cahuilla Indians of Southern California.* Berkeley: University of California Press, 1972.

Cook, Sherburne F. The Conflict between the California Indian and White Civilization. Berkeley: University of California Press, 1943.

Cook, Sherburne F. *Migration and Urbanization of the Indians in California.* Baltimore, n.p. 1943.

Cook, Sherburne F. *The Population of the California Indians, 1769–1970.* Berkeley: University of California Press, 1976.

Couro, Ted. *San Diego County Indians as Farmers and Wage Earners.* Ramona, CA: Ramona Pioneer Society, 1975.

Dippie, Brian W. *The Vanishing American: White Attitudes and U.S. Indian Policy.* Middletown, CT: Wesleyan University Press, 1982.

Du Bois, Constance Goddard. *The Condition of the Mission Indians of Southern California.* Philadelphia: Office of the Indian Rights Association, 1901.

DuBois, Cora. *Wintu Ethnography.* Berkeley: University of California Press, 1935.

Dutton, Bertha P. *The Rancheria, Ute, and Southern Paiute Peoples.* Englewood Cliffs, NJ: Prentice-Hall, 1976.

Gayton, Anna Hadwick. *Yokuts and Western Mono Ethnography.* Berkeley: University of California Press, 1948.

Heizer, Robert F. *Aboriginal California.* Berkeley: University of California Press, 1963.

Heizer, Robert F., ed. *The Destruction of California Indians: A Collection of Documents from the Period 1847 to 1865.* Santa Barbara: Peregrine Smith, 1974.

Heizer, Robert F., and A. J. Almquist. *The Other Californians: Prejudice and Discrimination under Spain, Mexico, and the United States to 1920.* Berkeley: University of California Press, 1977.

Heizer, Robert F., and Albert B. Elsasser. *The Natural World of the California Indians.* Berkeley: University of California Press, 1980.

Heizer, Robert F., and T. K. Whipple, eds. *The California Indians: A Source Book.* Berkeley University of California Press, 1971.

Hundley, Norris. *The Great Thirst: Californians and Water, 1770s–1990s.* Berkeley: University of California Press, 1992.

Hurt, R. Douglas. *Indian Agriculture in America: Prehistory to the Present*. Lawrence: University Press of Kansas, 1997.

Hurtado, Albert L. *Indian Survival on the California Frontier*. New Haven: Yale University Press, 1988.

Kroeber, A. L. *Handbook of the Indians of California*. Washington, DC: Government Printing Office, Smithsonian Institute Bureau of American Ethnology, 1925.

Lewis, David Rich. *Neither Wolf nor Dog: American Indians, Environment, and Agrarian Change*. New York: Oxford University Press, 1994.

Lewis, H. T. *Patterns of Indian Burning in California: Ecology and Ethnohistory. Ballena Press Anthropological Papers, No. 1*. Ramona, CA: Ballena Press, 1973.

Margolin, Malcolm. *The Ohlone Way: Indian Life in the Monterey–San Francisco Bay Area*. Berkeley: Heyday Books, 1978.

Margolin, Malcolm. *The Way We Lived: California Indian Reminiscences, Stories, and Songs*. Berkeley: Heydey Books, 1981.

McEvoy, Arthur. *The Fisherman's Problem: Ecology and Law in the California Fisheries, 1850–1980*. New York: Cambridge University Press, 1986.

Merriam, C. Hart. *The Dawn of the World: Myths and Tales of the Miwok Indians of California*. Lincoln: University of Nebraska Press, 1993.

Miller, Joaquin. *My Own Story (Or, My Life amongst the Modocs)*. Chicago: Belford-Clarke, 1890.

Powers, Stephen. *The Northern California Indians: A Reprinting of 19 Articles on California Indians, Originally Published 1872–77.* Berkeley: Department of Anthropology, University of California, 1975.

Shipek, Florence C. *Pushed into the Rocks: Southern California Indian Land Tenure, 1769–1986*. Lincoln: University of Nebraska Press, 1988.

Solnit, Rebecca. *Savage Dreams: A Journey into the Hidden Wars of the American West*. San Francisco: Sierra Club Books, 1994.

Turner, Allen C. *The Kaibab Paiute Indians: An Ecological History*. New Haven, CT: Human Relations Area Files, 1985.

Chapter 3

The Spanish and Russian Frontiers

Documents

A Spaniard Explores the Southern California Landscape, 1774

This mission [San Diego de Alcalá] was founded on the sixteenth of July, 1769. It is situated on a hill about two gunshots from the beach, facing Pebble Point and the mouth of the harbor named San Diego, which is in north latitude thirty-two degrees and forty-two minutes. Both the beach and the vicinity of the mission are very well populated with heathen, and in the district of about ten leagues there are more than twenty large villages, one of them being close to the mission.

In the beginning the Indians of this port showed themselves to be very bold and arrogant, even daring to attack the camp, seeing that there was but a small number of soldiers and most of them sick, when the expedition left to look for the harbor of Monterey. But [the Indians] were fright-

From Fray Francisco Palóu, *Historical Memoirs of New California*, ed. Herbert E. Bolton, 4 vols. (Berkeley: University of California Press, 1926), vol. 3, pp. 214–17, 220–24, 252–59. Reprinted by permission of the Regents of the University of California and the University of California Press.

ened away, three or four of them having been killed in the firing, and as many more of them wounded, while there was only one servant of the camp killed and two persons wounded, one of them being one of the missionary fathers, but they were not seriously hurt.

On account of this affair the Indians took offense, and for a long time did not come near the camp or mission. But since then they have gradually been subdued, and there are now baptized, among children and adults, eighty-three. Of the latter seven have died recently baptized and twelve have been married. They live in the village formed of their little houses of poles and tule and near them live also the heathen catechumens, who punctually attend catechism every day. . . .

Inside the stockade is the church, which consists of a chapel made of logs with a tule roof, and the dwelling for the two missionaries, with their corresponding offices, partly of adobe and partly of wood, with tule roof.

There is also inside the stockade a hut which serves as barracks for the soldiers of the guard and a warehouse in which to keep the provisions. And for defense there are inside the stockade two bronze cannons, one pointed toward the harbor and the other toward the village. At one side of the stockade they opened some trenches for the foundations of a church. . . . The missionary fathers are the foremen of the work and the neophytes the workmen. . . .

As this mission lacks water to irrigate the plentiful good land that it has, they must suffer want unless they succeed in raising seasonal crops. They have had experience of this in the first two years. The first year the river, which flows only in the rainy season near the hill on which the mission is situated, rose so high that it carried away all the seed. The second year they planted farther away from the river, and the rains failed at the best time and the seed was lost, excepting only five bushels of wheat, which they sowed about two leagues from the mission, having found by experience that rain fell more frequently at that place. Exploration has been made in the district around the mission for a radius of ten leagues, but no running water has been found for irrigating. But for the cattle there is sufficient water in several places, with a great abundance of pasture.

The heathen live on grass seeds which they harvest in their season, and which they make into sheaves as is usually done with the wheat, adding to it fish and game, hares and rabbits, of which there is an abundance. The missionary fathers have asked for a canoe and a net from San Blas, so that the new Christians may help by fishing, and if this is granted I do not doubt that it will be of great assistance to them.

Of the cattle that came by order of the illustrious visitor from California for these missions they gave eighteen head, large and small, to this

MAP 5. MISSIONS, FORTS, AND TOWNS DURING THE SPANISH PERIOD, 1769–1822. *Spanish exploration in California began with Juan Rodriguez Cabrillo's 1542 voyage, which sought a shortcut to the Orient. Between 1769 and 1823, twenty-one Franciscan missions dedicated to christianizing the Indians were founded in locations that had good water, soil, and timber and that produced crops and livestock using Indian labor. Forts and towns (pueblos) provided military protection, water, and additional food. From Warren A. Beck and Ynez D. Haase,* Historical Atlas of California. *Norman: University of Oklahoma Press, 1974, map 19. Reprinted by permission.*

mission, and in the beginning of last October it had forty head; of sheep it had sixty-four head; goats fifty-five, and swine nineteen; brood mares fifteen, four fillies and a colt; eight tame horses and one stallion; two breeding donkeys, one jack and one gelding; four saddle mules and eighteen pack mules equipped with leather saddle-bags, and two unbroken mules.

The mission has two ploughshares, and other tools and implements necessary for agriculture, carpentering, and bricklaying, and a forge for blacksmithing, although they are without workmen for those trades.

The mission of San Diego is followed by that of San Gabriel Arcangel, distant from it about forty-four leagues in a northwesterly direction. The road to it runs at first along the beach, and the rest of the way farther off, at a distance of eight to ten leagues.

The mission is situated on the slope of a hill in the valley called San Miguel, about half a league from the source of the river of that name. It is in latitude thirty-four degrees and ten minutes, and it has in sight that plain, which is very spacious, with plenty of land and an abundance of water. It runs through the plain in channels formed by the river, and it would be easy to take the water from them to irrigate all the land that they might wish for planting. . . .

A league and a half farther along the same road is the river of Nuestra Senora de los Angeles de Porciuncula, which carries all the year enough water to irrigate the large extent of land there, and other streams no less appreciable, all populated with so many heathen that it will be impossible to provide for all who may go to live at the mission. . . .

This mission of San Gabriel has thirty-eight head of cattle, thirty of sheep, twelve of goats, and twenty of swine, a filly, a stallion, five broken horses, two saddle mules, and fourteen pack mules equipped and furnished with everything, including leather saddle bags. It has six ploughshares supplied with everything for work in the fields, and all other tools necessary for farm work, as well as for carpentering; but it lacks a forge, and workmen for these trades, so necessary for the making of a new settlement.

The next mission is that of San Luis Obispo de Tolosa. It is distant from San Gabriel about seventy leagues. From San Gabriel to the beginning of the channel of Santa Barbara it is about twenty-seven leagues northwest, by a road somewhat apart from the coast. Along the channel it is as many more leagues to the west, which is the direction followed by the coast. . . . From the end of the channel to the mission of San Luis it is about sixteen leagues, part way along the beach and part way retired from it, although not very far.

This mission was founded September 1, 1772. It is situated on a hill on whose skirts runs a good stream of water supplying enough for the use of the mission and to irrigate a good field for crops. . . . In all four directions it is populated with many heathen and many very well-built villages. But there is no permanent village close to the mission, for in the neighborhood of the settlements no site was found with all the conditions required, while in this spot the mission has them in abundance, with the advantage of much good arable land, timber, firewood, and water. For, besides the stream of water mentioned, the mission has, a gunshot away, an arroyo with a little running water, with which it would easily be possible to irrigate another field; and it has in several places other abundant streams, with much land useful for planting as well as good pastures for every kind of cattle. . . . From its founding until the latter part of last October they only succeeded in baptizing twelve persons, all children, although at the time they were catechizing some adults.

The heathen of the neighboring villages harvest an abundance of very savory and nutritious wild seeds, and have game, such as deer and rabbits; and the beach Indians catch large quantities of fish. For this reason it will not be so easy to induce them to live at the mission, for they have the custom of building their towns in places where the seeds are, and as soon as they are all gone or gathered they move to another place, moving at the same time their little houses, which they make of tule mats. Hence it will be only through their interest in clothing, which they like and desire very much, that they can be reached.

They are in the habit of frequenting the mission and stopping there several days at a time, lodging in houses near the stockade in which live the relatives of the Christians, who have finally become permanent residents. It is expected that through their example others will do the same, and as they go on being baptized they will go on building houses to form a town.

At the present time they are building houses for the four married couples from California, and for five unmarried boys who remained there to put the fields in order. They are now ready to sow about eight bushels of wheat, and at the proper time they planted all they could of corn and beans, for which they had the aid of a small field of corn which they were about to gather. . . . With this crop they have enough seed to make larger plantings, and judging from the good crop they have yielded they expect to reap large harvests.

Inside the stockade they have their little church of logs and tule, and some rooms for a dwelling for the fathers, with the corresponding offices, a granary, and a dwelling house for the soldiers of the escort, all made of logs and tule.

The mission has forty-one head of cattle, four brood mares, one stallion, four broken horses, two saddle mules, fourteen pack mules, equipped and provided with everything, with leather saddle bags, and five heads of swine for breeding. It has six ploughshares, and all the tools necessary for farm work, as also some tools for carpenters and masons, for the time when they succeed in securing workmen.

November 23 [1774].—The commander having named the afternoon of November 23 for the departure from the royal presidio of Monterey, after the blessing was received from the reverend father president and farewell taken of the father companions, I set out from the mission of San Carlos at eleven in the morning, with a youth for service and a boy sacristan to aid me in the Mass. I arrived before twelve at the presidio, where I found the captain and the soldiers preparing for the departure. . . . I, with the greater part of them, and the pack train loaded with provisions for forty days, set out from the presidio at that hour, taking the road to the southeast, over hills and in sight of the beach. At half-past five we arrived at the Santa Delfina [Salinas] River, known as the Monterey because it empties into that bay about five leagues from the royal presidio on the road by the beach. . . .

November 25 [1774].—We set out from the camp about seven and took the road to the north-northeast, but after traveling a short distance, when the valley ends and one enters a spacious plain, our passage was stopped by a large marsh full of tule. . . . It is all as level as one's hand, of good arable land, with good pastures. It was named by the expedition which crossed it the Valley of San Pascual Bailon. . . .

This place has the advantage of plenty of firewood in the valley of San Benito, as well as from the large number of live oaks growing on the hills at the entrance of the valley, among which we saw many smokes, which are a sign of villages. It also has at hand timber for building, and in the valley of San Benito, near the camping place of yesterday, there is plenty of stone, and good pasture for all kinds of cattle.

We crossed this plain and found that it ends with a medium-sized river which runs through a thick growth of cottonwoods, willows, and alders, with a good stream of water. But on account of the depth at which it runs it does not seem to me it would be easy to make use of it to water the plain, unless further up toward the east it should be possible to take the water from it. This river, the soldiers said, is the one that the expedition of 1769 crossed on the beach, and they called it Santa Ana alias El Pájaro [Pajaro River].

After crossing the river we ascended some hills of earth well covered with grass, but with no trees except a live oak here and there. . . . After ascending the hills we took the road to the northwest, and by it we descended to a spacious valley called since the last expedition through it San Bernardino de Sena [east of Gilroy]. This valley runs from southeast to northwest. To the southeast no one knows its terminus, and to the northwest, the soldiers say, it reache[s] to the great estuary of San Francisco. Its width at the place where we descended must be four leagues, all land as level as the palm of the hand, and good, although in parts one finds spots of bad alkaline soil, without grass or trees.

After two hours' travel in this valley we came to a large grove of trees, cottonwoods, alders, willows, and blackberries, and inside it there was a large village. We stopped opposite it, about a gunshot away, and as soon as they saw us many Indians armed with bows and arrows came out. When we called to them they came at once, and many of them gave me arrows, which among them is the greatest demonstration of friendship. To this we responded by giving them some strings of beads, which they appreciated very much. We stayed with them a little while, and they showed great friendliness toward us. I made the sign of the cross on every one that came up, and not one resisted, being very attentive to the ceremony, as though they understood it. They gave us some baskets of atole, pinole, and seeds, and a surronato of the skin of a wildcat. The women and children did not come near, being more timid.

The men went about totally naked, like the rest of the heathen, but here and there one carried a little cape of skins or grass, protecting the body from the cold as far as the waist, but leaving all the rest of the body, and the parts that they ought especially to cover, exposed. Some of them I noticed were heavily bearded and most of them had good features and were corpulent. The women go covered with skins of animals and grass in place of skirts, and wear on their shoulders their little capes of skins. Judging by the people that permitted themselves to be seen, there were no fewer than three hundred souls of both sexes, young and old. Near the village we saw a large pool of water, and, judging by the course of the growth of trees, it may be part of a running arroyo.

After taking farewell of this village we continued our way in the same direction and valley. . . . We halted at half-past twelve near the water, the day's march having covered, at a good pace, five hours and a half, with the short delay at the village, a site which was called The Wounds of Our Father San Francisco.

A Russian Sailor on the Sea Otter Trade, 1813

Sea-otters abound in the harbour [of San Francisco] and in the neighboring waters. Their fur is too valuable for them to be overlooked by the Spaniards. An otter skin of good size and of the best quality is worth $35 in China. The best grade of skins must be large, of a rich colour, and should contain plenty of hairs with whitish ends that give a silvery sheen to the surface of the fur.

Russians from Sitka (Norfolk Sound), the headquarters of the Russian-American colony, are established at Bodega Bay, thirty miles north of San Francisco. Their chief in this new settlement is M. Kuskof, an expert fur-trader. They are thirty in number and they have fifteen Kadiaks with them. They have built a small fort which is equipped with a dozen cannon. The harbour will admit only vessels that draw eight or nine feet of water. This was formerly a point for the selling of smuggled goods to the Spaniards. M. Kuskof actually has in his settlement horses, cows, sheep, and everything else that can be raised in this beautiful and splendid country. It was with great difficulty that we obtained a pair of each species from the Spaniards because the government had strictly forbidden that any be disposed of.

M. Kuskof, assisted by the small number of men with him, catches almost two thousand otters every year without trouble. . . . The otter skins are usually sold to American fur-traders. When these fail of a full cargo, they go to Sitka where they obtain skins in exchange for sugar, rum, cloth, and Chinese cotton stuff. The Russian company, not having a sufficient number of ships, sends its own skins to China (or only as far as Okhotsk) as freight on American ships.

Two hundred and fifty American ships, from Boston, New York, and elsewhere, come to the coast every year. Half of them engage in smuggling with enormous profit. No point for landing goods along the entire Spanish-American coast bathed by the Pacific Ocean, from Chili to California, is neglected. It often happens that Spanish warships give chase to vessels, but these, being equipped with much sail, having large crews, and

From Louis Choris, *Voyage pittoresque autour du monde* . . . , trans. Porter Garnett in *San Francisco One Hundred Years Ago* (San Francisco: A. M. Robertson, 1913), pp. 16–20.

having, moreover, arms with which to defend themselves, are rarely caught.

The commodities most acceptable to the Indians of the coast of Northwest America are guns, powder, bullets, and lead for their manufacture, knives, coarse woolen blankets, and mother-of-pearl from the Pacific which they use to make ornaments for the head and neck.

Ships are often attacked with the very arms that they themselves sold, and even on the same day that they were delivered. Most of them, however, carrying from eight to fourteen guns, are able to defend themselves. Such occurrences are frequently turned to profit, for, should they carry off one of the chiefs, they are certain to get a great deal of merchandise as ransom, and gain greater facilities for trading.

A Historian Chronicles the Return of the Sea Otter, 1938

Down the San Simeon coastal highway, about fourteen miles below Carmel and just north of "the largest single-arch concrete bridge in the world," there is perched on the edge of a steep bluff overlooking the ocean a wayside Tea House. The proprietress, Mrs. Howard G. Sharpe, wife of a retired naval officer, on peeping over the porch rail early the morning of March 19, 1938, was overwhelmed to observe a populous group of strange animals swimming and sporting in the waves just beneath her lookout, a most astonishing sight to behold. State Commissioner Captain William Lippincott happening by at the time they examined the herd through a telescope and concluded it must be a body of sea otters.

Phoning the startling information to Monterey the authorities were quite skeptical, arguing, "Impossible—mistaken—you are confusing seals with sea otters as the latter disappeared decades ago, have practically become extinct" and they were only convinced after a personal inspection by professor Harold Heath of the Hopkins Marine Laboratary, who confirmed the decision.

From Augustin S. Macdonald. *Pacific Pelts: Sea Otters Choose California Coast*. (Oakland, CA: n.p., 1938), pp. 1–2, 5–8.

It is some twenty-two years since even an individual specimen was sighted in the vicinity and over one hundred years ago that they were seen in such numbers; it looked like staging a come-back to their native lair. The sudden, Phoenix-like appearance was a phenomenal occurrence which astounded everybody and from whence they came a puzzle which remains as yet unsolved.

In this particular drove there were counted about ninety-six otters. There are two other herds not far away, one off the outlet of Torres Canon and another group at Point Sur with a total variously estimated between three hundred and four hundred mammals.

The first assemblage was headed for Hurricane Point just beyond where the erosion by the waves has left standing in the surf tall pillars of rock, like sentinels, forming a semi-circle and it was within this sheltering cove they sought refuge shielded from enemies, which evidently was quite necessary as shown by two incidents that soon occurred. A thrasher whale making a dash in after them became stranded in the shallow water and only released at high tide; shortly thereafter a single animal strayed out beyond the kelp only to be immediately pounced upon by two immense sharks and literally torn apart. . . .

Much excitement has been stirred up by the unexpected visit of this local school of sea otters which has aroused the curiosity of naturalists everywhere. Just why they should suddenly arrive after a century's absence is a matter of conjecture, evidently returning by instinct, lured back by divination to their native habitat; perhaps for breeding purposes or in a longing for their ancient birth place as they wait sleeping the hours away wrapped in the kelp. Their remarkable reappearance is an historical event, one of the wonders of the scientific world, as so little is actually known of their habits, customs, or marine life.

My own speculative suggestion is drawn from coastal geological formations, food conditions, water temperatures, and marine migrations.

It is the general theory of scientists that the original outlet of the Sacramento and San Joaquin rivers in California was at Monterey bay off of which is a deep oceanic basin swarming with fish and about the shores crustacean life in abundance; while along the coast line there is a continental shelf with irregular boundaries extending out into the sea.

When the few sea otters which survived escaped to the north [and] became a herd large and strong enough to move, they emerged in a body from their hiding place near the Arctic and drifted southward with the Japanese current along the edge of this continuous submarine plateau until they reached the fault opposite Monterey where the break in the ledge occurred caused by the cutting of the ancient river channel. Here they turned by instinct landward to their former feeding and breeding grounds. That the littoral prolongation was favorable to this inference is shown by the oceanographic investigation of the Coast and Geodetic survey which proved that there exists a long under-sea valley and submarine mountains to the north and south of this point forming a natural gateway. . . .

How long these peaceful, migratory creatures will remain is problematical and no one can predict their future movements. A similar instance of the kind occurred on the southern Kuril Islands in 1872 where after disappearing for many years they were rediscovered off Yetorup Island by a party attempting to locate whaling wrecks and three hundred caught hiding amidst the kelp.

The decline of this industry on a large scale began about 1830 and as a profitable venture, with the gradual extinction of the game, constantly dwindled to such small proportions that the big concerns finally retired and abandoned the business, although the archives of Monterey show a permit issued to hunt otters in 1841 and as late as 1851 a boat load of Kanakas from Honolulu were hunting in and about the islands of the Santa Barbara channel after which the search, petered out with only occasional individual attempts without success.

It was a long call from the early days of disposing of them with wooden clubs to that of shooting with high power rifles with which weapons they had no chance of escape, no wonder they became frightened and sought seclusion. The seriousness of their kill may be understood by the authentic records stating for five years from 1812 it amounted to fifty thousand pelts by the Russians alone and five thousand annually thereafter to 1831 and during the flush period of California hunting there was an aggregate of over two hundred thousand skins captured.

Now that a pack has returned the only method of keeping them from extermination by poachers or again drive them away to sea is to make trapping or killing a felony with a heavy fine of say $2500 or more. Such penalties would make the sport unattractive, not worth the risk and in that way afford the animals an opportunity of properly propagating with, in course of time, a restoration to their original position and utility to the civilized world.

They were the innocent cause of much adventure, not alone of ships sailing the high seas but the long haul and trek across the desolate wilds of Siberia, to say nothing of Indian fights and encounters with the northern natives as well as indirectly responsible for the earliest and most famous love romance of California; that of the Russian Count Rezanov from Fort Ross with Concepion Arguello, daughter of the Spanish Commandante stationed at San Francisco. These are incidental tales traceable to the advent of hunters in search of sea otters.

The prominent part they played in the forward foundation of the California Commonwealth touching its compulsory colonization is not altogether appreciated until one realizes that virtually not alone the initiation of the commercial advancement but their continual contribution at the start was evidence of their basic value and a resourceful factor of its growth.

They are a concrete example of the mysterious ways of nature; the remnant of a dying species which should be preserved and to whom we, at least in California, owe a deep debt of gratitude.

ESSAYS

Hispanic Water Rights

Norris Hundley

The Spanish Crown moved quickly to secure its claims in America by replicating the familiar: a Hispanic society living primarily in cities, towns, forts, and other communities (some created in response to unique New World conditions) dependent on agriculture and cattle, sheep, goats, horses, and other animals—and all relying for survival on nearby water sources. In settling Alta California and the northern Spanish borderlands generally, Spain predominately used three community forms: the fort, mission, and town. A form of private property, the rancho, also played a key role and sometimes took on many of the characteristics of a modest town. For some colonists on other frontiers there had been special inducements, such as precious metals or pearls or trade, but for those destined for California there were no illusions of wealth. Few, except missionaries, wanted to go to such a forlorn place where nothing even remotely valuable had been found. . . . National security, or at least its pretext, finally sent the colonizers to California so no other nation would be tempted to occupy the region and possibly threaten the really important Spanish possessions further south. . . .

The colonists had hardly entered Alta California when they found in San Diego an attractive site for the first mission and presidio (fort) as well as rendezvous point for other expedition parties coming by land and sea. There was a splendid, protected harbor, "a large river," "good arable land," impressive stands of "willows, cottonwoods, and alders," and "many large villages of heathen." Though natural harbors were scarce,

From Norris Hundley, Jr., *The Great Thirst: Californians and Water, 1770s–1990s* (Berkeley: University of California Press, 1992), pp. 29–30, 31–32, 33, 34–35, 37–38, 39–40, 41, 43, 46, 47, 48, 58–59, 60–62. Reprinted by permission of the Regents of the University of California and the University of California Press.

the presence of the other features determined the location of most settlements, especially the missions that became the most numerous expressions of Spanish presence. Typical was Mission San Antonio de Padua which rose in a "large [interior] valley full of oaks" with "many heathen in the vicinity" and "near a river which was running with a good deal of water" that could "irrigate the . . . extensive land." Francisco Palóu's glowing description of another locale—"good streams of water," "fertile land," "heathen" in "neighboring villages," "cottonwoods, willows, and other trees, thickets of blackberries, and innumerable wild grape vines"—prompted the founding of San Gabriel, eventually among the most productive of all the missions. San Luis Obispo emerged in a place "having two small arroyos of water, with plenty of land which could be irrigated with a little labor" supplied by nearby Indians noted for their "docility" and "affection." With such attention to site and Indian settlement patterns, the Franciscans eventually established a chain of twenty-one missions over a period of fifty-four years. . . .

In time, three pueblos were established: at San Jose (1777), Los Angeles (1781), Branciforte (1797). The instructions for the founding of all of them made expressly clear the importance of sound water planning. "For the foundation of the pueblo of La Reina de los Angeles," stipulated Governor Felipe de Neve in August of 1781, "all the lands should be ascertained which can have the benefit of irrigation" and "the pueblo is to be located . . . so that the whole or the greater part of the lands for cultivation . . . should be opened." . . .

Spain's close attention to water supply and physical planning did not automatically guarantee success. Erratic precipitation patterns, then as now, suggested an abundance of stream flow in some areas that longer experience proved were more akin to deserts than oases. Part of the explanation for the inability of presidio residents to meet their food needs was the lack of sufficient water for irrigation. The military's understandable concern about defensive considerations in selecting a site sometimes clouded early appraisals of water supply and led to miscalculations.

Missions as well as presidios occasionally suffered because of faulty judgments about water availability. A half dozen of them had to relocate at least once because of floods or drought. Such was the fate of the first permanent European settlement in Alta California at San Diego. There the fathers established the mission near a broad fertile valley which they sowed with wheat that heavy rains and floods then destroyed. A lesson seemed learned, for the next crop was planted at a higher elevation, but it too was lost when reduced rainfall lowered the level of the river to a point where the irrigation system became ineffectual. The colonists eventually overcame the problem, but only after relocating the mission, which became the supply depot for the nearby presidio, at a site six miles upstream where different topography and better engineering saw San Diego eventually emerge as the fifth most productive mission in the province. Also among those missions relocated because of an inadequate water supply was Junipero Serra's own headquarters at San Carlos Borromeo. Originally founded in 1770 at Monterey, a year later it was shifted about four miles south to Carmel where, Francisco Palóu noted, "water flows the whole year through" and "it would be easy to retain enough . . . to irrigate as much as might be wished of the plain that is in sight."

Relocation was not confined to missions. The pueblo at San Jose had to be moved to another site shortly after its founding when heavy rains, poor drainage, and a broken dam turned the original locale into a quagmire. Continuing problems with marshy soil would require a second move. As with San Diego, residents learned that too much water could be as serious a problem as too little. . . .

To the missions of California, the grants of land and water, though gargantuan in size, were always strictly temporary and usufructuary. Title remained in the crown with the missionaries serving as trustees for the Indians. Responsibilities of trusteeship included not only Christianizing the natives and training them as agriculturalists and herdsmen but also providing them, as the viceroy instructed California's military commander in 1773, with "sufficient . . . water to drink and for irrigation of the fields" they farmed in common as well as for any individual plots they cultivated on their own. This must be done with an eye to the future, admonished the viceroy. "As the mission settlements are hereafter to become cities, care must be taken in their foundation," including building "the houses . . . in line, with wide streets and good market squares." To make the missions more attractive to Indians, the viceroy empowered the commander to designate the future pueblo's common lands and to issue individual land grants "to such Indians as may most dedicate themselves to agriculture and the raising of cattle." In this way, the viceroy underscored the mission's importance in the colonization scheme even though his hope for future Indian pueblos would go unfulfilled when the missions were finally secularized in the 1830s. . . .

When [water] was in short supply, as was often the case, Spanish legalisms took on great significance. Since the water was "for the common benefit," no one had a superior right that could be exercised to the detriment of others. As in the irrigation communities of Spain, this meant that no person or family had a right to a specific volume of water. The quantity varied according to individual and community needs and according to the available supply. The goal, as the Plan of Pitic stipulated, was to allocate water with "equality and justice," and the authority responsible for assuring such fairness was the local *ayuntamiento* or town council whose members were elected by the residents. . . .

As might be expected, the *ayuntamiento* outlawed the wasting of water and levied especially severe fines on those who allowed their irrigation ditches to overflow and damage roads, alleys, or other property. If the break should suddenly occur in the *zanja madre*, or main irrigation ditch supplying the community, the "person nearest . . . [the] breakage must immediately repair" it. Such action served to affirm that water, from the earliest days of settlement in the towns of both Spanish and Mexican California, remained what Iberian custom and law had been long before determined—a community resource and responsibility. . . .

A community's responsibility for a water system extended to the quality as well as the quantity of the supply. Water channeled through the *zanja madre* had to satisfy many purposes before being abandoned—watering animals, irrigating crops, and satisfying a variety of domestic household needs. Such multiple uses have potential for great harm since the water for animals, laundry, and sewage and garbage removal was also needed for drinking. Detailed ordinances imposed by the central government as well as local authorities sought to prevent pollution, especially of the water destined for human consumption. The task then, as today, seemed never-ending, with the difference being no threat from industrially generated toxic wastes. . . .

In making grants of land and water in coastal California, the Spanish did not have to contend with precontact Indian agricultural villages as in New Mexico, but the steady growth in the number of missions, presidios, and pueblos nonetheless increased the potential for conflict over water. Officials recognized the threat. The dispersed location of the missions reflected in part a deliberate attempt to minimize disputes and ensure success. Presidios and pueblos, however, tended to be near missions where they could secure a priest for special ceremonies, solicit labor from the attached Indian community for onerous tasks, obtain supplies in times of shortage, and offer defensive aid when needed. Though pueblos and presidios were few in number, this proximity to missions was a source of constant concern to religious, military, and civil authorities.

Ranchos posed a special concern. Ultimately numbering more than

800, but amounting to only a couple dozen or so prior to secularization of the missions in the 1830s, ranchos represented grants of private property and consisted of vast amounts of land. . . .

The rights of rancheros (and by implication other individuals) were ordinarily inferior to those of communities—presidios, pueblos, missions, Indian villages. . . .

Devoted primarily to cattle raising and to the hide and tallow trade, these rancho communities also frequently planted extensive fields with fruit orchards, vineyards, corn, beans, onions, potatoes, hemp, and other crops. Springs, artesian wells, and nearby streams and rivers irrigated the tracts and watered the cattle and other domestic animals. . . .

The California experience with water reflected the general Hispanic commitment to *bien procumunal*—the common good—a concept that defied precise definition, encouraged flexibility, and was incompatible with monopoly. . . . The concept of "common good" was . . . invariably ethnocentric when transformed into action, a lesson learned early by the native peoples of the New World. Their land and water rights were often reduced to little more than a cipher, while their labor was commandeered on behalf of God, country, and other men's profits. Shielded somewhat from this process were those Indian communities, like the Indian pueblos of New Mexico, with long-established agricultural traditions that the Spaniards went out of their way to safeguard but not to the extent of giving them a prior and paramount right to water. They, like all communities, were obligated to share supplies, though they frequently complained that their share was too small. In California, where there were no such preexisting Indian agricultural settlements in the areas settled by the Spaniards, the opportunities for abuse were great.

Playing a powerful role in the exploitive process was the Hispanic system of water control and management. From the beginning at California's missions and almost that early at the presidios, pueblos, and ranchos, Indians erected the hydraulic works on which the new colony depended for survival. They hauled the rock for dams, dredged the canals, and performed the never-ending drudgery of keeping ditches open and free of weeds, silt, and other debris. Thus, the water system became a mighty vehicle and reason for mobilizing and controlling the labor of native peoples. The missionaries' call for souls was doubtless sincere, but so too was their desire for a vast work force that would gain experience for constructing the missions and other buildings by first transporting water to parched fields and then bringing in crops to sustain the laborers and residents elsewhere in the colony. Indians eventually tilled perhaps 10,000 acres at the missions in addition to laboring at their other tasks as builders, herdsmen, and artisans. . . .

The situation changed in the 1830s when secularization was decreed by the Mexican government and when Indians either abandoned or were defrauded of mission lands intended for them. Many now crowded into the towns. Some refugees from the mission villages flatly refused to work, not wishing to exchange the regimen of the padres for that of others, while most of the remainder discovered that their numbers exceeded the capacity of local economies to absorb them. . . .

The Indians of California found little comfort in the Hispanic society they labored to maintain. Lofty ideals about the common good accompanied by incarceration, forced labor, disease and death, and a hydraulic system as an instrument of coercion took on a relevancy best understood by the victims. Still, while no society can boast about its policy toward indigenous peoples, Spain and Mexico did develop a remarkably equitable arrangement for allocating scarce water resources, at least among non-Indian inhabitants. Higher authorities set the guidelines with the preeminent emphasis on community needs and responsibilities, though not to the exclusion of individual rights. Water was a resource to be shared—shared in common among the residents of a settlement, shared among the various kinds of communities (missions, presidios, pueblos, Indian villages) along a common water source, and shared between communities and such private landholders as rancheros, although community rights generally took precedence over those of individuals.

Viewed from the vantage point of the twentieth century, Hispanic principles contrast sharply with the individualism and monopolistic impulses of those who flocked to California following the American conquest in 1846. Admittedly, Spain and Mexico's imprint on the waterscape differed significantly from that of aboriginal Californians, but it paled in comparison with what was to come.

Indians Encounter Spaniards

Albert Hurtado

The large willow trees that stood in the valley of the San Joaquin provided lookouts for Yokuts Indians who watched for intruders. In the autumn of 1819 they saw a party of Spanish soldiers and their mission Indian allies enter the valley near the now dry Tulare Lake. They were expecting the Spanish, since two Wowol Yokuts had returned from a September fiesta at the Mission San Miguel with a rumor that a force would come to the valley to capture all the runaway neophytes and all the gentiles, as well. Moreover, the Wowols believed that the Spanish intended to kill anyone who resisted. When the Indians spotted the expedition from the coast, they fled to hiding places deep in the great tule swamps. To the north, another Spanish expedition went after runaway neophytes in Miwok country near the modern city of Stockton.

These expeditions represented the waning power of the Spanish Empire in America. Spain's colonies were in revolt and in two years Mexico would become independent. In 1819 California was on the periphery of these changes, ruled by loyal government officials and Franciscan missionaries who were more concerned about the immediate problem of controlling the numerous Indian population than about the outcome of events in distant Mexico. East and north of the missions, presidios, and pueblos lay thousands of square miles of country that remained in the hands of native people. Occasionally Spanish forces entered the interior to explore, capture runaway neophytes, and to reclaim the livestock that mission fugitives had taken.

In theory, Spaniards sought to incorporate Indians peacefully into frontier society by training them in Franciscan missions. In practice, the results were not ideal. Sadly, violence was common, for Indians often resisted alien invaders and the Spanish fought to protect their small, isolated settlements. Demographic decline and the friars' coaxing convinced many Indians that the missions provided a safe haven, but oth-

From Albert Hurtado, *Indian Survival on the California Frontier* (New Haven: Yale University Press, 1988), pp. 32–36, 47–48, 50, 52–54. Reprinted by permission.

ers—especially Yokuts and Miwoks in the interior—believed that resistance was the key to survival.

Whatever strategy Indians chose, survival depended on adapting to the Spanish presence on the coast. The development of Indian livestock raiding illustrates this point. Soon after Spanish colonization began, native people started stealing and eating mission livestock, but the seriousness of livestock raiding increased after missionaries began to search out interior Indians to replace the coastal neophytes who had died from infectious diseases. To prevent raiding, the Spanish had considered building some inland missions that would have given the Franciscans a new field while opening church lands on the coast to private settlement, but Indian resistance and lack of adequate government support prevented the construction of any institutions in the interior. Without permanent Spanish defenses in the interior, the Indians became a troublesome and then a formidable foe. Interior neophytes who learned the skills of Spanish horsemanship could avoid capture by taking horses and driving them swiftly to their interior homeland. In the central valley, horses proved useful to Miwok and Yokuts Indians, who used them to hunt antelope and elk. . . . In addition, the interior Indians commonly ate horse meat, thus making complete use of their newfound beast of burden. In 1819 Father Mariano Payeras claimed that all the valley Indians rode horses and even held horse fairs where they traded stolen stock. Payeras may have exaggerated, but his concern reflected the importance of Indian horse raiding to Spanish Californians.

The 1819 expeditions were, as usual, looking for livestock thieves and new mission sites. Lieutenant Jose-Maria Estudillo commanded the force that the Yokuts had spotted, while Sergeant Jose Sanchez directed the northern contingent. Estudillo proceeded to the Wowol Yokuts rancheria, where he was surprised to find the community deserted. From two captured Indians he learned that the others had fled because they were afraid to go to the missions, although they were willing to give up the runaway Christians. Estudillo assured them that he wanted only to reclaim stolen horses and return Christians to the missions.

The lieutenant and his troops spent nearly a month reconnoitering the rancherias in the valley and in the Sierra Nevada foothills near modern Fresno. . . . Some foothill people—probably Western Mono—who had not met Hispanic people before, said that they were anxious to trade, so Estudillo gave their headman a pass to go to the missions. . . .

While Estudillo saw the bones of many slaughtered horses and recovered one alive, he did not report seeing any mounted Indians or feral horse herds in the San Joaquin Valley. The southern Yokuts referred to the "horse-killers of the north," possibly to encourage Estudillo to take his expedition elsewhere. In the meantime Sanchez's operations in the north and an unrelated expedition farther south, headed by Lieutenant Gabriel Moraga, gave native people reasons to doubt Estudillo's bland promises of peace. Moraga, searching for escaped Santa Barbara neophytes, surprised them in a canyon and captured nine. One neophyte who escaped walked for four days and nights to report the news to the rancheria where Estudillo was camped.

On the same day that Estudillo heard about Moraga's foray, he learned that Sanchez had fought the Muquelemne Miwoks on the Calaveras River, killed twenty-seven gentiles, wounded twenty more, and captured sixteen. Sanchez also recovered forty-nine stolen horses, but the Miwok killed a neophyte ally and wounded five soldiers. Alarmed at the news of Spanish casualties, Estudillo sent an Indian messenger to tell native people that no soldiers had been injured, fearing that knowledge of the army's losses might inspire Indian resistance. . . . In the middle of November the small force returned to Monterey presidio. . . .

[In 1838] John A. Sutter, asked permission to establish a permanent settlement in the interior. A German-born Swiss, Sutter was a man of good appearance and gentlemanly bearing. . . . He had made two trading trips from Missouri to New Mexico, been an Indian trader, gone to the American trappers' rendezvous, seen the Hudson's Bay Company's operation at Fort Vancouver, sailed to the Sandwich Islands, then to the Russian-American post at Sitka, and on to California.

Sutter had asked Governor Alvarado for permission to plant a colony on the Sacramento River at a propitious time. Besides discouraging Indian raiding, the governor hoped Sutter's settlement would reduce the power of Alvarado's ambitious uncle, Mariano Guadalupe Vallejo, who had become commander of the northern frontier. Operating out of Sonoma, Vallejo assaulted interior rancherias and sent captured native workers to californio ranchos. Thus, Vallejo kept Indian society in turmoil and built a loyal following among rancheros who needed Indian labor. Alvarado authorized Sutter's settlement and subsequently gave him official civil authority and a land grant. As a result, Sutter and Vallejo became bitter rivals for power on California's northern frontier. . . .

Sutter tried to stop Indian livestock raiding, but he met with indifferent results and Mexican contemporaries did not always appreciate his efforts. Soon after his arrival, he returned some stolen animals to their Mexican owners. . . . It was in Sutter's interest to eliminate horse raiding. After all, he had his own vast herds to protect. To secure New Helvetia and extend his power, Sutter set up an army of 150 Indian infantrymen and 50 native cavalry, who were supervised by several white officers. The troops wore Russian green-and-blue uniforms with red trim that came from Fort Ross, which Sutter purchased from the Russians in 1841. . . .

The Mexican War stalled Sutter's plans to stop Indian raiding, and meanwhile the growth of huge feral horse herds in the San Joaquin Valley testified to these unrelenting depredations. . . . The herds were principally on the west bank of the San Joaquin, where the plains were "alive with immense droves of wild horses." . . . Since wild horses had little value, Indian raiders rode past them to risk their lives stealing from the californios. Two years after Fremont's visit, William Robert Garner reported "forty thousand wild horses and mares" in the San Joaquin Valley, which was a "complete nursery of horses for California." At the end of the Mexican era, rancheros who wanted to replenish their herds would ride to the valley to round up the progeny of their own stolen horses.

Indian livestock raiders dealt to Mexican California a crippling blow, but not a mortal one. Rancheros may have lost thousands of horses, but during the Mexican era the californio population doubled and the total white population more than tripled, while the number of Indians dropped sharply, though they still substantially outnumbered whites.

The years of Mexican rule brought momentous changes to the interior and the Indians who lived there. Once an isolated Indian country that Hispanic people visited only rarely, the interior became part of an international fur trading system. Mountain men provided a market for stolen horses and so presented new opportunities to the Indians. . . . Unintentionally, fur traders imported malaria and other infectious diseases. Trade, whether for pelts or horses, made the interior Indians dependent on the outside world for their livelihood. . . . Whatever time and energy Indians devoted to these new pursuits had to be taken from traditional subsistence activities. . . . The fur trade impoverished the environment because the formerly bountiful wetlands harbored the malaria-carrying mosquito. Previously a rich source of food for Indians, the lowlands [became] a perilous place. . . . Finally, raiding, trading, and native labor linked California's resources and the interior Indians to outside forces—nationalist ambitions, imperial rivalries, and an expansive market economy. . . . Thus, Indians became dependent and vulnerable to a new set of dangers in a primitive frontier economy.

Sea Otters Encounter Russians

Adele Ogden

Aleksandr Baranov, for three decades Lord of the Pacific North, was acting as host at a gay festal board in Novo-Arkhangelsk, modern Sitka, in 1818. Before him was his honored guest and friend, Captain Vasilii Golovin. But as the former governor entertained the imperial inspector his inner thoughts were bitter. He could never completely realize his cherished dream of a Russian Pacific empire. He was an old man and had been asked to resign. The great Baranov even as he presided in the midst of high festivities was overcome by a sweeping sense of uselessness. The hilarities only probed his wounded feelings more deeply. He called for music. The singers who entered glanced at their respected master, and then in deep, lusty voices burst forth with the song of the sea hunt. The old man became young again. Before him in swift review passed Russia's dauntless seamen of the past and those yet to be. From the Aleutian chain to the island fringes of Lower California they moved, and he saw them standing guard on rocky outposts along that distant strand. He arose, and, radiant with hope and enthusiasm, he triumphantly proclaimed their new world deeds as he joined his chorists in singing his favorite air, "The Spirit of the Russian Hunters."

The southward advance of Russian seamen to the California coast was in the beginning a cooperative move. Baranov joined forces with his greatest rival in northern otter waters, the Anglo-American, in order to hunt in a marine area which belonged to neither. Negotiations for a commercial agreement were opened in 1803 by an experienced Boston merchant, Joseph O'Cain. The New England captain had been among the advance guard of those Anglo-Americans who were leaving the North Pacific, just as it was beginning to be depleted by ruthless otter-killers, to hunt in southern fields. He was no doubt the first of all his Countrymen to realize that the California fur trade could not be continued by the risky method of bartering. If Boston men could hunt the sea otter themselves,

From Adele Ogden, "Russian Sea-Otter and Seal Hunting on the California Coast 1803–1841," *California Historical Society Quarterly*, (1933): 217–39, excerpts on pp. 217–19, 221, 226–33, 235–36, 239.

they could obtain more and could keep clear of populated centers watched over by law-enforcing Spaniards. However, equipment and expert otter-pursuers they did not possess.

The Russians were the rulers of the only skilled otter hunters of the world. For years their Aleutian Indian vassals had been braving the elements in their little skin-covered canoes for the fur wealth of the far north. Tireless and sea-hardened, these "marine Cossacks" would remain motionless for from ten to twelve hours, either kneeling or sitting with their feet extended in front of them. The waves might dash over the light framework speeding through the water, but it mattered not to the rowers. They were clothed in the skin of the sea lion, and each one had fastened securely the lower edge of his jacket to a ring around the small opening in which he sat, so that not a drop could enter the boat. From five to fifteen bidarkas usually went forth to hunt together. The moment an otter was seen floating on the water, the alert Indians fixed their eyes upon it, and trembled "like the eager dog at the sight of game." Swiftly but very silently the canoes approached from the windward. When within shooting distance, while the man at the stern guided the boat, the nearest bowman raised his dart, and with sure aim threw his pronged bone spear with incredible accuracy. For twenty minutes or so the otter would remain submerged, his course being marked by a bladder attached to a long cord. When the animal arose to the surface for air, a ruthless hunter was on hand to finish him.

With expert Aleutian spearsmen, Captain O'Cain was certain he could hunt along the California coast. He turned to the rulers of the North Pacific for help. In October, 1803, he and Governor Baranov met in conference on the otter hunters' rendezvous at Kodiak Island, south of the Alaskan Peninsula. The captain's plan was simple. If Baranov would furnish Aleutian Indians and bidarkas, he would provide transportation with his vessel, the O'Cain, and would conduct a hunting expedition into the new fields. The skins which were collected would be divided equally and large profits could most certainly be realized.

Baranov considered carefully before he made his decision. Many worries pressed upon him. Just the year before, the new settlement at Novo-Arkhangelsk had been wiped out by the hostile Kolosh. Hunting parties sent to the northern mainland had been set upon by other treacherous Indian tribes. Foreign encroachments were becoming more and more annoying. To climax all, one of those intruders came to tell him of the discovery of new hunting grounds. Baranov, the indomitable, recalled the secret orders which he had received from his company the year before to push settlements southward. Baranov, the dreamer, in thought linked

MAP 6. TRADE DURING THE SPANISH–MEXICAN PERIOD.
Sea otters and whales offshore, along with cattle hides and tallow inland, were traded by Russian, Spanish, and American ships off the California coast. From Warren A. Beck and Ynez D. Haase, Historical Atlas of California. *Norman: University of Oklahoma Press, 1974, map 41. Reprinted by permission.*

the new proposition with his vision of a greater Russia on the Pacific Coast. O'Cain's idea might be the means of making that dream come true. He would venture. The first contract was signed.

Preparations for the California hunting experiment occupied old and young during the next few weeks. Aleutian women made waterproofs for their husbands. Old men contentedly whittled away at canoe paddles and frames. Small Indian boys helped their fathers by cutting sticks and beams, or by smearing whale oil in chinks and seams of the completed bidarkas. Russian-American Company agents collected provisions palatable to northern hunters—youkala or dried fish, whale meat, and whale oil. Hooks and lines were supplied for catching fish to supplement food stores. At last twenty bidarkas and twice as many hunters under the command of an able Russian, Shvetsov, were ready. As pledge of good faith, O'Cain left behind 12,000 rubles of merchandise. . . . In the latter part of March, O'Cain completed his hunt. On the return voyage, near present-day Ensenada, he was refused wood and water. Arrived at Kodiak in June, he turned over to Baranov the half of 1,100 furs. In addition, O'Cain had 700 skins which he had managed to purchase on his own account from Spanish officials and missionaries.

The contract system proved satisfactory to both parties. The problem of obtaining furs along the California coast seemed to be solved for the Boston men. . . . Three Boston vessels under Russian contracts were along the California coast in 1807.

Baranov made two arrangements for California hunting voyages in the latter part of 1811, one with William Lanchard of the Catherine who was given fifty bidarkas, and another with Thomas Meek of the Amethyst who received fifty-two canoes. From June to August, 1812, both vessels were hunting along the Lower California coast. In 1812 the Charon, also under a contract, and commanded by Captain Whitmore, left a sealing party at the Farallone Islands and then cruised along the coast to San Quintin. Each of the above captains shared with the Russians from 1,400 to 1,800 skins.

Both Boston man and Russian . . . profited from the contract system of trade. The Angloman, in the face of Spanish law enforcement and due to his own lack of hunting equipment, could not have remained at that time in the California otter fields. The Aleut made it possible for him to establish a line of hunting bases, all removed from Spanish population centers. Bodega Bay, Drake's Bay, the Farallone Islands, the Santa Barbara Channel Islands, the bay of San Quintin, Todos Santos Island, and Cedros Island were such points where seals were killed or from which the coast was scoured in all directions for the otter. The Boston vessels were safe, always aiming to keep plenty of sea room. The Aleutian Indian, the tool of the Angloman, paid the penalty by seizure, searching examination, imprisonment, and life itself.

For the Russians the contract system was the means of hunting along the California coast at a time when they lacked transportation and when the finding of new otter grounds was of utmost importance. Furthermore, considering the Russian desire to keep the fur wealth of the Pacific from the Anglo-American, it was undoubtedly a means of preventing their rival from having the exclusive run of the new fields. . . .

Baranov planned early for independent otter-hunting in California. He determined "not to divide the profits of this business with anybody" and "waited only for a chance to go there in person on some sea-otter expedition." Looking to the future, the governor had given specific instructions in 1803 to Shvetsov and in 1806 to Slobodchikov, Russian commanders of the Aleuts sent on the [Boston vessel] O'Cain, to make detailed observations of the quantities and habitats of valuable marine animals. Further information concerning the wisdom of engaging in the California otter trade was obtained as a result of the Rezanov expedition. . . .

The first independent hunting expedition to California arrived in 1809. Kuskov, Baranov's trusty "co-laborer," was in command. Two vessels started from Novo-Arkhangelsk in October, 1808, but one was shipwrecked near the Columbia. The Kodiak, Captain Petrov, touched at Trinidad to look for sea otters, but finding none, sailed on to Bodega Bay. Temporary buildings were erected. According to the Spanish record, 150 Indians, including twenty women, and forty Russians . . . [arrived] on the vessel.

A new and safer approach was found to the rich otter field of San Francisco Bay. Marin Peninsula became a portage. Landing at Point Bonita on the Pacific side, the Aleuts shouldered their canoes and tramped across the country to the bay. In February, 1809, about fifty canoes were seen landing at the northern end of the harbor in order to make the portage west. As soon as the forbidden waters were entered troubles began. Early in February skin craft were moving around Angel Island. One Aleut who landed was seized by San Francisco neophytes and brought to the presidio. Bidarkas were skulking around the southern shores of the bay during the last of March. On the 26th twenty canoes came ashore and seventeen men landed. A Spanish sergeant and eight soldiers hurried to the spot. Firing occurred, and as the hunters fled four were killed and two wounded. Careful treatment was given the latter, but the troops were ordered to be in readiness in case the crew of the large

Russian vessel should attempt to take any means of revenge. In April canoes were still in the bay near Yerba Buena. The Kodiak did not leave Bodega until August, when it took back to Baranov over 2,000 skins.

A second otter expedition was supervised by Kuskov in 1811. The hunters who had been brought on the Chirikov found very few skins along the coast near Bodega. Consequently, in May, while Kuskov was making observations in connection with the future settlement, twenty-two bidarkas were sent to San Francisco. These canoes with those of the three Boston contract vessels which were then anchored at Drake's Bay made a total of about 140 canoes hunting in the port at one time. Spanish officials employed a most effective method of handling the situation. Sentries were stationed at all wells and springs where the Aleuts obtained water. As a result, "the party was compelled to go away." Aleuts were sent to the Farallones for a supply of sea-lion meat, and then the Chirikov sailed for the north in June with the skins of 1,160 otters and seventy-eight yearlings.

The next step in the Russian advance into California otter fields was the establishment of a hunting base. Dependence upon foreign vessels had become undesirable and unnecessary. Hunting in or south of San Francisco Bay was both impossible and unpolitical. . . . As the historian Kiril Khlebnikov explains, "At the harbor of San Francisco the Spaniards kept strict watch and allowed no hunters to enter and in order to retain their friendship we had to abstain from all attempts to visit that place for the purpose of hunting." Pursuing the otter north of the Spanish settlements from a permanent California base was the only form of independent hunting which could continue. Fort Ross and her southern outpost, the Farallone Islands, served as the needed bases.

Fort Ross, therefore, was a center for sea-otter hunting as well as for supplies. Shortly after the construction of the new establishment was begun in the spring of 1812, Kuskov sent out Aleuts in the forty bidarkas brought on the Chirikov. Whenever Indians could be spared they left on hunting expeditions going up the coast as far as Cape Mendocino but rarely farther south than Drake's Bay. . . . There is little doubt that hunting was carried on in San Francisco Bay in 1813 and 1814 when canoes were allowed to enter for supplies. Khlebnikov states that whenever the Russians came to trade they always took Aleutian Indians and sometimes secured a few skins. In 1813 four Aleuts were detained, according to Kuskov "without the least cause" at the presidio. Otter hunters and sealers from Fort Ross must have been kept busy for the first two years, or otherwise William J. Pigot, captain of the Forester, one of Astor's vessels, could not have purchased 3,400 skins from Kuskov in March, 1814.

The Farallone Islands became a permanent base for Russian sea-lion hunters and sealers in the same year that Fort Ross was established. For two decades Baranov's seamen held this southernmost outpost of Russia's Pacific empire. From six to thirty Aleuts and several Russians continuously occupied those lonely rocks. Five or six times a year bidarkas would fight their way through the open sea from Bodega Bay to bring food, wood, water, and kegs for sea-lion meat and oil. Once a year or so one of the large company boats would stop to land a new gang of men and take away the others. The sea lions on that rocky outpost supplied the Aleutian Indians hunting along the California coast with the meat, blubber, and oil so essential to their diet. Every year from 100 to 200 *puds* (3,600 to 7,200 pounds) of sea-lion meat were salted down in barrels and boxes. At Fort Ross bidarkas and waterproof clothing were made from the skins and bladders of Farallone sea lions. More valuable than the sea lion, however, was the fur seal. California seals were smaller than those of the more northern regions and their black fur was coarser and therefore less valuable. However, each year the Russians killed from 1,200 to 1,500, and between 1812 and 1818, 8,427 seal skins were obtained at the Farallone station. . . .

The fur seal became for the Russians an important medium of obtaining supplies from American ships. As repeated orders were sent from St. Petersburg to Novo-Arkhangelsk forbidding trade with foreign vessels, needed supplies were exchanged for seal skins at the Russian California base. From 1823 to 1825 William Hartnell, of the English firm, McCulloch, Hartnell and Company, received goods for seals, and was much interested in making the seal trade with the Russians an important branch of his company's business. In 1823 he obtained 3,276 fur seals at Bodega. The American vessels, the *Mentor*, *Washington*, and *Arab*, took on seal skins at the Russian settlement in 1823 and 1824. . . . In 1825 after touching at the Farallones and other islands as far north as Cape Blanco, he wrote, "On these islands or keys I expected to find fur-seals; whereas I found them manned with Russians, standing ready with their rifles to shoot every seal or sea-otter that showed his head above water." . . . In June, 1828, when Auguste Bernard Duhaut-Cilly touched at Bodega he was unable to do any business because an American ship had just taken away all the skins. The Russians did not give up their Farallone outpost until years of wholesale killing exterminated the seal. Whereas the number taken in 1825 was 1,050, only fifty-four were received from the Farallones in 1833.

In contrast to sealing on island outposts, Russian otter hunting along the Spanish coast was not easy to continue upon an independent basis.

After 1813 Russian otter hunters began to try to extend their activities farther from their California base. In 1814 the Ilmen came to Fort Ross with supplies and with fifty Aleuts under Tarakanov. A hunting trip to the north was unsuccessful because of hostile natives. The vessel then sailed down the coast. For two days the Aleuts hunted around the Farallone Islands. Then the captain ordered them to slip into the bay at night. Tarakanov tells what happened:

> The Aleuts did so and hunted all day, killing about 100 sea otter, but when we went to the beach on the south side to camp for the night we found soldiers stationed at all the springs who would not allow any one to take any water. At this the Aleuts became frightened and started back toward the ship which had remained outside. It was dark and some wind was blowing and two bidarkas were capsized and the men, being tired with their day's work, could not save themselves.

The Aleuts who were capsized must have been seized by the Spaniards. In July, 1814, Kuskov was petitioning Governor Jose Arguello for Kodiaks held at the presidio and explained that "They have done no wrong but were only compelled to save themselves from the surf in the bay at the port of San Francisco where they were captured." The Ilmen continued to hunt southward but the crew had learned to be on the lookout for the Spaniards. One night some Aleuts who were sent ashore for water were frightened at the sight of some soldiers, and returned with empty kegs. Around the Santa Barbara Channel Islands 160 sea otters and some fur seals were taken.

The day of reckoning was coming. Some Russian sailors, eleven Aleuts and Tarakanov were ordered ashore to get fresh meat. Seeing some cattle grazing in the hills near San Pedro, the men landed, but almost immediately they were surrounded by Spanish soldiers on horseback. The frightened sailors pushed off without waiting for Tarakanov or the Aleuts, who were tied together with ropes and taken on a two days' march to Santa Barbara. There they were destined to work for two years and more before being released. The Ilmen proceeded on down the coast trading and hunting as it went. Upon returning to Fort Ross, the vessel brought, besides provisions, almost 400 otter skins. . . .

After Mexican independence the question of Russian otter hunting received the immediate consideration of the new government. In December, 1821 . . . the final decision was that since the Mexican government did not have sufficient ships to prevent clandestine otter hunting, it would be best to make an agreement for one year only, on condition that no settlement be established on the coast and that none of the mission Indians be employed in Russian canoes. As a result an agreement was signed at Monterey on December 1, 1823.

The first Russian-Mexican contract for otter hunting in California was to last for only four months. Russians were permitted to hunt otter and other marine animals anywhere from San Francisco to San Diego. Skins were to be divided into two equal parts. The Russians were obliged to sell their share to the Mexican government at forty-five pesos for each skin and were to receive in return wheat at three *pesos* per *fanega*. During the term of the first contract, the Russians agreed to pay the Aleuts and to keep their bidarkas in repair. Provisions for the hunters were to be supplied by the California government although the expenses were to be borne equally by both parties. If the contract was renewed, Californians were to pay the Aleuts at the rate of two piastres for large skins and one piastre for small ones.

San Francisco Bay was the field for the first hunt. Here otter seemed to be thicker than ever before. The Indians claimed that these animals had come to the California coast to escape the ruthless hunters of northern waters. Juan Bautista Alvarado describes how they entered the bay in great shoals at about the time of the first contract. Jose Fernandez writes that along the shore of the harbor from San Francisco to the estuary of Santa Clara "the ground appeared covered with black sheets due to the great quantity of otter which were there."

By January 4, 1824, one month after the contract was signed, 455 skins were taken in the bay. . . . In January, 1824, the bidarkas returned to Ross to be repaired. The latter part of the same month Captain John Cooper came to Bodega on the Rover, which had been purchased by Arguello, to receive half of the skins taken at San Francisco that season. Evidently Cooper had some difficulty in obtaining his cargo. He wrote to Khlebnikov that, " . . . it is a plain and evident fact too notorious to be contradicted that there was taken by this last expedition upwards of seven hundred otter skins in and about San Francisco the last information was that there was about 600 dried and on the stretch. It appears very mysterious to me that Don Lewis part is but 112 out of upwards of 700." The *Rover* finally sailed across to Manila with 303 otter skins and 300 otter tails. Other canoes came from Bodega to carry on the first contract. In February, 1824, eleven bidarkas arrived at San Francisco and eight more were expected. All were sent immediately to the San Pablo side of the bay "in order not to miss the calm days." At the same time, Aleuts hunted in Monterey Bay where, by March, 429 skins had been secured. . . .

Decreasing returns from black pelts had much to do with the final withdrawal of the Russians from the southern coast. Legal restrictions and the gradual extermination of the otter meant the end of one important means of support for the Ross settlement. It also meant that there was no further need for a southern hunting base. Sir George Simpson when reporting to the Hudson's Bay Company in 1842 told of the end of Russian otter hunting in California, and added, "After the loss of that profitable branch of trade . . . the Russian American Company withdrew."

The sea otter had brought the Russians to California. It had accounted greatly for their continued interest in the southern coast, and decreasing returns from sea-otter hunts was one of the reasons for their withdrawal. As a means of entering California seas, the last hunting ground of the Pacific, the Russians had resorted to the method of contracting with Boston sea captains. After a decade, the Angloman had been shaken off and independent hunting had been tried. However, Spanish opposition had been prohibitory. Russia then had become a petitioner for the privilege of being the business partner of Spain. The privilege had never been given. The Mexican government during the first years of its control in California had given the concession, for a gainful purpose. Finally, the Russians had become merely experienced employees of licensed Mexican merchants. Thus, for four decades, "the spirit of the Russian hunters," so nobly expressed in the indomitable Baranov, had carried Slavic seamen and their Aleutian crews through three trying stages of activity along the California coast.

Further Readings

Adams, Kenneth C. *California Missions*. Los Angeles: California Mission Trails Association, 1948.

Almaguer, Tomas. *Racial Fault Lines: The Historical Origins of White Supremacy in California*. Berkeley: University of California Press, 1994.

Alvarez, Robert, Jr. *Familia: Migration and Adaptation in Baja and Alta California, 1800–1975*. Berkeley: University of California Press, 1987.

Archibald, Robert. *The Economic Aspects of the California Missions*. Washington, DC: Academy of American Franciscan History, 1978.

Avina, Rose Hollenbaugh. *Spanish and Mexican Land Grants in California*. New York: Arno Press, 1976.

Bancroft, Hubert H. *California Pastoral, 1769–1848*. San Francisco: The History Company, 1888.
Bolton, H., ed. *Historical Memoirs of New California by Fray Francisco Palou*. Berkeley: University of California Press, 1926.
Chapman, Charles E. *A History of California: The Spanish Period*. New York: Macmillan, 1921.
Chapman, Charles E. *The Founding of Spanish California: The Northwestward Expansion of New Spain*. New York: Macmillan, 1916.
Cowan, Robert G. *Ranchos of California*. Fresno, CA: Academy Library Guild, 1956.
Crosby, Alfred W. *Ecological Imperialism: The Biological Expansion of Europe, 900–1900*. New York: Cambridge University Press, 1986.
Crosby, Alfred W. *The Colombian Exchange: Biological and Cultural Consequences of 1492*. Westport, CT: Greenwood, 1972.
Gibson, James R. *Imperial Russia in Frontier America*. New York: Oxford University Press, 1976.
Gordon, Burton L. *Monterey Bay Area: Natural History and Cultural Imprints*. Pacific Grove, CA: Boxwood Press, 1977.
Harlow, Neal. *California Conquered: The Annexation of a Mexican Province, 1846–1850*. Berkeley: University of California Press, 1982.
Hundley, Norris. *The Great Thirst: Californians and Water, 1770s–1990s*. Berkeley: University of California Press, 1992.
Hurtado, Albert L. *Indian Survival on the California Frontier*. New Haven: Yale University Press, 1988.
Jones, Oakah L. *Los Paisanos: Spanish Settlers on the Northern Frontier of New Spain*. Norman: University of Oklahoma Press, 1979.
Mackie, Richard Somerset. *Trading beyond the Mountains: The British Fur Trade on the Pacific, 1793–1843*. Vancouver, BC: University of British Columbia Press, 1996.
Martinez, Oscar J. *Border People: Life and Society in the U.S.–Mexico Borderlands*. Tucson: University of Arizona Press, 1994.
Monroy, Douglas. *Thrown among Strangers: The Making of Mexican Culture in Frontier California*. Berkeley: University of California Press, 1990.
Morgan, Dale L. *Jedediah Smith and the Opening of the West*. Indianapolis: Bobbs-Merrill, 1953.
Ogden, Adele. *The California Sea Otter Trade, 1784–1848*. Berkeley: University of California Press, 1941.
Pitt, Leonard. *Decline of the Californios: A Social History of the Spanish-Speaking Californians, 1846–1890*. Berkeley: University of California Press, 1966.

Reff, Daniel. *Disease, Depopulation, and Culture Change in Northwestern New Spain, 1518–1764*. Salt Lake City: University of Utah Press, 1991.

Robinson, W. W. *Land in California: The Story of Mission Lands, Ranchos, Squatters, Mining Claims, Railroad Grants, Land Scrip, Homesteads*. Berkeley: University of California Press, 1979.

Rolle, Andrew F. *Los Angeles: From Pueblo to City of the Future*. San Francisco: Boyd and Fraser, 1981.

Royce, Josiah. *California: From the Conquest in 1846 to the Second Vigilance Committee in San Francisco*. New York: Knopf, 1948.

Smith, Wallace. *Garden of the Sun*. Los Angeles: Lymanhouse, 1939.

Treutlein, Theodore E. *San Francisco Bay: Discovery and Colonization*. San Francisco: California Historical Society, 1968.

Weber, David J. *The Spanish Frontier in North America*. New Haven: Yale University Press, 1992.

Weber, David J., ed. *New Spain's Far Northern Frontier: Essays on Spain in the American West, 1540–1821*. Albuquerque: University of New Mexico Press, 1979.

Chapter 4

Environmental Impacts of the Gold Rush

Documents

A Traveler Reports on the Gold Country, 1849

SAN FRANCISCO (Upper California), June 1, 1848,

Sir: —I have to report to the State Department one of the most astonishing excitements and state of affairs now existing in this country, that, perhaps, has ever been brought to the notice of the Government. On the American fork of the Sacramento, and Feather river, another branch of the same, and the adjoining lands, there has been discovered, within the present year, a placer, a vast tract of land containing gold in small particles. This gold, thus far, has been taken on the bank of the river, from the surface to eighteen inches in depth, and is supposed deeper, and to extend over the country.

On account of the inconvenience of washing, the people have, up to this time, only gathered the metal on the banks, which is done simply with a shovel, filling a shallow dish, bowl, basket, or tin pan, with a quantity of black sand, similar to the class used on paper, and washing out the sand by movement of the vessel. . . . I presume nearly $20,000 of this

From Fayette Robinson, *California and Its Gold Regions* (New York: Stringer and Townsend, 1849), pp. 16–20.

gold has already been so exchanged. Some 200 or 300 men have remained up the river, or are gone to their homes, for the purpose of returning to the Placera, and washing immediately, with shovels, picks, and baskets; many of them, for the first few weeks, depending on borrowing from others. I have seen the written statement of the work of one man for sixteen days, which averaged $25 per day; others have, with a shovel and pan, or wooden bowl, washed out $10 to even $50 in a day. There are now some men yet washing, who have $500 to $1000. As they have to stand two feet deep in the river, they work but a few hours in the day, and not every day in the week.

A few men have been down in boats to this port, spending twenty to thirty ounces of gold each—about $300. I am confident that this town (San Francisco) has one-half of its tenements empty, locked up with the furniture; the owners— storekeepers, lawyers, mechanics, and laborers—all gone to the Sacramento with their families. Small parties, of five to fifteen men, have sent to this town and offered cooks ten to fifteen dollars per day for a few weeks. Mechanics and teamsters, earning the year past five to eight dollars per day, have struck and gone. . . .

FIGURE 2. CALIFORNIA MINERS.
Together, Indians, blacks, Chinese, Anglo Americans, and miners from around the world panned the forks of the Sacramento River for gold. From the Honeyman Collection, Courtesy, The Bancroft Library, University of California, Berkeley.

One American captain, having his men shipped on this coast in such a manner that they could leave at any time, had them all on the eve of quitting, when he agreed to continue their pay and food; leaving one on board, he took a boat and carried them to the gold regions—furnishing tools and paying his men one-third. They have been gone a week. Common spades and shovels, one month ago worth $1, will now bring $10 at the gold regions. I am informed $50 has been offered for one. Should this gold continue as represented, this town and others would be depopulated. Clerk's wages have risen from $600 to $1000 per annum, and board; cooks $25 to $30 per month. This sum will not be any inducement a month longer, unless the fever and ague appears among the washers. The *Californian*, printed here, stopped this week. The *Star* newspaper office, where the new laws of Gov. Mason for this country are printing, has but one man left. A merchant, lately from China, has even lost his Chinese servants. Should the excitement continue through the year, and the whale ships visit San Francisco, I think they will lose almost all their crews. How Col. Mason can retain his men, unless he puts a force on the spot, I know not.

I have seen several pounds of this gold, and consider it very pure—worth, in New York, $17 to $18 per ounce. $14 to $16 in merchandise is paid for it here. What good or bad effect this gold region will have on California, I cannot foretell. It may end this year, but I am informed that it will continue many years. Mechanics now in this town are only waiting to finish some rude machinery, to enable them to obtain the gold more expeditiously and free from working in the river. Up to this time, but few Californians have gone to the mines, being afraid the Americans will soon have trouble among themselves, and cause disturbance to all around. I have seen some of the black sand, as taken from the bottom of the river (I should think in the States it would bring 25 to 50 cents per pound), containing many pieces of gold; they are from the size of the head of a pin to the weight of the eighth of an ounce. I have seen some weighing one-quarter of an ounce ($4). Although my statements are almost incredible, I believe I am within the statements believed by every one here. Ten days back, the excitement had not reached Monterey. I shall, within a few days, visit this gold mine, and will make another report to you. Inclosed you will have a specimen.

I have the honor to be, very respectfully,

Thomas O. Larkin

Hon. James Buchanan, Sec'y of State, Washington.

P.S. This placer, or gold region, is situated on public land.

A Writer Recounts the Difference a Year Makes, 1851

Preface

The author became convinced of the propriety of publishing a work of the character here proposed, soon after his arrival in California, in August, 1850. All those who reached there in that year, found a state of things existing very different from what they had previously anticipated, and were almost universally disappointed in the yield of the gold mines. Those mining districts, which had heretofore been very productive, were now becoming exhausted, and new ones discovered in other parts of the country. But all were claimed and occupied. Every gulch and ravine in the whole country were, by the first of October, completely filled, while every road and by-path were crowded with a floating population, or itinerant miners, who were continually wandering from place to place, and district to district, in search of some untenanted spot, in which to try the realities of their golden dreams. The man who had spent the previous season in that country, who had returned home, and was now making his second advent into the mines, was perfectly lost and bewildered, with the changes that had taken place in his absence. Those mines which he now found the richest, were then undiscovered. And where then the foot of civilized man had scarcely made an impression, were now to be seen thriving and business towns. But the country had become too full of people, the wealth of the mines had to be divided between too many. The result of which was almost universal dissatisfaction. Consequently, many vague, uncertain and contradictory reports reached the States, in regard to California as it was in 1850.

The very rapid transition of California from the rude, wild and almost uninhabited wilderness of 1848, to the rich, the great and populous State of 1850, affords a theme of itself from which a writer might glean facts sufficient to fill a volume with interesting matter. . . .

From Franklin Street, *California in 1850, Compared with What it Was in 1849* (Cincinnati: R. E. Edwards, 1851), pp. 5, 16–19, 37–42.

Feather River

This is the largest tributary to the Sacramento, and the most northward of any on which gold has yet been discovered. It is a handsome stream, with a bold, rapid current and is navigable for steamers of a small size, a considerable distance, at all seasons of the year. Its bottoms in many places are wide, and contain some excellent farming land, and are also well adapted to grazing, and pasturage, producing the year round a luxuriant coat of grass. Captain Sutter has a large farm, or rancho, on this stream, a short distance above its mouth, where he is raising several hundred head of hogs, and a great number of horses, mules and cattle. His hogs live and fatten chiefly on the oak acorns, with which the uplands abound; whilst his other animals are fat at all seasons, without the trouble and expense of feeding them grain. . . .

Feather river is formed by three branches, of about equal size, all of which head far in the mountains. Their general course is southwest with very rapid currents, forming in some places abrupt and precipitous falls. Along these streams, in 1850 were probably developed more of the two extremes of fortune, that follow the mining operations in that country, and of the precariousness and uncertainty of the business, than were to be found in any other portion of California. Some men realized very large amounts in a short time, while others were entirely broken up and ruined in fortune, and were compelled, late in the fall, after having spent a whole summer's work, and a hard-earned fortune of the previous year, to take their pan, shovel, and pick, and seek for other diggings to get money to carry them home. These were generally the result of damming operations.

Gold was discovered in 1849, deposited in the beds of these streams. . . . Incredible hardships were encountered in getting there. All the provisions were packed on mules, and sometimes a great distance through snow. But these were cheerfully endured, and at the commencement of the dry season, in the spring, hundreds were found in their tents, who had been there many weeks, anxiously awaiting the close of the wet season, to commence operations. Some of these were fortunate enough to sellout in the spring to others, for big prices, who came too late to get a claim in any other way than by purchase. But how fickle is the Goddess of Fortune. A few realized their most sanguine expectations whilst much the largest number, (perhaps four-fifths) after having spent an immense amount of labor and money in building, dams, and digging ditches never realized a cent in return. Why the gold should have been so choice in the

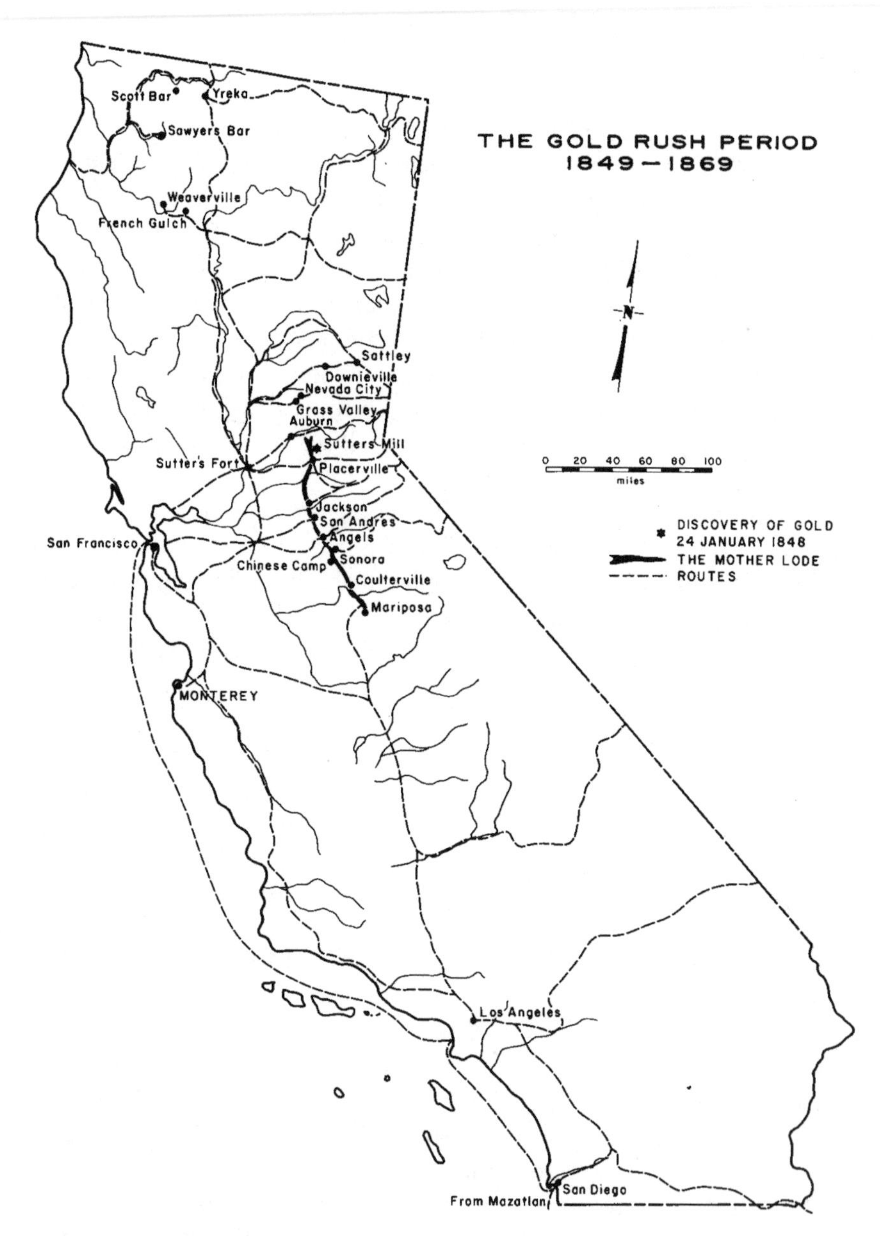

MAP 7. THE GOLD RUSH PERIOD, 1849–1869.
Beginning in 1848, miners converged in the Mother Lode country, especially along the Feather, Yuba, and Tuolumne Rivers, to seek their fortunes. Mining soon progressed from simple placer mining to capital-intensive hydraulic and quartz mining. From Warren A. Beck and Ynez D. Haase, Historical Atlas of California. *Norman: University of Oklahoma Press, 1974, map 50. Reprinted by permission.*

selection of its hiding places, in the beds, and along the banks of these streams, is a question hard to settle. But the strongest probability is that the channel has in many places changed, since the deposits were first made; and many of those places where it is now found, several feet below the surface, were then occupied as the beds of the rivers.

Gold has been discovered in nearly all the gulches and ravines, in the vicinity of these streams, and in some places in large quantities; but that near the surface is becoming pretty well exhausted. But new discoveries, a greater depth under ground, are daily being made, which will continue to be the case while Americans have existence in California. . . .

Mining Operations

Under this head I will first consider the general character and durability of the mines; secondly, the different processes of mining now in use, and thirdly, the amount that a man may expect to realize from the mining operations.

In the first place, then, I will lay it down as an axiom, that the mines of California are inexhaustible. The scope of country along the western slope of the Sierra Nevada, in which gold has been discovered, occupies in extent, from north to south, near five hundred miles, and varying in width from fifty to a hundred, extending in some places far into the mountains, and westward into the upland that borders on the valley of the Sacramento and San Joaquin rivers. As yet but a comparatively small portion of this great extent of country has been worked. In many places rich discoveries have of late been made far below the surface, and no good reason can be given why thousands of other places do not contain deposits of gold similarly situated. But seeking for it at the depth of from twenty-five to one hundred feet, is the work of chance and requires capital. But such is the character of the American people, that they will never stop their researches until every foot of earth in this vast El Dorado has been prospected to the bed rock, and two hundred years from this time will find the Yankees digging for gold on the western slope of the Sierra Nevada. . . . The appearance of the gold, both that which is found in fine particles and in larger lumps, proves conclusively that it has once been melted and thrown in scattered fragments over the earth by volcanic eruptions, where it is now discovered, buried under ground which the washing rains of perhaps a thousand winters have heaped upon it, and which gathered as it moved onward to its resting place, other particles of the precious metal which became generally diffused through the whole earth, by the action of the water. The earth on the mountains, on the hill sides, and in most of the low grounds, is of a very loose, washy nature, and has consequently been, through all time, continually changing its location. . . .

The process of digging the dirt is the same throughout the mines, and with which every man who is acquainted with the use of the shovel, pick and crowbar, is familiar. But the process of washing the dirt varies to suit the location of the mines. The common rocker being used where water is scarce, while along the streams where it is more plenty the long tom is used, and in many places the plan of sluice washing is adopted, which is a great improvement on either of the other modes. And where water is plenty, men can make good wages washing gold in this way, where, by any other process, it would not pay their board. It is a very easy, cheap and simple mode of washing. For this purpose a ditch is dug, some two or three feet wide and about the same depth, with a sufficient fall to give the water a rapid current, and may be of any length. The water is then let in at the upper end, when the men commence throwing in the dirt, which is carried off by the action of the water, leaving the gold deposited in the bottom. This is continued for several days, when the water is stopped, and the dirt and sand are scraped from the ditch, which is washed through a common rocker, and the gold goes through the usual process of panning, drying, etc.

The long tom is simply a trough or box, ten or fifteen feet long, about twenty inches wide and six inches deep, with a riddle at one end, while the other is so fixed as to receive a current of water which passes rapidly through the tom, into which the dirt is thrown, while two or three men stand by with shovels, and keep the dirt in motion until it passes through the riddle at the lower end, under which a box, about five feet long, with ripples or cleats across the bottom is placed to receive the gold, which, being much heavier than the dirt, sinks to the bottom, while most of the dirt and sand pass off over the ripples. The gold, together with the dirt and sand, with which it is deposited, is taken from the box and go through the usual processes attending the common rocker.

The common rocker is the means that was used in California, when the mines were first discovered, for separating the gold from the dirt. It is now made of plank, and requires the skill of a mechanic to manufacture. It is usually about four feet long and about two feet broad at top, with a ripple about an inch and a half across the bottom at the middle; and an opening about an inch above the bottom at the lower end, for the water and dirt to pass through; a riddle, about two feet square and six inches deep, is placed on the top at the upper end, into which the dirt is thrown, while a man sits by, with one hand keeps the rocker in motion and with the other keeps constantly pouring water into the riddle with a large dip-per, made for the purpose. The gold and fine sand are deposited in the bottom above the ripple, from which it is taken in a large tin pan, holding about six quarts, in which the sand is separated from the gold, by the simple process of dipping water and pouring from the pan, the sand following leaving the gold in the bottom.

The quartz is ground to powder, by steam mills erected for this purpose, after which the gold is extracted by quicksilver. . . .

Much of the dissatisfaction that prevailed in California, in the latter part of 1850, arose from the fact that there were too many people in the country; it was not prepared for such an immense emigration. The mines were crowded to overflowing, while much the largest number of those engaged in them, found the yield of gold, fall far short of their previous anticipations. These, added to the overdone condition of the trading and business operations generally, rendered the country, for a few weeks previous to the setting in of the rainy season, a perfect scene of confusion. Consequently, thousands started for home, as soon as they got money enough to take them there. And hundreds came who had to work their passage on the vessels that brought them. And they will continue leaving until the number left will be no greater than the country will require to work the mines, and cultivate the soil, and to carry on commerce and trade with a properly regulated competition. California will then be a desirable country to settle in. . . .

A Federal Agent Assesses Mining's Impact on the Indians, 1853

Diamond Springs, El Dorado County, December 31st 1853

The Indians in this portion of the State are wretchedly poor, having no horses, cattle or other property. They formerly subsisted on game, fish, acorns, etc., but it is now impossible for them to make a living by hunting or fishing, for nearly all the game has been driven from the mining

From "A letter from E. A. Stevenson, Special Indian Agent to Hon. Thomas J. Henley, superintendent of Indian Affairs, San Francisco," in Robert Heizer, ed., *The Destruction of the California Indians* (Lincoln: University of Nebraska Press, 1993), pp. 15–16, by permission.

region or has been killed by the thousands of our people who now occupy the once quiet home of these children of the forest. The rivers or tributaries of the Sacramento formerly were clear as crystal and abounded with the finest salmon and other fish. I saw them at Salmon Falls on the American river in the year 1851, and also the Indians taking barrels of these beautiful fish and drying them for winter. But the miners have turned the streams from their beds and conveyed the water to the dry diggings and after being used until it is so thick with mud that it will scarcely run it returns to its natural channel and with it the soil from a thousand hills, which has driven almost every kind of fish to seek new places of resort where they can enjoy a purer and more natural element. And to prove the old adage that misfortunes never come singly, the oaks have for the last three years refused to furnish the acorn, which formed one of the chief articles of Indian food.

They have often told me that the white man had killed all their game, had driven the fish from the rivers, had cut down and destroyed the trees and that what were now standing were worthless for they bore no acorns. In their superstitious imaginations they believe that the white man's presence among them has caused the trees (that formerly bore plentifully) to now be worthless and barren. In concluding this brief report I deem it my duty to recommend to your favorable consideration the early establishment of a suitable reservation and the removal of these Indians thereto, where they can receive medical aid and assistance which at the present time they so much require.

All of which is very respectfully submitted.

E. A. Stevenson, Spec. Indian Agent

To Hon. Thos. J. Henley, Supt. of Indian Affairs, San Francisco, Cal.

A Newspaper Reporter Describes North Bloomfield Mine, 1879

We stand on the brink of the mine and try to fix the salient points in thought and memory before we descend into the great amphitheatre, vaster in its circle than the stony base of the Coliseum. Around us are naked rocks and well-scraped furrows, piles of pine wood blocks for use

From "A Great Gravel Mine," *The Daily Transcript*, Nevada City, CA, July 30, 1879.

in the flumes, rusting joints of condemned water pipe, and shops where soot-covered men are riveting joints of new pipe, sharpening drills at glowing forges, or in a thousand ways giving examples of the uses of iron. . . .

As we turn to descend, a measured succession of sounds begins. Far down, under the highest cliff, on the slope bedrock, and half hid in shadow, are a multitude of men. The water has done its work here, and washed out all the loose earth and smaller rocks. Now the next thing is to get rid of the large boulders, often weighing tons. They must be blasted into fragments so small that when the water is turned on here again they will be swept down and out through the tunnel. They need not be very small for that. A boulder of six or eight hundred pounds weight goes out like a flash. So here are thirty or forty men, busy with drills, in a great hammering company. It is, at this instant, wild music. After from fifty to a hundred holes have been drilled, and loaded with giant powder and properly prepared with fuses of exactly the same length, the men will take irons of about three feet long, made red hot at one end, and run from boulder to boulder touching the ends of fuse. This being done, they will scud hastily into a grim looking "block house," a couple of hundred yards distant, built of old flume blocks and roofed with logs. Then, after a brief space, there will be wild artillery, and much smoke. . . .

There is a real pleasure, very distinct, but hard to describe, about this gigantic force. This is the water which left the Bowman reservoir a few hours ago, and has been worried and tumbled and beaten into foam until one might easily believe that it comes out with not merely the force of so much gravity, but also with a wicked, vicious, unutterable indignation. The black pipe, three feet in diameter, leads down the cliff, and across the mine. It becomes smaller, and ends in a jointed, elbow-like pipe, with a movable nozzle. By laying the weight of a hand on the lever, this rim-like nozzle enters the edge of the stream, and the weight of the water turns the machine to any angle desired. . . . It is not hard work to manage one, but it requires much experience and judgment to know how to use the stream to best advantage, and with greatest safety. Large boulders and lumps of pipe-clay are slowly washed down to the bed-rock for the blasters to handle, but rocks two feet in diameter fly like chaff when struck by the stream. The actual work of tearing down the cliff is hard to see, for there is a cloud of red foam hanging over the spot. You hear little rattling and slipping noises through the incessant roar, and a stream which seems ten times greater than could come out of the pipe flows down the dripping pile, and so into the rock-channels which lead to the tunnel.

When a portion of bed-rock has been cleared off, there is a general "clean-up," coming every two or three weeks. The blocks in the flume are taken up, the undercurrents are cleaned, the bed-rock is scraped, and those few flakes for which so much earth was dissolved, are obtained. . . . The gold, even when lying on the bedrock, is not so easily found. It slips into tiny crevices; it sticks to lumps of yellow clay, which are supposed to carry off a good deal of gold—and it is not so much after all. People who think that the whole bottom would be shining with flakes, sparkles and nuggets of gold, might easily spend half an hour just before a "clean-up" without finding a bit, unless they knew just where to look for it.

But the most interesting portion of the mine is at the outcome of the main tunnel, which extends under Humbug Creek. . . . The stream of water is so powerful that no man could stand against it a moment. The water after leaving the tunnel is carried half a mile or so in a flume, so as to allow a chance for undercurrents to collect more of the gold . . . and then the thick muddy stream is allowed to find its own way down without hindrance.

PIPING, OR HYDRAULICING.
NORTH BLOOMFIELD HYDRAULIC MINE, NEVADA COUNTY, CAL.
Photographed by C. E. Watkins, San Francisco.

FIGURE 3. HYDRAULIC MINING.
The North Bloomfield Hydraulic Mine, Nevada County, California, depicted in this famous Carleton E. Watkins photograph, became a source of controversy between miners and farmers in the 1870s. Hydraulic mining, which funneled water through a hose to a nozzle created enough pressure to bring down entire mountainsides, causing debris flows and flooding in towns below. Courtesy, The Bancroft Library, University of California, Berkeley.

Judge Sawyer Halts Hydraulic Mining, 1884

Sawyer, J. This is a bill in equity to restrain the defendants, being several mining companies, engaged in hydraulic mining on the western slope of the Sierra Nevada mountains, from discharging their mining debris into the affluents of the Yuba River, and into the river itself, whence it is carried down by the current into Feather and Sacramento Rivers, filling up their channels and injuring their navigation; and sometimes by overflowing and covering the neighboring lands with debris, injuring, and threatening to injure and destroy, the lands and property of the complainant, and of other property owners, situate on and adjacent to the banks of these water-courses. . . .

Hydraulic mining, as used in this opinion, is the process by which a bank of gold-bearing earth and rock is excavated by a jet of water, discharged through the converging nozzle of a pipe, under great pressure, the earth and debris being carried away by the same water, through sluices, and discharged on lower levels into the natural streams and water-courses below. Where the gravel or other material of the bank is cemented, or where the bank is composed of masses of pipe-clay, it is shattered by blasting with powder, sometimes from 15 to 20 tons of powder being used at one blast to break up a bank. . . . For example, an eight-inch nozzle, at the North Bloomfield mine, discharges 185,000 cubic feet of water in an hour, with a velocity of 150 feet per second. The excavating power of such a body of water, discharged with such velocity, is enormous; and, unless the gravel is very heavy or firmly cemented, it is much in excess of its transporting power. At some of the mines, as at the North Bloomfield, several of these Monitors are worked, much of the time, night and day, the several levels upon which they are at work being brilliantly illuminated by electric lights, the electricity being generated by water power. A night scene of the kind, at the North Bloomfield mine, is in the highest degree weird and startling, and it cannot fail to strike strangers with wonder and admiration. . . .

The Yuba River is a tributary of Feather River, entering it at Marysville, 30 miles above the mouth of the Feather, where the latter

From "Woodruff versus North Bloomfield Gravel and Mining Co.: The Sawyer Decision of 1884," *The Federal Reporter*, 18, no. 14 (1884) 753–818, pp. 756–57, 759–60, 763, 764, 765, 767, 770, 782–83, 784, 786, 792–93, 797, 800, 802, 808.

joins the Sacramento. It is the fourth river in size in the Sacramento valley, and drains about 1,330 square miles of the western slope of the Sierra Nevada mountains, comprising portions of Sierra, Nevada, and Yuba counties, its extreme breadth being about 36 miles, and its extreme length about 60 miles, excluding the 12 miles of its lower course from the foot-hills to its junction with Feather River at Marysville. . . .

The portion of the valley here referred to as covered with sand is that portion of the borders of the Yuba River extending across the Sacramento valley from the foot-hills to its junction with Feather River at Marysville, a distance of about 12 miles. Formerly, before hydraulic mining operations commenced, the Yuba River ran through this part of its course in a deep channel, with gravely bottom from 300 to 400 feet wide, on an average, with steep banks from 15 to 209 feet high, at low water, on either side. From the top of the banks, on each side, extended a strip of bottom lands of rich, black, alluvial soil, on an average a mile and a half wide, upon which were situated some of the finest farms, orchards, and vineyards in the state. Beyond this first bottom was a second bottom, which extended some distance to the ridge of higher lands, the whole constituting a basin between the higher lands on either side of from a mile and a half to three miles wide. Not only has the channel of the river through these bottoms been filled up to a depth of 25 feet and upwards, but this entire strip of bottom land has been buried with sand and debris many feet deep, from ridge to ridge of high land, and utterly ruined for farming and other purposes to which it was before devoted, and it has consequently been abandoned for such uses. . . .

The waters of the Yuba are so charged with debris that they are wholly unfit for watering stock, or for any of the uses, domestic or otherwise, to which water is usually applied, without being first taken out of the stream and allowed to stand in some undisturbed place and settle. As is comes down to Marysville it is so heavily charged with sand as to render it unfit even for surface irrigation. . . .

The North Bloomfield Mining Company, defendant, has constructed a dam to impound its debris, 50 feet high, near the junction of Humbug canyon with the south Yuba. The dam, not having been carried higher as it filled up, is now full, and the debris that has passed over the dam has filled the canyon and the south Yuba below the dam to a level with the debris above, so that now the debris passes along down the canyon over the dam without obstruction, as though no dam at all existed at that point. . . .

About 1868 the people of Marysville found it necessary to build levees around the city and along the north bank of Yuba River to protect it from the rapid encroachment of the debris coming down the Yuba; and levees were built. It has been found necessary to increase these levees in height and thickness from year to year ever since. In 1875 the levee on the north side of the Yuba broke, some three or four miles above the city, and the city and other lands were not only flooded, but a large amount of debris was deposited. This was the first time Marysville was over flooded. . . . So, in 1881, with much less water than at the great flood, it rose to a higher point at Marysville than ever before. . . .

The defendants have attempted to show that much of the danger from overflows results from the acts of the people themselves, in consequence of the improper system of levying adopted, and the cutting off by such means of some outlets of water, available at high water. There is, as might be expected, some conflict in the testimony of experts and others on these points; but it is probable that they have not in all instances adopted the wisest plan possible in their efforts to protect life and property. These works are always erected on the judgment of engineers, or other men presumed to be competent, and rarely without some difference of opinion, and it is scarcely possible that any plan wholly unobjectionable to all could be adopted. . . .

Defendants allege that both congress and the legislature of California have authorized the use of the navigable waters of the Sacramento and Feather Rivers for the flow and deposit of mining debris; and having so authorized their use, all the acts of defendants complained of are lawful, and the results of those acts, therefore, cannot be a nuisance, public or otherwise. . . .

The conditions . . . imposed upon California by the act of congress admitting her into the Union cannot be lawfully violated by obstructing, much less destroying, the navigation of her rivers and bays for purposes having no relation to facilitating navigation or commerce. The power of congress to regulate commerce between the states would also, doubtless, enable it, by proper legislation, independent of these conditions imposed by the act of admission, to prevent the state from destroying or obstructing, or authorizing the destruction or obstruction of, the capacity for navigation of her navigable waters. . . .

[Edwards] Woodruff's interests involved are by no means insignificant, no matter how much may have been said to belittle them. His block of stores, built on one of the most eligible business locations in Marysville, at a cost of at least somewhere between $40,000 and $60,000 his nearly 1,000 acres of farming land—among the best in the state—in Sutter county, called the Hock Farm, and his Eliza tract of over 700 acres on the opposite side of the river, in Yuba county and upon which a little settlement, embracing business houses and a public regular steamboat-landing, once existed, of which 125 acres in the aggregate on the two tracts are conceded to have been already destroyed certainly constitute an estate of no inconsiderable value. . . .

The brief flood occasioned by the breaking of the English dam, in June last, afforded a striking illustration of what is liable hereafter to occur. This enormous deposit of debris in the Yuba, and near Marysville, and in the streams in the mountains above, is a continuing, ever-present, and, so long as hydraulic mining is carried on as now pursued it will ever continue to be, an alarming and ever-growing menace, a constantly augmenting nuisance, threatening further injuries to the property of complainant, as well as the lives and property of numerous other citizens similarly situated. Against the continuous and further augmentation of this nuisance the complainant must certainly be entitled to legal protection. . . .

The supreme court of California has never recognized the validity of any custom to mine in such a manner as to destroy or injure the property of others, even in the district or diggings where the local customs and usages of miners are sanctioned by the statutes. But the California reports are full of cases where the principle has been enforced in the mines that every one must so use his own property as not to injure another. . . .

After an examination of the great questions involved, as careful and thorough as we are capable of giving them, with a painfully anxious appreciation of the responsibilities resting upon us, and of the disastrous consequences to the defendants, we can come to no other conclusion than that complainant is entitled to a perpetual injunction. But as it is possible that some mode may be devised in the future for obviating the injuries, either one of those suggested or some other, and successfully carried out, so as to be both safe and effective, a clause will be inserted in the decree giving leave on any future occasion, when some such plan has been successfully executed, to apply to the court for a modification or suspension of the injunction.

Let a decree be entered accordingly.

Joaquin Miller on Environmental Deterioration in the Gold Country, 1890

As lone as God, and white as a winter moon, Mount Shasta starts up sudden and solitary from the heart of the great black forests of Northern California. You would hardly call Mount Shasta a part of the Sierras: you would say rather that it is the great white tower of some ancient and eternal wall, with nearly all the white walls overthrown. . . .

Ascend this mountain, stand against the snow above the upper belt of pines, and take a glance below. Toward the sea nothing but the black and unbroken forest. Mountains, it is true, dip and divide and break the monotony as the waves break up the sea: yet it is still the sea, still the unbroken forest, black and magnificent. To the south the landscape sinks and declines gradually, but still maintains its column of dark-plumed grenadiers, till the Sacramento Valley is reached, nearly a hundred miles away. Silver rivers run here, the sweetest in the world. They wind and wind among the rocks and mossy roots, with California lilies, and the yew with scarlet berries dipping in the water, and trout idling in the eddies and cool places by the basketful. On the east, the forest still keeps up unbroken rank till the Pitt River Valley is reached; and even there it surrounds the valley, and locks it up tight in its black embrace. To the north, it is true, Shasta Valley makes quite a dimple in the sable sea, and men plow there, and Mexicans drive mules or herd their mustang ponies on the open plain. But the valley is limited, surrounded by the forest, confined and imprisoned.

Look intently down among the black and rolling hills, forty miles away to the west, and here and there you will see a haze of cloud or smoke hung up above the trees; or, driven by the wind that is coming from the sea, it may drag and creep along as if tangled in the tops.

These are mining camps. Men are there, down in these dreadful cañons, out of sight of the sun, swallowed up, buried in the impenetrable gloom of the forest, toiling for gold. Each one of these camps is a world of itself. History, romance, tragedy, poetry, in every one of them. They are connected together, and reach the outer world only by a narrow little

From Joaquin Miller, *My Life amongst the Indians* (Chicago: Morril, Higgins & Co., 1892 [1890]), pp. 18–22, 54–55.

pack trail, stretching through the timber, stringing round the mountains, barely wide enough to admit of footmen and little Mexican mules, with their apparajos, to pass in single tile.

But now the natives of these forests. I lived with them for years. You do not see the smoke of their wigwams through the trees. They do not smite the mountain rocks for gold, nor fell the pines, nor roil up the waters and ruin them for the fishermen. All this magnificent forest is their estate. The Great Spirit made this mountain first of all, and gave it to them, they say, and they have possessed it ever since. They preserve the forest, keep out the fires, for it is the park for their deer.

This narrative, while the thread of it is necessarily spun around a few years of my early life, is not of myself, but of this race of people that has lived centuries of history and never yet had a historian; that has suffered nearly four hundred years of wrong, and never yet had an advocate.

Yet I must write of myself, because I was among these people of whom I write, though often in the background, giving place to the inner and actual lives of a silent and mysterious people, a race of prophets, poets without the gift of expression—a race that has been often, almost always, mistreated, and never understood—a race that is moving noiselessly from the face of the earth; dreamers that sometimes waken from their mysteriousness and simplicity, and then, blood, brutality, and all the ferocity that marks a man of maddened passions, women without mercy, men without reason, brand them with the appropriate name of savages.

I have a word to say for the Indian. I saw him as he was, not as he is. In one little spot of our land, I saw him as he was centuries ago in every part of it perhaps, a Druid and a dreamer—the mildest and tamest of beings. I saw him as no man can see him now. I saw him as no man ever saw him who had the desire and patience to observe, the sympathy to understand, and the intelligence to communicate his observations to those who would really like to understand him. He is truly "the gentle savage"; the worst and the best of men, the tamest and the fiercest of beings. . . .

A singular combination of circumstances laid his life bare to me. I was a child, and he was a child. He permitted me to enter his heart. . . .

All this city [Sacramento] had been built, all this country opened up, in less than two years. Twenty months before, only the Indian inhabited here; he was lord absolute of the land. But gold had been found on this spot by a party of roving mountaineers; the news had gone abroad, and people poured in and had taken possession in a day, without question and without ceremony.

And the Indians? They were pushed aside. At first they were glad to make the strangers welcome; but, when they saw where it would all lead, they grew sullen and concerned. . . .

I hurried on a mile or so to the foot-hills, and stood in the heart of the placer mines. Now the smoke from the low chimneys of the log cabins began to rise and curl through the cool, clear air on every hand, and the miners to come out at the low doors; great hairy, bearded, six-foot giants, hatless, and half-dressed.

They stretched themselves in the sweet, frosty air, shouted to each other in a sort of savage banter, washed their hands and faces in the gold-pans that stood by the door, and then entered their cabins again, to partake of the eternal beans and bacon and coffee, and coffee and bacon and beans.

The whole face of the earth was perforated with holes; shafts sunk and being sunk by these men in search of gold, down to the bed-rock. Windlasses stretched across these shafts, where great buckets swung, in which men hoisted the earth to the light of the sun by sheer force of muscle.

The sun came softly down, and shone brightly on the hillside where I stood. I lifted my hands to Shasta, above the butte and town, for he looked like an old acquaintance, and again was glad.

ESSAYS

Mining on Trial

Robert Kelley

During the last quarter of the nineteenth century, farmers pouring into the valleys of California created an agrarian empire and set in motion years of controversy. The older economic interests were dismayed at this rapid rise of farm power, and fought bitterly to protect their position. San Francisco, railroads, the cattlemen, and the dwindling ranks of the miners combined to resist agrarian encroachments. As a state senator from the San Joaquin Valley hotly exclaimed in the legislature of 1878: "It seems that when anything is introduced in the interest of the farmers of the State, it is jeered down and sneered at by some of the Senators from San Francisco." . . .

Of these clashes of economic interest, none was more remarkable than the long controversy which raged in the Sacramento Valley over the fate of hydraulic gold mining in the northern Sierra Nevada. By the seventies this industry had become a multimillion-dollar giant. Scores of mines, situated on vast deposits of gold-bearing gravels, poured forth a steady stream of wealth. Roaring jets of water, fed by hundreds of miles of ditches and flumes, hollowed out great pits in the mountains and filled nearby ravines and canyons with mud, sand, and gravel. During high water, these tailings washed downstream and spread out on the flat Sacramento Valley floor, burying vast areas under mining debris.

Thousands of acres of some of the richest farm lands in the state became barren wastelands. Orchards, grain fields, and houses were buried out of sight. As their beds filled with silt, the Sacramento and its tribu-

From "The Mining Debris Controversy in the Sacramento Valley," *Pacific Historical Review* 25 (Nov. 1956): 331–346, pp. 331–34, 336, 339, 341, 342, 344–45. Reprinted by permission.

taries became sources of terror and widespread destruction. By the seventies, the drainage system of the lower Sacramento Valley was so overloaded with mining debris that it was unable to carry off even normal rainfall. Increasingly devastating floods became almost annual occurrences. Marysville, Sacramento, and the other river towns had to build miles of expensive levees, and tax rates mounted higher by the year. Property values dropped, river boats had to cease operations, the flow from city hydrants became a turgid gruel of mud and water, and townsmen joined farmers in a desperate counteroffensive. . . .

With each passing year, the floods grew more destructive. In 1875, heavy rains sent brown waters swirling over the flatlands again. The Yuba River overran Marysville's levees and the town filled like a bowl. Desperate valley residents besieged the legislature of 1876 with pleas that an investigation of their problem be launched. But the miners attacked the proposal, San Francisco and most of the rest of the state were indifferent to it, and the farmers returned to their threatened valleys with only the solace that a memorial had been sent off to Congress by the Legislature. . . .

Thus rebuffed, the farmers turned again to the courts. A group of men whose farms along the Bear River were already partly buried by mining debris asked the Yuba District Court to issue an injunction prohibiting all mines upstream from using the river as a dump. . . . Meanwhile, the San Francisco entrepreneurs who had created and directed most of the hydraulic mines took alarm. In September of 1876 they held a meeting in San Francisco, and from these deliberations emerged the Hydraulic Miners Association. Dominated by the men of great wealth who had the most to lose, in the following years it led the campaign against the farmers. . . . In late 1878 farmers from the valleys of the Bear, the Yuba, and the Feather met in Marysville and created the Anti-Debris Association of the Sacramento Valley. The battle lines were drawn, and through the next fifteen years these opposing groups were almost continually at war with one another.

In the initial skirmish, it looked as though the farmers had won quick victory. Early in 1879 the District Court of Sutter County ruled that miners had no right to discharge their debris into the rivers. The farm and river towns broke into happy rejoicing, while the miners grimly secured a stay of proceedings and appealed the decision to the State supreme Court. . . .

In the latter part of that year, 1879, the California Supreme Court . . . rul[ed] for the miners on the ground that it was not equitable to join all of the mines in one watershed together in a single suit just so that all possible sources of damage would be encompassed. . . . When this news flashed over the telegraph, the mountains went wild with joy. Bells pealed in mountain towns, cannon were fired, exploding fireworks rent the air, and all available steam whistles screamed. "It was one of the biggest jubilees Nevada City has ever had," exulted Calkins of the Transcript. "Men were going up and down some of the streets adding to the racket by hurrahing like mad." The flatlands, sunk in gloom, contemplated the ruin of their hopes. Individual suits for damages had been shown to be impossible in 1873, and thousands of dollars poured into the fight to get an injunction in equity against all miners in a single watershed had also met only frustration.

But all was not lost. Many farmers . . . had felt that the problem was only one of building adequate public works; that some flood control system based upon levees and brush dams could be worked out to everyone's satisfaction, retaining the profitable relationship between the mountain towns and the valley farms. . . .

To solve [the] problem, whose implications for both miner and farmer were catastrophic, the State Engineer recommended that a comprehensive flood control system be built as rapidly as possible. Dams should be placed near the mouths of mountain canyons to catch and hold coarse debris, and levees should be built at strategic points to constrict the rivers and give them sufficient power to scour out their own beds. . . .

In the face of furious objections, the Drainage Act of 1880 was passed. It created a reclamation program almost identical to that recommended by the State Engineer. . . . In the following months of 1880, hundreds of men toiled along the banks of the Sacramento and its tributaries, building miles of levees and great brush dams stretching across the wide beds of the Bear and Yuba rivers. . . .

Meanwhile, a series of storms buffeted central California. January and February of 1881 saw floods even more extensive than their disastrous predecessors. Suffering and misery walked the land; forbearance turned to rage. Farm families huddled on wind-swept levees watched in bitter frustration as their hard-won flood control system proved powerless to hold back the swollen rivers. As soon as the brown waters receded, farmers and townsmen gathered in Marysville at the call of the Anti-Debris Association to shout approval of heated speeches demanding the prohibition of hydraulic mining. A petition to the legislature to this effect was overwhelmingly adopted . . . [and a] mass rally in Sacramento lent new force to Marysville's request, but even this pressure was insufficient. The legislature refused even to consider the proposal. . . . Since most of the largest mines drained into the Yuba hundreds of miners were thrown out of work.

Meanwhile, farmers were faced with an infuriating situation. After an initial period of drastic unemployment, miners along the upper Yuba began ignoring the Marysville injunction. Various stratagems were developed to evade the service of injunctions, and the mines were soon busier than ever before. The mountains echoed to the roar of scores of hydraulic monitors, sluice boxes ran full day and night, mining towns bustled with prosperity, and rivers turned brown and greasy with mud and sand. . . . Where miners would scorn an injunction from a court in Marysville, they would hardly be willing to do the same to an injunction secured in the name of the people of California.

In June of 1882 . . . Judge Jackson Temple ruled that miners had no right to discharge what he termed "coarse" debris into the rivers of the state. . . .

The farm counties were delighted. The mayor of Marysville and the president of the local anti-debris association called upon all mines in the Yuba watershed to cease operations. If they did not, suits seeking injunctions . . . would be instituted, thus forcing them to shut down until costly dams were built. . . .

By this time, farmers and miners were once again locked in judicial combat. Shortly after the Gold Run decision, the anti-debris groups realized that they had not gotten what they wanted. Judge Temple's decision simply embodied the principle of the unlamented Drainage Act. The mines would continue to operate, and scores of dams in the mountains would store debris for future floods. Furthermore, only one mine at a time could be enjoined, which was prohibitively expensive. Consequently, the anti-debris movement began seeking a more comprehensive injunction. Edwards Woodruff, a citizen of New York and a property owner in Marysville, entered the Ninth United States Circuit Court in San Francisco in September of 1882 and filed suit against the North Bloomfield and all other mines in the Yuba watershed, asking for perpetual injunction. Thus began a suit which was to last for the better part of a year and a half. Not until January of 1884 did Judge Lorenzo Sawyer—who had been a miner in the Nevada City area during the gold rush period—render his decision. Hundreds of witnesses were called, and thousands of pages of testimony were taken. Judge Sawyer made several inspection trips along the rivers and into the mountains, trailed by anxious representatives from both sides. Meanwhile, the hydraulic industry continued to operate at peak capacity.

By late 1883, it was apparent that his decision was soon to be reached. Each Monday, the day on which Sawyer customarily rendered his decisions, men gathered at telegraph offices throughout the mountains and plains, waiting for news. Finally, in early January of 1884, word came that he would give his decision on the following day. Marysville prepared hopefully for a celebration. . . .

The following day, Marysville had good reason to celebrate. Judge Sawyer handed down an injunction which denied to the mines any right whatsoever to discharge any type of debris into the river, whether it be "tailings, boulders, cobble stones, gravel, sand, clay, debris or refuse matter. . . .

The flatlands broke into joyous celebration; even Bakersfield and Merced sent congratulations to their northern colleagues. In the mountains, there was profound gloom. As a Dutch Flat resident sadly remarked, "Most of us will pack our grip sacks." In San Francisco, a new and significant point of view was expressed. The passing of the mines was seen to be perhaps not at all a bad thing. As the *San Francisco Bulletin* observed, though mining had been dealt a crushing blow, it was, after all, a transitory industry, while "the wheat field produces year after year, and wine and oil and wool are perennial."

The hydraulic industry did not die without a struggle. The big corporations took their losses, which ran to hundreds of thousands of dollars and ceased operations. But their facilities were available for any man to use. The succeeding years were marked by a kind of guerrilla warfare where injunctions were bullets, and stealth became a way of life in the mountains. Farmers found that continued vigilance was the price of liberty from hydraulic mining; they had to carry on offensive operations for eight more years through their anti-debris associations. Spies were hired, patrols toured the mountains, and contempt proceedings were resorted to when men could be caught. In the spring of 1885 the distraught Yuba Board of Supervisors even called unsuccessfully upon the governor to order out the militia to enforce injunctions. . . .

Appeals to the federal government finally brought that agency into the affair. In June of 1886, the Attorney General of the United States, acting upon instructions from Congress, announced that he intended to seek injunctions against all hydraulic mines in order to protect navigable rivers. Federal marshals were sent up all the streams in the Sacramento Valley. But even these found their task not an easy one; the first marshals to visit the upper regions of the Feather River were frightened off in the middle of the night when they overheard a group of men outside their hotel window actively debating the question of whether or not to tar and feather them. But by 1892 the Marysville Anti-Debris Association was able to decide that its task was finished, and disband what was one of the most successful organizations in the history of California.

In the following year, a major attempt was made to revive the industry. Following recommendations made by the Army Corps of Engineers, Congress in 1893 passed the Caminetti Act. It provided that all mines could operate once again if they built debris dams which were approved by a commission created for the purpose. The result was a fleeting revival of the whole controversy, with overjoyed miners on the one side, and enraged farmers on the other creating a new Anti-Debris Association in Sacramento.

But the excitement was unwarranted. The industry did not have a chance to revive. Bad winters had wrecked the vast ditch and flume system in the Sierra, and no company could undertake the enormous expense of repairing this essential basis for operations. Furthermore, debris dams turned out to be prohibitively expensive and inefficient. This short revival of interest was hardly more than a dying gasp. By 1895, the hydraulic mining industry in the northern Sierra Nevada was no longer a major element in the state's economy. Bustling mining towns died, schools closed, and much of the mountain country lapsed into drowsy somnolence, a region of ghost towns and quiet forest. The long struggle was over, and the farmer was dominant in the Sacramento Valley.

Mining's Impact on the Land

Randall Rohe

The importance of mining to the development of the West . . . contains few studies that take an environmental approach. What effect did mining . . . have on the land? How did mining alter the physical landscape in terms of the destruction of natural vegetation, the alteration of original terrain and drainage conditions, and the formation of landforms, and how permanent were these changes? Were the effects produced by man the same as those produced by "natural agencies"? Did mining invalidate

From Randall Rohe, "Man and the Land: Mining's Impact in the Far West," *Arizona and the West*, 28 (Winter 1986): 299–388, reprinted by permission, and Randall Rohe, "Man as Geomorphic Agent: Hydraulic Mining in the American West," *Pacific Historian*, 27, no. 1 (1983): 5–16, reprinted by permission.

any geomorphic principles or did man merely change the magnitude of certain variables? Are there distinguishable differences between the landforms created by man and those by "natural agencies"? . . . To what degree did the land rebound from the effects of mining and what remains in the present landscape from this past era?

Placer and Lode Mining

At first, gold mining in the West consisted almost entirely of placer operations. The discovery of placer gold characteristically marked the first step in the development of a mining industry. Working placer deposits in turn often led to the discovery of vein or lode deposits. In fact, placer operations frequently financed the initial development of lode mining and provided the necessary capital to begin production.

Placer mining remained the chief source of gold until 1873, when lode mining surpassed it in importance. . . . Placer mining continued as an integral part of western mining for nearly a century, and the activities inherent in a century of placer and lode mining provide a basis for analysis of the impact of an intensive human activity on the land.

Destruction of Vegetation

Throughout much of the American West, the vegetation of many river valleys remained largely unchanged until the advent of mining. Placer mining typically was concentrated in these valleys and resulted in the alteration of relatively small areas. These alterations of vegetation, however, proved significant enough to elicit comment from many contemporary writers: "It (Nevada City) is beautifully situated on the hills bordering a small creek, and has once been surrounded by a forest of magnificent pine trees, which, however, had been made to become useful instead of ornamental, and nothing now [1851] remained to show that they had existed but the number of stumps all over the hillsides." . . .

As the scale of mining increased, a corresponding increase occurred in the area of vegetation altered. Many contemporary photographs indicate the almost wholesale destruction of vegetation that accompanied hydraulic mining. A description of the diggings near Nevada City, California, in 1854 compared the scene with that four years earlier and noted "the immensity of the work that has been done." . . .

All forms of mining, even the small-scale methods, entailed the removal of adjacent vegetation. This, of course, did not alone account for the great destruction of vegetation described by contemporary writers.

The need for building material probably played an even greater role. Typically, early photographs and sketches of mining towns indicate a scarcity of timber. Often, in fact, they have only a few scattered trees with no undergrowth and little or no dense growth nearby. Mining, itself, used great quantities of lumber for construction purposes. In fact, the demand for lumber generated by hydraulic mining in California reached such proportions that some mills almost exclusively produced blocks (riffles) for the bottoms of sluices.

Lode mining, too, had a tremendous impact on vegetation. . . . Besides lumber for mine props, railway ties, mine buildings and the like, lode mining consumed quantities of timber to fuel mining machinery, stamp mills, and smelters. Oak, juniper, piñon, and mountain mahogany were commonly utilized as fuel or manufactured into charcoal. In many parts of the West, the needs of mining soon exhausted the local supplies of fuelwood. On the Comstock of Nevada, for example, the supply of wood seldom met the demand. The pinyon and juniper of the nearby ravines and hills were removed in an ever-expanding circle. By 1864 several hundred American laborers were constantly cutting and hauling firewood from nearby woodlands. Chinese followed the woodcutters, pulling up the brush, stumps, and roots from the cutover hills. An estimated 120,000 cords of firewood were used in the district in 1886. The pinyon and juniper on the neighboring hills was rapidly exhausted, and woodcutters moved to the eastern slopes of the Sierra Nevada.

To satisfy the Comstock's demand for wood, gigantic drives of lumber and cordwood up to four miles or more long took place on the Carson River each spring. More than 150,000 cords of wood were floated down the Carson in a typical season. The use of the Carson and other rivers, however, did not prove wholly satisfactory and in the 1870s expensive V-flumes were built, some a dozen miles long, to move logs to the mills. More than seven hundred cords (500,000 ft) of mining timber were transported down the Carson and Tahoe Lumber Company's flume daily. It has been estimated that over a thirty-year period, the Comstock lode consumed eight hundred million feet of lumber—enough to build fifty thousand ranch-type houses, each with two baths and a double garage. . . .

Revegetation

The multiplicity of factors involved preclude any detailed, irreproachable statements on the revegetation of mined areas. Differences in geographic location, climate, age, composition, slope, and adjacent extant vegetation prevent the delineation of a sequential pattern of revegetation. Despite these differences, however, some generalizations are possible. Typically, revegetation most quickly occurs in areas mined by the traditional methods. Panning, ricking, and tomming altered the original soil to a limited depth and over only a small area. As a result, these methods produced the least disturbance in the original conditions. Today, vegetation covers, if not obscures, most areas worked in these traditional ways. Even in the arid Southwest sagebrush partially covers many of the tailing piles. Revegetation took longer in areas worked by sluicing, which not only resulted in greater changes in the original soil conditions but over a larger area. In some areas sluicing stripped the soil to bedrock. Today, many of the areas mined by sluicing in the last century still lack a complete vegetation cover. Most of these locales, however, contain notable evidence of revegetation.

Compared to the traditional methods, the revegetation of areas where hydraulicking constituted the chief form of mining took notably longer. Hydraulic mining removed soil to greater depths, commonly to bedrock, over large areas, and left steep slopes. Characteristically, revegetation occurred most rapidly in the bottom of the pits left by this method. In California, most hydraulic mining ended in the middle 1880s. By the turn of the century, vegetation in some of these pits already almost equaled the adjacent forest which remained untouched. Hydrophytes normally appeared quickly, too, around the margins of the pits common to many hydraulic mines. The bottom of some hydraulic mines, however, remain only slightly revegetated to the present. Revegetation occurs least quickly on the slopes of the hydraulic pits. Hydraulicking often produced nearly vertical slopes that greatly impede any revegetation. Subsequent erosion usually reduced these slopes, but some even to this day contain little or no vegetation. Others, however, show evidence of revegetation less than fifty years after mining ceased.

Of the placer methods, dredging perhaps most altered riverine vegetation. Dredging removed all vegetation, completely overturned the soil and left in its place large parallel rows of gravel with steep slopes. The additional factor of time (the recency of most dredging operations) has resulted in the almost complete absence of vegetation in some dredged areas. Characteristically, vegetation first reappears in the low areas between the parallel rows of gravel. Plants appear especially quickly around the margins of the ubiquitous dredge ponds. These ponds are in various stages of fill. Once filled, vegetation quickly covers them. In some cases the revegetation between the gravel ridges almost completely masks the tailing from ground level. In other cases, areas worked by dredging show little revegetation. [Frank] Scott noted the varying degrees of revegetation encountered in different dredge areas of California. Near Hornitos the dredge tailings were only slightly overgrown with grass; south of Jacksonville the tailings were overgrown with Digger pine, scrub oak, bushes of several kinds, and berry vines; and in the vicinity of Jenny Lind the tailings were heavily overgrown with cottonweeds, grass, and berry vines. In almost all areas worked by dredging, even some where mining stopped less than thirty years ago, the discerning eye sees evidence of revegetation.

Without doubt, placer mining affected the distribution of different plant species. In many areas of California originally covered by dense Ponderosa pine forests, mining removed almost all vegetation. Chaparral and Digger pine often spread into such areas. Even today some still support Chaparral or Digger pine. The latter originally occupied a narrow zone at the base of the Sierra foothills but spread upward rapidly as a result of the complete removal of timber during the mining period. In several areas in California mining increased the extent of Digger pine to lower elevations as well. Mining debris washed down from hydraulicking destroyed much of the original vegetation. Seeds of the Digger pine, however, brought downstream with the mining debris quickly established themselves.

Undoubtedly, the movement of great numbers of people occasioned by the gold rushes resulted in the introduction and spread of various plants. Little conclusive proof exists, however, of plant introduction, either intentional or accidental, that accompanied mining. Apparently, only the Chinese, to any degree, purposely introduced and spread a particular plant species. The Chinese, especially in California, planted "Trees of heaven" (*Ailanthus altissima*) widely. The present distribution of *Ailanthus* in California corresponds closely to the Chinese mining activities of the nineteenth century.

In many lode mining areas considerable recovery has occurred and "forest cover may now be much the same as it was before active mining and settlement." . . . Almost immediately following the end of mining, revegetation started. The initial vegetation, typically, consisted of the expected pioneer species for the locale. In most cases, the present vegetation either closely resembles the original vegetation described by contemporary observers or appears to be approaching it. In many areas, vegetation now covers mined areas sufficiently enough that few people realize the presence or extent of past mining operations. . . .

Geomorphic Effects

Hydraulic mining . . . enabled a few miners to accomplish in weeks what formerly required a hundred men months to do. Even early hydraulicking removed upwards of fifty to a hundred cubic yards of material daily. The resulting effects on the landscape left an indelible imprint on several contemporary observers:

> It is impossible to conceive of anything more desolate, more literally forbidding, than a region which has been subjected to this hydraulic mining treatment. . . . The whole vista is one of extreme desolation and ruin. Certainly by no other means does man more completely change the face of nature than by this method of hydraulic mining. Hills melt away and disappear under its influence, . . . whole valleys are filled with clean-washed boulders of quartz. The desolation which remains . . . is remediless and appalling.

Hydraulic mining left a noticeable impact wherever it took place. However, hydraulicking reached its apogee in California and there exerted its greatest impact. In some mining districts, wherever one looked, the effects of hydraulic mining stood out "in the form of a desolate waste of tailings, [and] mammoth excavations in the flanks of the pine-clad and round-topped hills." The hydraulic mines of California commonly consisted of "an open cut of huge dimensions." In the mid-1880s an observer described the North Bloomfield mine as a "barren amphitheater, so vast that it could contain a whole settlement and so deep that a high church steeple could hardly reach to the ledge." The North Bloomfield mine did not represent an isolated example. Dozens of similar mines operated in California during the height of hydraulic mining.

In some localities, hydraulicking cut back hillslopes for considerable distances. In the process it created long vertical cliffs. . . . Hydraulic mining so altered some river valleys that a reconstruction of their original appearance seems impossible, for at times, only small unworked "islands" of the original valley floor remain.

In some areas, hydraulicking removed the soil almost completely and exposed large areas of bedrock. Around Columbia and Springfield, California, especially, hydraulic mining produced such landscapes, so different from the original "as to be unrecognizable." . . . Here hydraulicking produced a landscape covered with many great masses of white limestone in bizarre shapes and "the earth, torn up everywhere, resembles a battlefield of the antediluvian giants and monsters."

In hydraulic mining, redeposition of part of the material excavated, especially the larger size, occurred nearby. These tailings formed deposits ranging in size from a few cubic yards to thousands of cubic yards. Often the tailings accumulated in fan-shaped heaps or dumps. Many times nearby ravines served as dumps and eventually became "choked with the large masses of tailings." In some areas of the West, especially in parts of California, bedrock tunnels served for disposal of tailings. These tunnels led from a shaft at the bottom of the hydraulic pits and discharged the tailings into adjacent ravines and valleys. The longest and greatest number of bedrock tunnels occurred in California. The longest of these, the North Bloomfield, consisted of a main tunnel eight thousand feet long with a branch nine hundred feet long.

Weathering and erosion, of course, modified the features produced by hydraulic mining. Many slopes left by hydraulicking are still unstable. A moderate building of talus, however, is common. It seems that these features are fairly permanent. Time coupled with weathering and erosion simply make them more a part of that natural landscape.

The material removed and redeposited by mining constitutes, perhaps, the most visible and lasting effect placering left on the landscape, but the effect of mining on waterways proved equally if not more important. All forms of placer mining required water and diversion of water for mining purposes occurred early in most mining districts. Construction of the first mining ditch in California occurred in 1850 in the Nevada City district. Two years later the state contained only six ditches with a total length of fifty-five miles. The establishment of hydraulic mining, however, created a gigantic system of canals, ditches, dams, and reservoirs that increased in size each year.

During 1855 more than 1,159 miles of mining ditches were built in California. In 1882 the total reached six thousand miles of main ditches, another thousand miles of subsidiary lines, and an unknown length of small distributors. None of the other Western states approached the development of water facilities for mining that took place in California but the pattern of development elsewhere roughly approximated that of California. . . .

Early mining ditches diverted either directly from streams or from natural lakes. As demands for water increased, mining companies frequently increased the capacity of these lakes by dams. Other times, these companies built dams across stream valleys and created wholly new lakes. The reservoirs created by these dams ranged in size from forty or fifty to as much at five hundred acres in size. In 1892 the storage capacity of mining reservoirs on just the Yuba, Bear, Feather, and American rivers aggregated over six billion cubic feet. . . .

Traditional mining methods introduced relatively small amounts of debris into the streams, but even these mining methods sufficiently affected the streams to evoke comment by numerous contemporary observers, such as: "The bed of the creek, which had once flowed past the town [Nevada City], was now [1851] choked up with heaps of 'trailings'—the washed dirt from which the gold has been extracted—the white colour of the dirt rendering it still more unsightly."

From available evidence the relatively small amounts introduced into streams by traditional methods probably little affected stream regimes. As late as 1853 some streams of the California goldfields were said to still run clear. Wherever hydraulic mining took place, however, on almost any scale, it noticeably affected the streams. In California hydraulic mining took place on a gigantic scale and the streams of California received huge quantities of debris. Indeed, these waterways came to exhibit in the extreme the effects of hydraulic mining. The Sacramento system received by far the greater amount of mining debris, as compared to that of the San Joaquin. A conservative estimate places the amount of debris dumped into tributaries of the Sacramento at 1.3 billion cubic yards.

The mining debris introduced into the streams of California changed the whole character of these rivers. It converted clear, high-banked streams into sluggish, turbid, erratic watercourses flowing on elevated beds between artificial banks. The water became yellow with mud, and bars formed all along the Feather and in the Sacramento for many miles. Before hydraulic mining, the beds of the upper rivers consisted largely of bedrock and larger boulders, with only a few reaches of coarse gravel. The normal load of these streams fell below their capacity. During the height of hydraulicking, however, their loads greatly exceeded their capacity and deposition resulted. The Bear River, for example, filled its bed to a depth of nearly eighty feet in the center and tall pine trees, formerly far above the stream, were gradually engulfed until by 1870 only the top branches were above the current. Between 1870 and 1873 mining debris raised the bed of parts of the lower Bear River ninety-seven feet. Like the Bear River, other streams that drained the California goldfields received great amounts of mining debris.

Graphs of low-water records of the Yuba River at Marysville and the Sacramento River at Sacramento for 1843–1913 reveal the general trend of the deposition of mining debris. Figures show that during these years the bed of the Yuba rose about .31 foot per year and the Sacramento .25

foot per year. The greatest deposition occurred near the edge of the foothills where the greatest reduction of slope and velocity occurred. Along the lower Yuba River from Smartsville to Marysville, the debris spread out on adjoining plains to form the Yuba Debris Basin. A fan-like deposit with a width of one to three miles covered twenty-five square miles and contained 360,000,000 to 600,000,000 cubic yards of mining debris.

The overloading of streams by mining debris caused formation of numerous sand bars, occasionally even islands, and streams "spread at will in many shifting channels." The width of some streams expanded to five and six times their original size. Occasionally mining debris so choked the riverbeds that the streams cut new channels, sometimes shifting as much as half a mile. At times tributaries even changed mainstreams.

Contemporary writers describe the Feather River in 1848–1849 as a clear stream bottomed by sand and gravel, with well-defined banks. Even at low water the river consisted of a succession of pools eight to fifteen feet deep. During the period of hydraulic mining the descriptions by contemporary observers characterized the stream as "very muddy, the channel . . . crooked and narrow." Instead of a few bad shoals during low water, continuous miles of shoal water characterized the stream. The once characteristic deep pools became the exception rather than the rule. . . .

When streambeds rose above the level of the surrounding land a levee system was required for flood protection. The levees confined the water but continued deposition of mining debris caused the beds to rise still higher. The streets of Marysville, once twenty to twenty-five feet above the bed of the Yuba, by 1879 were below it. Eventually the levees "were finally overtopped by the floods, and the bottom lands were submerged from rim to rim."

Floods have always been a normal periodical natural phenomenon of the California rivers. Mining debris, however, exaggerated flooding on the lower rivers. The periodicity, the destructiveness, and the area of inundation all increased. Mining debris so overloaded the lower rivers that they proved unable to carry off even normal rainfall. Increasingly devastating floods became almost annual occurrences. In the process, mining debris covered and destroyed thousands of acres of agricultural land and likewise reduced navigability on the lower rivers. Steamboat Slough, a subsidiary channel of the Sacramento River, furnishes an excellent illustration. Originally, Steamboat Slough was the deepest and most important channel for steamboat travel on the lower Sacramento.

By 1880, however, the accumulation of mining debris in the channel almost completely curtailed its navigability. . . .

Accumulations of mining debris even reached the tidal waters of San Francisco Bay and its dependencies. Mining debris noticeably reduced the depth and area of these tidal waters and caused the formation of shoals in Suisun Bay and larger deposits in the Straits of Carquinez. The filling of the bays reached such proportions, especially in Suisun Bay, that some expected the eventual shoaling of San Francisco Bay and destruction of its harbor. All these disastrous effects of dumping mining debris into streams provoked ever greater opposition to the practice, and in 1884 a series of injunctions finally stopped this method of tailing disposal.

Almost immediately the streams began eroding the deposits of debris and moving "toward the normal profile" and a "restoration of the river conditions of the early fifties." They trenched the deposits of upland creeks and largely removed those of the river canyons. The streams removed deposits lodged in river canyons traversed by larger streams especially rapidly. . . . Shortly after the turn of the century, the canyon portions of most California rivers again resembled their pre-mining conditions. Once again these streams at nearly all points rested on the rock bottom of the canyon and were engaged in deepening them. What debris remained took the form of terraces along the canyon walls.

After 1884 the streams likewise attacked deposits of tailings along upland creeks. Here, too, erosion of these deposits by streams to reach their pre-mining base level produced terraces. In some mining districts the recutting left successive terraces ten, twenty, and up to thirty feet above the present stream. Streams also eroded many piedmont deposits of debris into terraces. Typically, this erosion resulted in well entrenched streams with successive terraces that mark the stages of cutting.

In 1902 tidal pulsations again occurred in the Sacramento River at the capital city. They became quite pronounced by 1910 and continued to increase in height. By 1913 the daily fluctuations reached a height of almost twenty inches, a figure nearly that of the tides before mining.

Surveys and observations after the cessation of hydraulic mining indicate that streams generally moved to regain their pre-mining condition. Presently, the remaining debris may be regarded as fairly permanent. Further erosion of the terraces seems unlikely. Human influences other than mining prevent the streams from ever exactly duplicating pre-mining conditions. However, conditions have reached or will reach the point where the effects of mining are indistinguishable from those of nature.

As a rule, the geomorphic effects produced by placering proved the same as those produced by natural agencies. Placer mining simply changed the magnitude of certain variables, resulting in no more than acceleration or deceleration in basic geomorphic processes. Mining did not abrogate existing geomorphic principles. In fact, it often provides an excellent illustration of some of the basic geomorphic processes. Some landforms that resulted from mining are almost indistinguishable from ones produced by nature. It appears, too, that the artificial origin of these landforms became successively less and less clear with time. As a result, with time, man-induced alterations of the landscape are accepted as results of nature. Today, the artificial origin of some landscapes would be difficult to determine except for the existence of documentary evidence.

Dredging for Gold

Clark Spence

Among the many innovations in mineral technology in the late nineteenth and early twentieth centuries, several would truly revolutionize the industry. Gold dredging, [in particular], would apply the mass production of Henry Ford's America to placer deposits, again enabling profitable working of ground heretofore untouchable. It was the American West, California in particular, that would develop a highly sophisticated dredge technology, one that would be disseminated in ever widening concentric circles for use throughout the globe.

According to a contemporary, a gold dredge consists of a floating hull with a superstructure, a digging ladder, endless chain of digging buckets, screening apparatus, gold-saving devices, pumps and stacker. It could be described as a floating mill with the addition of apparatus for excavating and elevating the ore. . . .

This type of machine, . . . the connected-bucket or bucket-ladder dredge, [is] the most common and at the same time the most complex, expensive, and efficient of the larger dredge family, which has several collateral branches. The development and use of this specific technolo-

From Clark Spence "The Golden Age of Dredging," *Western Historical Quarterly* (October 1980): 401–14. Reprinted by permission.

gy and the impact of the dredge on the landscape and the environment are of particular interest.

Full-blown, the bucket dredge was awesome and spectacular: ugly, graceless, megalosaurian. It might cost half a million dollars and be as large as an ocean freighter. Huge, tireless, it clanked relentlessly along, creating its own dirty pool and ripping pay dirt from bedrock fifty or even a hundred feet below the surface.

"A squat hightailing monstrous water widgeon diving its chain of spoonbills down and under," a Colorado poet called one. Unsightly and grotesque in an often scenic setting, it stank of oil and muck and filth and scum. Round the clock, its greed seemed insatiable, its presence at once fascinating and dreadful.

It wallowed in its water-bed; it burrowed, heaved and swung;
It gnawed its way ahead with grunts and sighs;
Its bill of fare was rock and sand; the tailings were its dung;
It glared with fierce electric eyes.

To some, there was a touch of romanticism about the dredge. It reminded one of the steamboats on the Mississippi. Journalists and promoters spoke of the "Modern Gold Ship" and of "Flagships of the Gold Fleet," or they likened them to grain harvesters: "Farming for Gold." The dredge was an instrument for conquering nature, "the acme of scientific mining and inventive genius." It was "a wonderful modern invention," able to do the work of 2,000 men and to make profits "where a cooly would starve." . . .

"The California-type dredge, known all over the world, is so efficient that it is being used on every continent where large quantities of low-grade metals are found," said an observer in 1938.

[But] another part of the story concerns the environment and the havoc wreaked upon it by the gold dredges. There would be far less criticism of their impact than there had been of hydraulic mining. . . . Like other kinds of mining, dredgers were not beautifiers of the scenic West. They left an unsightly wake—"a rocksnake of cold blue cobbles mounded," a genteel Colorado poet put it. . . . Topsoil lay buried deep under stones and cobbles in conical piles, long windrowed hills, or rough, irregular heaps with pools between.

Dredge leavings were indeed offensive to the eye, apologists acknowledged, but no more so than automotive junkyards, the charred remains of forest fires, or miles of cut-over stump land. Tailings were part of the price of progress, of resource development, jobs, and prosperity. Moreover, they argued, much dredging was done "Over the pits and scars of the first deflowering"—over land already once mined by hand or by hydraulicking. This was especially true around Marysville, California. . . .

Because of possible flood or physical damage, dredge towns themselves were sometimes at cross purposes with the industry on which they were dependent. Both Folsom and Oroville, for example, had problems with dredge companies which brought their mammoth machines into the city limits, buying rights to dig up cemeteries as well as streets. Sacramento was particularly concerned about flooding from debris attributed to dredges upriver.

A third complaint was that dredging despoiled good farm land. . . . Dredge men argued, and perhaps correctly, that the bulk of dredge land was marginal, unusable, or low-value grazing land from the many "rattlesnake ranches" of California; that some of it had already been mined over; that some of the vineyards suffered from phylloxera; and that even where good orchard land had been destroyed that was more than offset in benefits to people in the dredge communities. . . . Overgrazing and soil erosion were far more destructive, the placer men claimed, and a strip one foot wide taken from all of California's roads and railroads would conserve more land than that taken out of cultivation by all the dredges combined.

Not that all agreed that the land was ruined: under proper circumstances, some contended, it could be reclaimed by recoiling—by techniques that left the ground level and its soil on top. Auxiliary sluices, as used in New Zealand, were advocated, as were suction equipment, sand pumps, silt elevators or stripping and resurfacing with steam shovel or electric scraper, but none proved practically effective.

The few limited recoiling or replanting efforts that were actually made received far more attention than they warranted, especially the few acres restored by James Leggett at Oroville. According to a fawning press, Leggett "simply poked the roots with a crowbar, tickled the rocks, and willing Mother nature did the rest." His eucalyptus trees did indeed thrive, but descriptions of his alfalfa, figs, and citrus fruit were grossly exaggerated. Natomas Consolidated had equipped two of its eight dredges with special stackers and double-deck sluices to level and leave soil on top, and by 1919 had resoiled some 500 acres east of Sacramento and planted some in fruit and hay. But resoiling required additional power and another dredge hand and was not applicable to dredging deeper than about twenty feet. Estimates in 1919 were that it added from $200 to $250 per acre to dredging expenses; in the mid-1940s the figures ranged from $484 to $1,127 per acre. Unless the land were extremely rich, resoiling simply was not economically feasible.

Realizing this, dredge operators argued that it was the responsibility of the original owners when they sold their dredge rights to take less and to require recoiling. And they soon pointed out that dredged-over land was not unproductive—that several companies, including Natomas Consolidated, sustained a thriving industry in the sale of crushed rock, indicating again the American talent for utilizing everything but the squeal.

Unlike New Zealand and parts of Australia, any compulsion from the government to resoil was lacking. Businessmen contended that forced resoiling would only drive capital into other fields. But beginning in 1905, when the California Federation of Women's Clubs spearheaded the drive, the legislature would hear a number of bills to prohibit dredging on farmlands unless the soil was left "tillable for agricultural purposes." At least seven such measures were introduced down to 1953, all easily defeated after lobbying by the California mining interests. In 1945, when Merced County sought to achieve the same ends by use of local police powers, the federal courts intervened.

Thus, until very recent times, westerners evidenced only limited and lackadaisical concern for the environmental impact of gold dredging. The California Debris Commission played no major role, and there were no comparable agencies elsewhere. So long as tailings could be land-locked, complaints were minimal. And western citizens had few anxieties about arable land as long as it was abundant; in the twenties and thirties, crop surpluses made it clear there was no problem. Even in 1945, in a dredge case, a federal judge saw no impelling need to preserve farmland. Where state police power had in the past been used to touch on fertility of the soil, he said "it has been exerted to protect land owners and the public from the economic consequences of excess production."

Occasionally a skeptic might question the pursuit of precious metal. In 1919, engineer W. H. Gardner, in "Song of the Gold Dredge," weighed gold—"the senseless dross"—against the importance of steel in modern society:

> Oh! this the whine of bucket line—the plaint of grinding gear—
>
> The song that lifts through graveyard shifts by hour, day and year—
>
> The anguished grind as buckets bind in stress and strain and shear: "Gold—Gold—futile gold!

Of wasted steel a wealth untold,

Of useful steel the ransom made,

Of squandered steel the purchase paid! For Gold—Gold—Gold!

But this was an unusual attitude. In World War II, the federal government may have deemed gold mining unessential, but rarely did the average citizen question the vitality of gold to the world monetary system or the vast expenditure of energy needed to tear it from the land. Whatever the standards of Americans in the 1970s, contemporaries of the dredging scene in the golden years of the early twentieth century applied the standards of their own era and were seldom much bothered with the impact on the environment.

Further Readings

Bruff, J. Goldsborough. *"Gold Rush": The Journals, Drawings, and Other Papers of J. Goldsborough Bruff*. Edited by Georgia Willis Read and Ruth Gaines. New York: Columbia University Press, 1949.

Caughey, John W. *The California Gold Rush*. Berkeley: University of Califonia Press, 1975.

Caughey, John W. *Rushing for Gold*. Berkeley: University of California Press, 1949.

Clappe, Louise Amelia ("Dame Shirley"). *The Shirley Letters, Being Letters Written in 1851–1852 from the California Mines*. Salt Lake City: Peregrine Smith Books, 1985.

Farquhar, Francis P. *History of the Sierra Nevada*. Berkeley: University of California Press, 1966.

Goode, Kenneth G. *California's Black Pioneers: A Brief Historical Survey*. Santa Barbara: McNally and Loftin, 1973.

Greever, W. S. *The Bonanza West: The Story of the Western Mining Rushes, 1848–1900*. Norman: University of Oklahoma Press, 1963.

Heizer, Robert F., ed. *The Destruction of California Indians*. Salt Lake City: Peregrine Smith, 1974.

Kelley, Robert. *Gold vs. Grain: The Hydraulic Mining Controversy in California's Sacramento Valley*. Glendale, CA: Arthur H. Clark Co., 1959.

Lapp, Rudolph M. *Blacks in Gold Rush California, Yale Western Americana*. New Haven: Yale University Press, 1977.

Leshy, J. D. *The Mining Law: A Study in Perpetual* Motion. Washington, DC: Resources for the Future, 1987.

Levy, JoAnn. *They Saw the Elephant: Women in the California Gold Rush*. Hamden, CT: Archon Books, 1990.

Lingenfelter, R. E. *The Hardrock Miners: A History of the Mining Labor Movement in the American West, 1863–1893*. Berkeley: University of California Press, 1974.

Lydon, Sandy. *Chinese Gold: The Chinese in the Monterey Bay Region*. Capitola, CA: Capitola Book, 1985.

May, Philip. *Origins of Hydraulic Mining in California*. Oakland, CA: Holmes Book, 1970.

McPhee, John. *Assembling California*. New York: Farrar, Strauss, & Giroux, 1993.

Moynihan, Ruth B., et al. *So Much to Be Done: Women Settlers on the Mining and Ranching Frontier*. Lincoln: University of Nebraska Press, 1990.

Paul, Rodman W. *California Gold: The Beginning of Mining in the Far West*. Cambridge: Harvard University Press, 1947.

Paul, Rodman W. *Mining Frontiers of the Far West, 1848–1880*. New York: Rinehart & Winston, 1963.

Paul, Rodman W. *The Far West and the Great Plains in Transition, 1859–1900*. New York: Harper & Row, 1988.

Peterson, Richard H. *The Bonanza Kings: The Social Origins and Business Behavior of Western Mining Entrepreneurs*. Lincoln: University of Nebraska Press, 1977.

Pomeroy, Earl. *The Pacific Slope*. New York: Knopf, 1965.

Robbins, William G. *Colony and Empire: The Capitalist Transformation of the American West*. Lawrence: University Press of Kansas, 1994.

Robinson, Fayette, and Franklin Street. *The Gold Mines of California: Two Guidebooks*. New York: Arno Press, 1973.

Rohrbough, Malcolm J. *Days of Gold: The California Gold Rush and the American Nation*. Berkeley: University of California Press, 1997.

Smith, Duane. *Mining America: The Industry and the Environment, 1800–1980*. Lawrence: Kansas University Press, 1987.

Williams, Stephen. *The Chinese in the California Mines: 1848–1860*. San Francisco: R and E Research Associates, 1971.

Wyman, Mark. *Hard Rock Epic: Western Miners and the Industrial Revolution, 1860–1910*. Berkeley: University of California Press, 1989.

Young, Otis E. *Black Powder and Hand Steel: Miners and Machines on the Old Western Frontier*. Norman: University of Oklahoma Press, 1976.

Chapter 5
Forests Transformed

Documents

A Lumberman on Board-Feet Profits, 1884

Of all that has been told or written by travellers and correspondents concerning California scenery, its huge growth of beets, melons, squash, pears, and fruits of all descriptions, the least attention has been called to our grand forests of Redwood. This, however, is not much a matter of surprise, as the facilities for a careful inspection of this favorite building material are quite or nearly as primitive as during the early settlement of the State. Especially is this the case in the northern section of the State, where the redwood belt has greater width, and from climatic causes has developed a heavier growth of timber. Not only are the trees in this northern section larger in circumference, but they attain a much greater height, and withal give a product to the millmen that is far superior in quality to that obtained in the southern extremity of the redwood belt. . . .

The California Redwood Company (the largest in this line on the coast) has already taken the initiative step looking to a supply of clear seasoned lumber for the Eastern market. At Tormey Station just below Port Costa, and convenient for shipping both by rail and sea to all parts

From C. G. Noyes, *Redwood and Lumbering in California Forests* (San Francisco: Edgar Cherry & Co., 1884), pp. 3–4, 14–18, 34–35, 75.

of the world, they have built wharves and opened a yard covering some twenty acres, where their lumber can be seasoned properly before being offered to the markets abroad. . . .

Many will argue—and justly, too—that it would be better for the country that a demand which causes such a draft upon its lumber resources should, by some manner of means, be restricted. . . . Others can argue, however, that owners of timber lands can assist in reproduction by a slight effort in the way of timber culture, and thereby extend the supply to an indefinite period. . . . We have often thought that should the Government offer as great inducements in the reproduction of redwoods as it is doing to encourage timber culture in parts where it is unnatural for forests to thrive, that the redwoods would never become exterminated, as has so frequently been predicted. One must confess, however, that the matter of cultivating this tree with a view to growing timber like anything of its present size, would require a people possessing a higher regard for generations a hundred or more years hence than the mind of an average American can comprehend.

The Government map . . . shows that the really valuable portion of the belt (from Russian River to the northern limit) covers about two hundred and seventy miles from north to south. . . . The Government estimate (board measure) of timber standing in this belt in the census year 1880 was 25,825,000,000 feet. This was made up from estimates furnished by a few lumbermen. . . . But it is also true that many others, including millmen and lumbermen, estimate from 50 to 100 per cent higher, and taking the estimated area of the belt from Russian River to the Oregon line with the estimate of timber standing, we shall find even their figures largely increased. The 275 miles covered by this portion of the belt multiplied by the least estimated width (15 miles) gives 4125 miles. A square mile contains 640 acres, and the average yield per acre (according to government estimate) is 50,000 feet, which would give 32,000,000 feet to the square mile. This would give us a total for the 4125 square miles 132,000,000,000 feet of standing timber. . . .

Economy in the manufacture of redwood lumber is a matter in which the pioneer millmen of the Pacific Coast have taken but little interest until within the past three or four years. This could hardly be expected to have been otherwise, for the reason that the supply seemed to them unlimited and inexhaustible. The interest manifested by foreign investors and eastern capitalists in the timber reserves of America, however, has of late checked the inclination to waste which our old lumbermen inconsiderately indulged in for years. This check upon waste is commendable, more especially in the redwoods, because of its adaptability for building purposes, where white pine and the softer woods of the eastern forests are considered indispensable. And that the redwood is largely to fill the demand which has caused the almost entire destruction of the pineries of Maine, Michigan, Wisconsin, and Canada, there is not the least doubt among observant lumbermen of the eastern States. . . .

Almost the first thought passing in one's mind, as he enters a virgin forest of redwoods, is one of pity that such a wonderful creation of nature should be subject to the greed of man for gold. The same feelings of awe pervade one's being upon his first introduction to this apparently exhaustless army of giants, that impress the beholder of Niagara, Yosemite, and the near relatives of the redwoods—the Big Trees of Calaveras and Merced. . . .

When transportation facilities are complete, either by rail or by water, . . . as they certainly will be within a few years, it needs no prophet to predict that the California Redwood will, in the near future, have no rival in the lumber marts of the world.

Within a generation to come the question will be asked: "How long will the Redwoods last? A few years at most. But in that brief time men will build their castles and their thrones of power upon the mighty race of giants, with the one regret that there are no more to conquer."

An Advocate Calls for Forest Preservation, 1903

In Northern and Central California the creation of extensive forest reserves, particularly those "temporary" in character, has excited no little opposition, particularly among the sheep and cattle men, who have been accustomed to pasture their stock on the public lands in the forest area.

It has been objected that there is no need of reserving forests; that they renew themselves after cutting, the young growth being often more dense than the original forest; that sheep and cattle do no harm, and that the permanent withdrawal of so much timber land would result in serious checks to the development of this part of the State. It has been said, for instance, that the construction of new railroads into Northern

From Anonymous, "Should the Forests Be Preserved?" (San Francisco: California Water and Forest Association, 1903), pp. 14–15, 17, 20–21, 48.

California from the East or the Northeast might be blocked or hindered by creation of the reserves. . . .

The forest reserve question has been argued at length at meetings of the Sacramento Valley Development Association, when the stock interests were much in evidence. In the first instance the Association adopted a memorial to Washington authorities appealing for the preservation of the forests through the creation of reserves. . . . After a long debate the Association passed a resolution favoring "the withdrawal of all timber lands from sale by the Federal Government, and placing all forests, whether under public or private ownership, under a competent forestry administration, in which the State shall co-operate with the general Government."

In addition to this the Association also passed at the same meeting a resolution opposing the establishment of forest reserves in Northern California "under the present laws governing the formation and control of such reserves," and inviting the National Government to investigate, with the view to preserving the forests, and at the same time permit the cutting of timber and the pasturing of stock. . . .

Pasturing Stock in Reserves

Many disinterested observers have declared that unrestricted grazing of stock in the Sierra is highly injurious to the young tree growth. This is particularly the case with sheep, which trample down small trees or strip them of their foliage, especially where the land is overstocked, which is said to be usually or often the case. Many fires in the forests are attributed to sheep men, sometimes from negligence and at other times for the deliberate purpose of improving the pasturage for the following season.

The regulations of the Department of the Interior governing forest reserves prohibit grazing by sheep and goats, save by special permit to be obtained through the General Land Office, when such grazing shall appear not to be injurious. The pasturing of other live stock is allowed so long as it appears that injury is not being done to forest growth and water supply, but permission must be had from the Land Office. . . .

However, the day when the inaccessibility of the forests was their protection is a thing of the past. The attention of the lumber interests is being centered in this State as never before, and already vast areas have been purchased by Eastern companies for lumbering purposes.

The timber interests of the State have a legitimate place, and a very large one, in our commercial life. Lumber is needed and must be supplied, and there is no reason why the home forests should not supply this need. But it is possible to do this and yet preserve those features of the forests which tend to make of them the great fountain-heads of the streams and underground channels that keep alive the industrial centers. . . .

The means of destruction may be briefly stated as the axe and fire. Small words with a far-reaching significance!

The first means of destruction to be impressed upon me was one that possibly few people know anything about, and that cannot be condemned too strongly. Even those engaged in the business, whom I could get to express an opinion, condemned the practice as a wanton waste of timber that ought to be stopped. I refer to the making of shakes. From the small area already mentioned there goes out each year about 2,000,000 shakes. Nothing but the finest of sugar pine trees are cut for this purpose. If the first length cut from the tree does not split perfectly straight, the tree is abandoned and another cut down.

The probable average production from the trees utilized, making no allowance for the ones thrown away entirely, is not more than 5000 shakes per tree, which represents but a small fraction of the available lumber in the tree. It can readily be understood, therefore, that the waste of timber is great.

Passing from the individual operator splitting shakes to the large company operating mills, we find the most important means of destruction. Within this same area there are two companies, one with mills in operation and the other preparing to erect mills, and controlling together about 30,000 acres of the cream of the timber belt. . . . Passing from the shade of heavy forests the traveler will emerge upon an open waste, with here and there a tree of inferior quality or small size standing to tell the story of what has been done. On the ground the way is blocked by vast numbers of saw logs prepared for the mill. . . .

President Roosevelt's address [1903] . . . may here be fittingly quoted. . . . "And now first and foremost, you can never afford to forget for a moment what is the object of our forest policy. That object is not to preserve the forests because they are beautiful, though that is good in itself; nor because they are refuges for the wild creatures of the wilderness, though that, too, is good in itself; but the primary objective of our forest policy, as of the land policy of the United States, is the making of prosperous homes. It is part of the traditional policy of the home-making of our country. The whole effort of the government in dealing with the forests must be directed to this end, keeping in view the fact that it is not only necessary to start the homes as prosperous, but to keep them so."

A Woman Recalls the Lumber Camps, 1931

One Sunday Evening in early April 1926, I stepped aboard a caboose-like railroad car at Korbel, California, a small mill town. I was on the way to my first logging camp job to work for the Northern Redwood Lumber Company. Young, naive, inclined to be bookish, I watched with interest as men of varying ages, clad in waist overalls and hickory shirts, crowded into the car. These men were loggers returning to camp after a Sunday off in Eureka. There were a few young choker setters and rigging slingers, but most were middle-aged with weather-beaten outdoor faces. These were the donkey punchers, firemen, hooktenders, fallers, and buckers, and even a powder monkey was aboard. . . .

The men settled back for a two-hour ride as the train jerked and slowly began picking up speed. We sat facing each other on two long benches against the walls of the car, and as the only young woman in a car full of rugged loggers, I was soon the object of good-natured banter. Noticing that I was too shy to respond, an older man responded to the teasing. I learned later he was the woods boss.

It was dark as we pulled into the cluster of wooden camp buildings, but the cookhouse and rows of bunkhouses were brightly lit by the camp generator. I followed the men into the kitchen where Christina, the head cook, had a bountiful smorgasbord of cold cuts, pastries, bread, and coffee for the returning loggers. Loggers, I soon found, were always hungry and had cast-iron stomachs besides. The next day I was to learn they could lay away a big meal in eight minutes or less. . . .

Morgan, a flunky, was serving the men thick slabs of bread and meat. "Hi," she smiled. . . . She was a slender young woman nicknamed Two-Speed because she was a very fast worker but very slow in speaking. . . .

Finding a seat by the woodbox, I began looking around the clean compact kitchen where I would be working. The long woodburning range with its large firebox and two big ovens caught my eye. On its 24 by 36 inch cast iron griddle, hot cakes, steaks, chops, fish, and eggs were

Anna M. Lind, "Women in the Early Logging Camps: A Personal Reminiscence," *Journal of Forest History*, 19 (February 1931): 128–35, pp. 129–31, 135. Reprinted by permission.

cooked. The pine floor was clean from scrubbing. Strong shelves on the bare walls were stacked with deeps (oval bowls about three inches deep) and platters. Gray graniteware tea and coffee pots were lined up neatly on one large shelf. The cookhouse was rugged and primitive, but I liked what I saw. The feeling of belonging in the woods was already putting its claim on me.

So far as I know, women first began working in the logging camps of the Pacific Northwest during the WWI years of 1917 and 1918. By their own endeavor and efficiency they proved themselves capable of good work under the hard conditions of camp life. Looking at the impassive cook, Christina, and the elderly Ida, I wondered if they had been among these pioneering women.

Christina walked over to where I was sitting. "Can you cook?" she asked abruptly. . . .

I knew I could cook, but waiting table was entirely new to me, and my first attempt proved very embarrassing. At breakfast the following morning I loaded up with as many plates of hot cakes as my inexperienced hands allowed. Entering the dining room, I was confronted with a roomful of seated men, their eyes turned toward me, and overcome by nervousness I stumbled over my own feet and fell, plates and hot cakes flying. I wanted to sink through the floor and disappear.

No one laughed. Loggers are said to be rough men, but they can also be gentle. An old windfall bucker helped me to my feet while others gathered up the plates and spilled hot cakes. "There, hon," the old bucker said kindly. "Just go back to the kitchen and get us some more hot cakes." His common sense eased my embarrassment, and somehow I managed to keep the food coming through breakfast. . . .

Cooking dinner in the dinner shed was easy, but walking up the steep hill to it made my young legs ache. Later, whenever the woods boss wasn't around, the men and I would ride up the steep slope on an empty logging car towed by a steam donkey.

Everything was new and fascinating to me. I watched the big redwood logs come in from the landing on cars shunted to the tracks at the top of the incline by a puffing steam locomotive, or "lokey" as we called it. There, the loaded cars, one at a time, were hooked onto a cable and lowered down the steep car would jump the track, and I would watch in awe as Red, the donkey puncher, would snatch off his hat, throw it on the ground, and systematically stomp it to pieces with his caulked boots, swearing continuously. Even more amazing was Red's uncanny skill at getting the load back on the track again.

The dinners I cooked in the dinner shed were big meals with meat, potatoes, gravy, salad, vegetables, and always pie. Loggers love pie, and I would have to hurry to get them baked by dinner time. There would also be a large pan of freshly made doughnuts, still warm, waiting for the men. After dinner there was always washing up, tables to reset, and the hike back to camp. . . .

Between work periods, Two-Speed and I enjoyed hiking among the tall beautiful redwoods. Besides the new logging grades cut in the virgin timber, there were game trails to follow. . . .

Times were good and camp jobs easy to find. After several years with the Northern Redwood Lumber Company, I pulled up stakes and paid my train fare to Seattle, Washington—alone in the world and on my own. I wanted to see more of the Pacific Northwest. . . .

It seems that women in the logging camps, whether working in the cookhouse or living in family houses, exerted a good influence over the men. I've worked in a number of logging camps and can truthfully say I was always treated courteously and like a lady. I've read or heard somewhere that Stewart Holbrook said "loggers are Nature's Gentlemen." And it's true.

Lizzie McGee Describes the Sequoia Mills, 1952

To get vast acreage [in the Sierras, San Francisco industrialists Hiram D. Smith and A.D. Moore] took advantage of an existing law that permitted a United States citizen to file on 160 acres of timbered government land. There were few requirements attached. You saw the land, you made a filing, and in due time could pay a flat sum and get your title. A few early settlers did this in good faith. But in this instance, most people considered an unfair advantage was being practiced. Agents scoured the nearby valley towns, looking for men who had not yet used their timber claim rights and who were willing to bargain for a few easy dollars. They gathered them up by the stage loads. . . . The men were taken to the vast beautiful timber belts where many of them had their first glimpse of the magnificent redwoods. They took a birdseye view of them—probably admired them. Large township maps were spread out, showing desirable

Lizzie McGee, *Mills of the Sequoias* (Visalia, CA: Tulare County Historical Society, 1952), pp. 1–2. Reprinted by permission.

quarter sections of vacant land. Each picked himself out a claim and they were taken back to Visalia where a U.S. Land Office was then located. A stop there, and with a scratch of a pen, they were in possession of a filing on a timber claim where some mighty big trees had flourished for some thousands of years. . . . In course of time the acreage price to the Government was paid off, his deed was delivered and the trees were his. But not for long. There was a wholesale transfer of deeds to the Company, for which each owner received a modest sum, in exchange for his timber right inheritance. All perfectly legal as far as the law was concerned, that is, if he didn't advertise his pre-arrangement concerning the deal. The Company was all set to begin slashing trees. . . .

Along about 1899 the Smith and Moore Millwood mill was moved to Converse Basin. Most of the freighting and moving operations was done over the railroad and up the hoist. A makeshift road permitted some teaming over the top and down to Converse. . . . Here would be cut the really big trees. Reducing a ten, twenty or more foot sequoia log into sizeable proportions for transportation to the mill, and for running through the saw, required cruel blasting. With long augurs, holes were sunk deep into the heart of the log, the hole was packed with black powder or dynamite and a long fuse attached. Sometimes the explosion split the log more or less evenly but many more times it was wastefully shattered. The loss of valuable timber was pitiful. In time a splitting saw was invented and the loss was not so great. Even so the woods were a sorry looking shambles when cutting operations were completed. About as much waste timber covered the ground as was cut at the mill. . . .

Nearing Maxon's Ranch (Trimmer Springs) the flume came down an elevated incline to the top of a high structure over a bridge that spanned the mighty Kings and crossed over to the north side. From here on the wooden ditch was set on high trestles that paralleled the river to several miles beyond Centerville where it turned westward into Sanger.

At suitable points along the flume route, where slackening waters permitted, stations or camps were maintained. There were fifteen in all. From these stations flume walkers patrolled the ditch on wooden catwalks attached to, and a part of, the side of the flume. These trouble shooters were constantly on guard against lumber jams, breaks in the flume, or obstacles that might obstruct free passage of lumber trains. . . . Overlooking steep mountain cliffs, deep ravines and dangerous rocky canyons, these flume walkers made their way cautiously and by some were dubbed "flume snakes." They each carried a picaroon, a sort of lumber tamer.

It is said that a man by the name of Bowell invented the picaroon idea

FIGURE 4. LOGGING REDWOODS.
A business extending from the California coast into the Sierra Nevada, felling huge old-growth trees was a profitable and dangerous enterprise. Here, men yard split sections of redwood near present-day General Grant Park in the Sierras, 1888. Harold G. Schutt collection, reproduced with permission from the collections of the Sanoian Special Collections Library, Henry Madden Library, California State University, Fresno.

but they were fashioned at the lumber company blacksmith shop from double bitted axes. One bit was only slightly reshaped and the other hammered to a curving point. With this contrivance the flume walker reached out and straightened into line a train if it got wild and tried to buckle up and block the stream. As a lumber train reached the several stations along the way, it was halted and the next down coming train was hooked on by the herders. By the time the valley floor was reached, great trains of lumber bundles moved gracefully and slowly into the Sanger lumber yard. They were slid off and piled up to dry, in readiness for finishing operations at the milling plant nearby. The flume had a capacity of 250,000 feet of lumber daily. . . .

On steep inclines there was considerable slopping out of water. To replenish this waste, feeder streams were run in by small flumes. At Millwood "Little Lake" on Mill Flat Creek above the mill, furnished the first supply. Rancheria and Cow Flat each furnished a supply, with probably others.

There seems to have been two types of boats. A long boat that formed to the sloping sides of the flume, and a dinkey, braced underneath in W-shape, each section appears as a V-shaped culvert. The dinkey accommodated two men, the front one with feet braced against a backstop, the second behind and braced against him. . . . It must have been a breathtaking experience on some of the steep tumbling sections. These boats were used by company inspectors, the supply crew, and in emergency by others. For use between Millwood and Maxon's, supplies were freighted up from Sanger and shipped down by water. Clamps and boats were sent back by freight.

A ride down the flume was a thrill to be remembered. At times it was dangerous. The load was balanced to dip the rear end slightly. Before starting a boat down, a poster was attached to a lumber clamp, signifying such an intention. No more lumber was sent down till the bundle passed all the stations and arrived at Maxon's. An "all clear" was sent to Millwood. The boat was released and away the passengers went on a wild, wild ride. At any of the camps they could, if they chose, crawl out and proceed on foot down the catwalk. It is not reported that one ever did so.

Coming down the rapids, the boat with the speed of fifty miles per hour, outran the water. When it scraped the bottom the speed was lessened. But if the rear end kicked up a bit, when the onrushing waters overtook the boat, an awful spill could result. Such an incident happened on a steep stretch of rapids below the lower mill. Two men with two suitcases were in the midst of a flume 'shoot the chutes' joy ride. The boat was winning the race. With a burst of speed the water bumped it from the rear. The whole outfit, boat, men and suitcases took a terrific spill. The suitcases kept on riding the stream. By the time the passengers righted themselves and got underway their luggage had been fished out at Rancheria Station where the owners picked them up. . . .

When the trees had all been slashed in the Converse Basin area, a track was laid to a saddle overlooking the upper Kings River Canyon. Here a hoist was installed. From there the track was laid down into Indian Basin (in the vicinity of the junction of Kings River Highway and the Hume Road). This was the Rob Roy Chute and was operated with a hoist. [But] in a year or so the Converse Mill partly burned down.

Judi Bari Recounts the Timber Wars, 1994

The environmental battle in the Pacific Northwest has reached such a level of intensity that even the press now refers to it as the Timber Wars. At stake is the survival of one of the nation's last great forest ecosystems. Our adversaries are giant corporations—Louisiana-Pacific, Georgia Pacific, and Maxxam in Northern California, where I live.

These companies are dropping trees at a furious pace, clogging our roads no less than 18 hours a day, with a virtual swarm of logging trucks. Even old timers are shocked at the pace and scope of today's strip-logging, ranging from 1000-year-old redwoods, one tree trunk filling an entire logging truck, to six-inch diameter baby trees that are chipped up for the pulp mills and particle-board plants.

One-hundred-forty years ago the county I live in was primeval redwood forest. At the current rate of logging, there will be no marketable trees left here in 22 years. Louisiana-Pacific President Harry Merlo put it this way in a recent newspaper interview: "It always annoys me to leave anything on the ground when we log our own land. We don't log to a 10-inch top, we don't log to an 8-inch top or a 6-inch top. We log to infinity. It's out there, it's ours, and we want it all. Now."

So the battle lines are drawn. On one side are the environmentalists, ranging from the big-money groups like The Wilderness Society and Sierra Club to the radical Earth First!ers and local mountain people fighting the front-line battles in the woods. Tactics being used include tree-sitting, logging road blockading, and bulldozer dismantling, as well as the more traditional lawsuits and lobbying.

On the other side are the big corporations and the local kulaks who do their bidding. Tactics used by them have included falling trees into demonstrators, suing protesters for punitive damages (and winning), buying politicians, and even attempting to ban the teaching at a local elementary school of a Dr. Seuss book, *The Lorax*, which the timber companies say portrays logging in a bad light.

But what about the timber workers? Where do they fit into this scenario? Their true interest lies with the environmentalists, because, of

Judi Bari, "Timber Wars," in *Timber Wars* (Monroe, ME: Common Courage Press, 1994), pp. 11–13, 16, 18. Reprinted by permission.

course, when the trees are gone, the jobs will be gone too. Logging is listed by the U.S. Labor Department as the most dangerous job in the U.S., yet the current speed-up has some loggers and millworkers working ten or more hours a day, six days a week.

Clearcutting is the most environmentally devastating logging method, and also the least labor-intensive. In the long run, the only way to save timber jobs in our area is to change over to sustained yield logging, where logs can only be taken in a manner and at a rate that doesn't destroy the forest. This is exactly what the environmentalists are asking for.

Yet in spite of all this, those timber workers who are organized at all have been organized by the companies against the environmentalists. There are a few noteworthy exceptions . . . but by and large timber workers around here are either doing the companies' dirty work or keeping their mouths shut.

A good example of this is the spotted owl campaign. Scientists and environmentalists have been trying to get the owl listed as an endangered species, as 90–95% of its habitat, the old growth forest, has already been annihilated. The timber companies have responded with a vicious campaign promoting the extinction of the owl so that it would no longer stand in the way of them destroying the last of the old growth. Loggers are the pawns of this game, wearing T-shirts that read: "Save A Logger, Eat An Owl" and "Spotted Owl Tastes Like Chicken." Recently a hearing on the owl's status was held in Redding, [California]. The timber companies closed the mills and logging operations for the day and bused 5,000 workers to the hearing, carrying anti-owl banners and cheering as speakers denounced environmentalists. . . .

Pacific Lumber is another of the "big three" timber companies in the area. Until recently, it was a locally based, family-run operation paying good wages and amazing benefits. Pacific Lumber also treated the forest better than most and, because of its conservative logging and avoidance of clearcutting, has ended up owning most of the privately owned old-growth redwood that's left in the world.

But in 1986, Pacific Lumber was taken over in a leveraged buyout by Maxxam Corp., a high-finance holding company owned by . . . Charles Hurwitz. Hurwitz financed the takeover with junk bonds, and is now liquidating the assets of the company to pay off the debt. But in this case, the assets of the company are the last of the ancient redwoods. Hurwitz has tripled the cut, instituting clearcutting, gutted the pension plan, and started working people overtime. . . .

Historically, it was the IWW who broke the stranglehold of the timber barons on the loggers and millworkers in the nineteen teens. The ruling class fought back with brutality, and eventually crushed the IWW, settling instead for the more cooperative business unions. Now the companies are back in total control, only this time they're taking down not only the workers but the Earth as well. This, to me is what the IWW-Earth First! link is really about. And if the IWW would like to be more than a historical society, it seems that the time is right to organize again in timber.

ESSAYS

Myth and Reality of the Humboldt Forests

Tamara Whited

Logging the redwoods with axes and the short, light saws much in use before the Civil War, took courage, perseverance, and a certain ruggedness of character. Mill owners faced uncertainty as well, though of a less lethal sort: long-term profits were never sure in an industry that was entirely unregulated, located in a most isolated part of the country, and subject to wild fluctuations in demand. It is this sense of heroism, epitomized by the will of the entrepreneur and the courage of the logger in the woods, that largely characterizes the historiography of the Humboldt Bay region of the mid to late nineteenth century.

Unlike the era that saw the establishment of redwood preserves and parks, the fledgling years of redwood logging have received scant critical attention. Historians have reveled in the perceived conjuncture between the status of Humboldt County as one of America's "last frontiers," the mightiness of and the challenge posed by the redwoods, the appearance of "superhuman" lumber entrepreneurs, the difficulties, in the early days, of communicating with the outside world, and the clear progression toward technical domination of the dense forests of Humboldt County. Assessed together, these elements form what might be called the heroic narrative of the region. Some examples are in order to impart some of its flavor. One of the early heroic narratives was Leigh Irvine's *History of Humboldt County*, California, written in 1915. Irvine sounded the note of conquest, which has been replayed by various historians since his work was published: "Owing to the great size and weight of the trees and their

Tamara Whited, "Humboldt County in Reality and Myth: Constructing the Redwood Belt from 1850 to 1890," by permission of the author, Department of History, Indiana University of Pennsylvania, Indiana, Pennsylvania, quoted sources in "Further Readings."

thick stand on the ground, redwood logging offered many problems not met with in other woods, but these conditions have been met and conquered, and now redwood logging moves along smoothly and systematically, conducted by men who know how."

For Irvine, the history of technical innovations recapitulated the history of redwood logging; in this way he inserted the substitution of machine for muscle into the heroic narrative, a curious theme in which later historians have rarely spotted contradictions. Irvine's list of technical improvements is actually quite useful in its completeness: the advent of the circular saw came in 1862, steam locomotives and thus logging railroads appeared in Humboldt County beginning in 1874, the "steam donkey" (also known as the "Dolbeer donkey" after John Dolbeer, a steam engine used to haul logs onto skid roads and load them onto cars) that was patented in 1882, and the band saw nearly completed the technological picture in 1886. . . .

Hyman Palais (1974), also enamored of the mystique of the redwood "challenge," affirms that the loggers were the true heroes:

> From ox teams to modern trucks, from sash saws to band saws, the history of redwood logging will always have a special aura around it because of the problems men faced in cutting and utilizing such massive trees. Brute strength was always required—first, that of men and animals, later, that of machines. But brute strength alone would have been helpless without the ingenuity, perseverance and inventiveness of the men who logged the tallest trees on earth. . . .

Given that historians are forever trying to complete the picture of past eras and phenomena, the heroic narratives cannot be too harshly condemned for what they leave out. Rather, the entire aura, with its overtones of conquest, Manifest Destiny, inevitability, and the infinite progress of technology, that they attribute to the early history of Humboldt County must be questioned for its historical truth. . . .

The region now called Humboldt County did not make for easy entrance by either land or sea, nor, as Oscar Lewis noted, was it particularly inviting to the average settler: it was marked by "extremely rough topography, dense forests, deep streams racing through precipitous canyons, a lack of natural passes, and an Indian population that was easily the most warlike and wily in California." Nevertheless, the impetus provided by the Gold Rush to explore and settle all of California quickly affected the Humboldt region. Early explorers of the area set out with plans based on knowledge of what settlers had done all over North America—they trapped, traded, and farmed, often in that chronological order. The redwood forests hindered their progress but did not change their minds.

The members of the Josiah Gregg party of explorers, credited with discovering Humboldt Bay for the Anglo-Saxon world in 1849, intended, according to Owen Coy, to secure claims for themselves around the bay, "provided the surrounding country was adapted to agricultural purposes." Other explorers viewed Humboldt Bay and its environs with an eye toward the local gold rush on the Trinity River. . . .

A state act of February 18, 1850, created Trinity County in Northwest California, comprising what is now Trinity, Humboldt, Klamath, and Del Norte Counties; in 1851 Klamath County was carved out of Trinity, and in 1853 what remained of Trinity County was divided into Trinity in the east and Humboldt County in the west. From 1853 to 1856 Union, now Arcata, held the honor of Humboldt's county seat, but after a fierce political battle Eureka became the permanent county seat in 1856. These organizational developments of the 1850's reflected growing interest and settlement in Northwest California. Much of this early interest stemmed, however, from speculation over the mineral resources of the area, not surprising in light of the Gold Rush of 1849. . . .

The first rush to grip the collective imagination was the short-lived "Gold Bluffs" gold rush of 1850–51. Centered around Trinidad Harbor, up the coast about 20 miles from Eureka, the Gold Bluffs rush soon petered out because no one knew of a method by which to extract particles of gold from the sand. A more profitable relationship developed with the Klamath River mines to the northeast and, slightly later, the Trinity mines to the east between 1853 and 1860. This inland mining region originally depended on coastal settlements for supplies; the site of Union had been selected largely because it was the closest point on Humboldt Bay to the Trinity mines. In 1855 the *Humboldt Times* claimed that Union's trade with the mines amounted to $500,000 annually. It seemed that the Humboldt Bay settlements would continue to focus far inland, but by the early 1860's, hostilities with native Californians had made the overland packing trade unprofitable and dangerous. Besides, traders in Shasta City had begun to build a road between Weaverville, the center of the Trinity mines, and the Sacramento Valley. The new trade route gradually eclipsed that between the mines and Humboldt Bay. . . .

A significant portion of the lands within the borders of Humboldt County lured settlers for their agricultural potential. The most promising agricultural lands lay in the Eel River Valley, about five miles below the southern end of Humboldt Bay. Walter Van Dyke, an early Humboldt pioneer and later California Supreme Court Justice, described the valley as "one of the finest agricultural regions in the world." The area around present-day Arcata and the flatlands south of Eureka also attracted a farming population.

Overwhelming evidence . . . shows that Humboldt County was a leading agricultural region in California during the last half of the nineteenth century. In 1859, 55% of the county's working population described themselves as farmers, farm laborers, stock raisers, and/or dairymen; in fact, not until three decades after the beginning of white settlement did the lumber industry surpass agriculture as the "bulwark of the Humboldt economy." Just as important, observers touted Humboldt's agricultural productivity even after the lumber industry had exceeded it; throughout the nineteenth century, spokespersons for Humboldt County sought to depict the entire region as a haven for the small farmer.

Variety, abundance, and rapid growth characterize the agricultural output of the chief farming areas. In 1854, 2500 acres were in cultivation and 1400 cattle were being raised in the sparsely populated Eel and Mad River Valleys; by the 1860's, Humboldt farmers exported grains, hay, peas and beans, potatoes, tobacco, and fruits in large quantities. After 1855, a stock industry developed along the south coast district of Humboldt County and through the redwood belt. Except for allowing their cattle to graze in the redwoods, agriculturalists avoided forests and even cutover lands.

Apparently enough farmers did cultivate cutover lands for A. T. Hawley, in a pamphlet largely extolling Humboldt agriculture, to write in 1879: "The forest lands, denuded of the mighty trees which have thrived upon their fatness for centuries, make splendid farms." But it can be assumed that removing redwood stumps posed too daunting a task for the average small farmer. For practical reasons, then, agriculture was not compatible with logging as it had been elsewhere in the United States. Instead, farmers stuck to the river valleys mostly to the east of the redwood belt.

The land laws of the 1840's and 50's demanded cultivation as a prerequisite to full ownership. Similarly, under the Homestead Act of 1862, anyone over 21 could acquire a quarter-section of public land for $1.25 an acre, paying as little as $10 at the time of entry, cultivate it for five years and then possess it if no one but the buyer had used it. In Humboldt County, over 9,700 acres were entered between 1863 and 1865. One could not obtain timberlands under the Homestead Act because they were considered uncultivable. Dubious land titles and depredations on public timberlands finally prompted passage of the Timber and Stone Act in 1878. The articles relating to forests authorized such lands "unfit for cultivation" in California, Oregon, Nevada, and Washington Territory to be sold for $2.50 an acre; no more than 160 acres could be sold to any person or association, and one hundred- to one thousand-dollar fines would be levied for removing trees from public lands. A neat system had been devised for the disposal of lands that were not of obvious use to farmers. . . .

Evidence from the *Humboldt Daily Times* (November 29, 1877) reveals the ways in which first-generation Humboldt residents conceived the optimal use of redwood forests. A number of editorials appeared in the months preceding passage of the Timber and Stone Act; all agreed with the proposed sales of 160-acre tracts to individuals at $2.50 an acre. The writers realized that the value of the forests had nothing to do with agriculture; one claimed that "their value is greater than ordinary tillable public lands." Undoubtedly aware of the impossibility of establishing a lumber industry on the basis of 160-acre holdings, they nevertheless perceived the value of the redwoods in extractive terms.

All of the above said, it remains true that a heavily capitalized redwood lumber industry emerged early on and, recovering now and then from an economic slump, steadily expanded over the course of the late nineteenth century. But it would be an abdication to historical "inevitability"—so easy to do when writing about the exploitation of nature under capitalism—to accept this development at face value. This would be especially dubious in light of the foregoing analysis, which suggests that some important early trends sought not to construct Humboldt County as logging country. The first step in explaining the unstated decision to systematically cut down the redwood forests of California lies in exploring reactions to the trees themselves, reactions which gradually turned into attitudes and values. . . .

There were those who envisioned redwoods only in terms of board feet; those who responded with differing combinations of love, awe, and pity; those who believed in the inexhaustibility of the forests (they could fall into either of the two former camps); and those for whom a sense of wonder and a desire to exploit were not contradictory.

Commentators such as C. G. Noyes reveled in conceptualizing the entire redwood belt of California in board feet. By Noyes's calculation, the area from the Russian River to the Oregon border represented 4,125 square miles of redwood forest, at 50,000 feet of lumber per acre; multiplied by 640 square mile and then by 4,125, the grand figure came to

132,000,000,000 board feet of redwood. Noyes scoffed at the government's estimate of 25,825,000,000 board feet.

A quite different set of reactions characterized groups who wrote descriptive tracts on Humboldt County. Titus Fey Cronise's voluminous *Natural Wealth of California* (1868) provided one of the earliest ecological descriptions of the redwood forests and spoke of value in other terms than board feet. Cronise's observations of redwood ecology emphasized the tenacity of *sequoia sempervirens*: redwoods had "imperishable" roots, could sprout from stumps, and resisted fire. Redwoods converted fog into rain and played an important role in protecting watersheds. Cronise forebodingly predicted the consequences for agriculture in the event of total redwood exploitation: "It will surely happen that if the redwoods are destroyed, and they necessarily will be if not protected by law, certain portions of California, now fruitful, will become comparatively a desert."

J. M. Eddy and John Muir were also much impressed by the redwood's ability to sprout from stumps. All three writers offered an alternative based on this phenomenon to the standard heroic narrative, substituting the trees themselves for the lumberjacks. But this narrative was unmistakably declensionist in tone, as captured best by Muir's words in *Our National Parks*: "The redwood is one of the few conifers that sprout from the stump and roots, and it declares itself willing to begin immediately to repair the damage of the lumberman and also that of the forest burner."

Beyond ecology, the redwoods' impression on the senses and the imagination could elicit reactions of awe and love. A remarkable essay by Dr. A. Kellogg, a member of the California Academy of Sciences, conveyed the possibilities of a spiritual response that truly did justice to the redwood forests. Perhaps the most remarkable thing about Kellogg's piece was that Noyes, who partially espoused the board-feet vision, included it as an appendix in his 1884 *Redwood and Lumbering in California Forests*. Kellogg's "Essay upon Redwood" deserves quoting at length. From him, watching redwoods emerge from morning fog elicited poetry and a vision of feminine beauty: "Soon the cambric night-curtain lifts, and vistas of grandeur and of glory, beauty unwonted! and still they rise, refreshed and charmed as a morning bride in her vail (sic!) but to dwell on these ever-varying visions would be to write an endless volume!" For Kellog, felling the redwoods evoked visions of doom and apocalypse: "The shocked earth doth groan and murmur her moan at the sound of their fall! Prone, and full oft crushed to the earth, they lie the lengthened ruin of a thousand years! . . . Canst witness the sad havoc, and feel no pained interest—nay, great solicitude?" Kellogg felt a deep sense of shame for his generation of Californians: "We seem doomed to be famous according as we have lifted up axes against the thick trees"—and he called explicitly for government protection of the remaining redwood forests. Kellogg was not the only one to sound this note in the early 1880's, nor the only one to view the forests as female.

The anonymous author of the *History of Humboldt County* (1882), extended the metaphor of the virgin forest to that of the raped forest: "Nothing can be more majestic and impressive than the land clothed with redwood forests, nor more naked, desolate, ragged and uncouth than the land after it is stripped of them. It is in the one case peace, beauty, plenty, virginity and bounty; in the other rags, fire, destruction, rapine, ghastliness and most unsightly death."

For the first several decades, redwood lumbering was closely tied to shipbuilding, due to the Pacific focus of the trade. The lumber companies themselves owned many sailing ships, steamers, and schooners; such vertical integration intensified over the course of the century. The first foreign shipments of redwood left San Francisco in 1854; only ten years later the South Pacific islands consumed more of Humboldt's lumber than anywhere outside of California. The domestic trade grew in importance only after 1896, when the Southern Pacific Railroad offered low rates on lumber shipped from San Francisco to the Midwest and promised rebates to mills to defray the expenses of shipping lumber from the mills to San Francisco. In 1915, a railroad finally connected Humboldt County to the rest of the country. . . .

Large firms dominated the redwood industry by the 1870's, and mill owners possessed not only ships but offices in San Francisco and supply stores as well. J. M. Eddy in his 1893 *In the Redwood's Realm* notes that the 77,000,000 board feet sawn in 1875 belonged mostly to half a dozen mills. The California Redwood Company became the largest, owning in 1884 two mills, two logging railroads, nine logging camps, and a machine shop; it employed around a thousand workers. Nevertheless, new mills continued to appear and force down prices until mill owners in Humboldt and Mendocino Counties attempted to form redwood cartels. From 1879 to 1894 at least five cartels formed and disbanded, failing for reasons varying from the differing qualities of wood between the Humboldt and Mendocino mills, to different interests between lumber dealers and dealers who were also manufacturers, to the persistence of independent agents, to fears that artificially high prices would invite competitors. Thus, overproduction continued to characterize the industry; no fewer than 26 mills existed in Humboldt County in 1893.

The emergence of the redwood industry is a story of about-faces and shifting values, not one of logical, foreordained development. For the first three decades of Humboldt County's existence, settlers did not conceive of their locale solely as "logging country." The fruitful environment supported small farmers who grew grains and legumes, fruit growers, raisers of livestock, and fishers in addition to the loggers.

This is not to say that agriculture in Humboldt dwindled; in fact it remained quite productive but is dwarfed economically and conceptually by the lumber industry. Few people think of Humboldt County when California agriculture comes to mind, not surprisingly since many single industries became identified with specific regions of the state during the twentieth century. In the end, the heroic narratives of Humboldt's history fail on many counts, doing violence not only to alternative values concerning the redwoods but to the entire, complex relationship that bound the county's early American settlers to their environment.

Cutting the Sierra Forests

Douglas Strong

Fishing and agriculture within the Tahoe Basin remained relatively small enterprises throughout the late nineteenth century, but the lumber industry came to dominate human activity. Although second growth timber eventually covered most of the scars left by loggers, little virgin timber survived. Both the vegetation and the eroding slopes on which the timber once stood underwent rapid change.

The earliest mills, established by 1860 at the south end of the lake, supplied building materials for local settlements and early trading posts. With the discovery of the Comstock Lode, however, the development of mines and of the neighboring towns provided an insatiable lumber market. As early as 1861, Augustus Pray and partners built a water-powered sawmill on the east shore at the site now known as Glenbrook. Here they cut as much as ten thousand board feet a day, which they transported by wagon over the summit to the east and down Clear Creek to Carson Valley. The same year, Samuel Clemens (Mark Twain) staked a timber claim nearby—then watched in fascination as it went up in smoke after his

From Douglas Strong, *Tahoe: An Environmental History* (Lincoln: University of Nebraska Press, 1984), pp. 22–33. Reprinted by permission.

unattended campfire jumped out of control. Clemens later called Tahoe "the fairest picture the whole earth affords," yet the consequence of his brief sojourn on its shores was the scorching of nearby mountain slopes.

Mining of the Comstock Lode increased the demand for wood. An immense quantity of timber was required for square-set timbering, a mining technique used to shore up the ceiling of underground excavations; quantities of wood were also used for building materials, as fuel for steam pumps in the mines, and for other purposes. At the end of the long winter of 1867, Chinese merchants sold the roots of previously cut trees near Virginia City for $60 per cord.

When the scattered piñon and juniper trees near Virginia City and the forest on the east side of the Carson Range fell quickly to loggers and the price of lumber skyrocketed, mining interests looked hungrily at the vast and essentially untouched forests within the Tahoe Basin. . . . The only problem was how to quickly and economically transport the timber across many miles of rough, steep terrain to where it was needed. . . .

The railroad, together with the V-shaped flume, provided the key to the possible exploitation of Tahoe's forests. By January 1870, trains operated by the Virginia and Truckee Railroad hauled ore from the Comstock to the stamp mills on the Carson River and returned with wood from the Sierra. By 1873 and the Big Bonanza strike, more than thirty trains per day were using the twenty-one miles of single track between Carson City and Virginia City.

To a considerable extent the history of the lumber industry at Tahoe is the story of Duane L. Bliss and the Carson and Tahoe Lumber and Fluming Company [CTLFC]. . . . [The company] purchased large tracts of timberland on the slopes of the eastern side of the basin and bought the mills at Glenbrook House and other properties. In time CTLFC held more than fifty thousand acres in the basin, including many miles of shoreline, some of which had cost as little as $1.25 per acre. The company purchased additional acreage from the Central Pacific Railroad in the northern and northwestern parts of the basin: alternate sections of land that originally had been granted by the federal government to promote construction of the first transcontinental railroad. . . .

With control of a vast acreage of timberland and a transportation system extending from Spooner Summit to Virginia City, the next task of CTLFC was to streamline logging operations within the Tahoe Basin. In 1874 survey crews inspected the steep grades between Glenbrook and Spooner Summit, and the following summer large crews completed construction of nearly nine miles of narrow gauge track over eleven trestles and through a 487-foot tunnel along the climb to the pass. Teamsters with the aid of powerful block and tackle hauled three locomotives and many log cars up the steep mountainside from Carson Valley to Tahoe—a noteworthy accomplishment.

Loggers cut trees in various parts of the basin and brought them to the lakeshore by flume, greased skids, or teams of oxen. There the trees were gathered in large booms and pulled across the lake by steamer to the Glenbrook mills, where they were sawed into timber. The newly completed Lake Tahoe Railroad then hauled the lumber to a large receiving yard at Spooner Summit, nearly a thousand feet higher than the lake. After being dumped into the flume, which was fed by feeder flumes from Marlette Lake to the north and from the mountain streams to the south, the wood made the twelve-mile journey with a three-thousand-foot drop in elevation to the Carson Valley. Finally, the Virginia and Truckee Railroad carried the lumber and cordwood twenty miles to the Comstock market.

As CTLFC exhausted the readily available timber near Glenbrook, it extended operations to the south shore and the west side of the lake. . . . Although Bliss's mills dominated the Tahoe lumber industry, several small operators cut timber throughout the basin. One competitor stood out from all the others. Walter Scott Hobart organized the Sierra Nevada Wood and Lumber Company and established a mill at Incline at the northeastern end of the lake. From there he shipped timber to the Comstock by an ingenious system. First a narrow-gauge railroad carried logs to the mill from timbered slopes to the northwest and from Sand Harbor, which served as a gathering point for logs from company land at the south end of the basin. The mill had a double-tracked incline railway. Powered by a stationary forty horse-power steam engine that operated a bull wheel, the railway hoisted the logs up a steep hill, gaining fourteen hundred feet in elevation. From the top of the railway the timber was moved via flume and railroad to the Comstock. . . .

Sometime later the California State Forester criticized the logging practices of the Sierra Nevada Wood and Lumber Company—practices not uncommon in the Tahoe Basin. Loggers cut all trees of marketable size, leaving the area nearly denuded. In the process, mostly because of carelessness in felling and skidding trees, they destroyed a large part of the young growth. Slash, consisting of broken timber, branches, and tree tops, remained in a jumbled mass where it fell—a hindrance to the reproduction of new trees and a fire hazard. . . .

When E. A. Sterling, a former assistant in the U.S. Bureau of Forestry, inspected the basin in 1904, he reported that the forest cover had been materially changed because of both lumbering and fires. Only on the inaccessible upper slopes did the original forest, mainly firs, remain untouched by lumbermen. In regard to the best timber close to the lake, he remarked: "The forest is much reduced in density; brush and reproduction are competing for possession in the openings; the sugar pine has disappeared almost entirely, and is scantily represented in the reproduction; the finest of the Jeffrey and yellow pine and white fir has been removed, fir production in general [is] replacing the pine; while considerable areas have reverted entirely to brush." . . .

Fortunately, loggers had left enough defective trees to provide sufficient seed for reproduction. As a result, in the absence of fire, a dense new forest, ten to twenty feet high by 1904, had risen. White firs held competitive advantage, producing an abundance of seeds, growing well in the shade, and surviving in relatively poor seedbeds. These trees grew from amid the chapparal and the old slash left by logging operators, providing a forest cover that promised to restore the scenic beauty, if not the quality of timber, that had once graced the slopes of the Tahoe Basin.

But because of the denseness of the new growth and the abundance of the slash, the new forest was vulnerable to fires as never before. In September 1889, for example, the *Sacramento Record Union* reported: "Witnesses fresh from the Lake Tahoe region inform us that almost that entire section is a scene of desolation on account of the destruction of fires." . . . The article noted the potential consequences of the loss of vegetation: floods, soil erosion, reduction in wildlife, destruction of valuable timber, and desolation of the finest scenery. . . .

Sheepherders, who intentionally set many of the fires to encourage the growth of new vegetation, also damaged the forests by bringing their sheep to the mountains early in the season. The animals destroyed more than they ate, sliding on the wet slopes and trampling the young grasses. By autumn no feed remained, and visitors to the basin had to carry hay and barley for their horses. . . .

The passing of the logging era marked the end of a period in which a relatively few men—farmers, sheepherders, miners, fishermen, and loggers—attempted to profit from Tahoe's natural resources. Prospectors never found the mineral wealth they sought. Fisherman gradually depleted the resources on which they depended. Loggers quickly decimated the forest and in the process eliminated the employment base it had briefly provided. By the turn of the century, Tahoe no longer offered attractive opportunities for the exploitation of natural resources. Future development, including the urbanization of parts of the basin, rested instead, upon the slowly growing tourist trade.

Resistance to Logging

Frederica Bowcutt

Capitalism is not deterministic; it does not drive all of human behavior towards nature even in a culture as enthusiastic about the market economy as the United States. Two primary forces pose significant resistance against the trend to commodify living and non-living entities of the natural world: social resistance and natural resistance. Social resistance can take the form of labor unrest, grassroots activism including civil disobedience, restrictive environmental policies, legal challenge, and political pressure. Natural resistance includes erosion, difficult access, tidal action, soil fertility, succession, time requirements for regeneration, and climatic change. These two forms of resistance were found on Sinkyone lands between 1850 and 1900 as well as a dialectic between the capitalistic community and nature.

Prior to settlement by Euro-Americans, the Sinkyone tribe of the California Athabascans lived on a portion of the Pacific Northwest coastline which straddles Mendocino and Humboldt Counties, extending from Usal Creek north of Fort Bragg to Scotia south of Eureka and inland to the South Fork of the Eel River. The tribe was subdivided into tribelets each of which possessed a small territory usually defined in terms of a watershed. Each tribelet addressed issues of land ownership and trespass as a unit. Redwood was logged using elkhorn chisels and stone mauls. Once fallen, planks were split off and used for constructing homes. Dugout canoes were also made from redwood logs via the combined use of elkhorn chisels and fires. The Sinkyone, based on extant stories, placed high value on conformance to social mores. Among northwestern Californians, wealth conserved within the family unit conferred status because it was a sign of the family's conformance to socially condoned behavior. Transgressions against others would result in the loss of wealth because compensation for the transgressions would be collected.

Settlement of the Mendocino coast by peoples of European decent

From Frederica Bowcutt, "Sinkyone: A History of Resistance to the Commodification of Nature," by permission of the author, Evergreen State College, Olympia, Washington.

began in the mid-1800s with the establishment of the first lumber mill at Big River in 1852; by 1855, settlers were living all along the coast. Before the Mendocino coast was surveyed in late 1869, Euro-Americans considered land not owned by the state or private grant holders to be part of the public domain owned by the federal government. The timber on these commons was considered by many on the western frontier to be public property available to anyone. Due to the agricultural focus of federal land policy prior to 1878, no legal means existed to obtain timber rights; therefore, timber was exploited primarily by fraud. Before passage of the Timber and Stone Act of 1878, two laws were the primary means used to gain title to timber lands. The Preemption Act of 1841 allowed those who settled on unsurveyed land legally to claim a 160-acre homestead. After proving residence on the land, the claimant had two years to file for title. Fraudulent claims were made by lumbermen who logged the land within two years and then abandoned it without having to pay the federal government. The second law, the Homestead Act of 1862, allowed any United States citizen to claim title to 160 acres after living on it and cultivating it for five years. This law was also used fraudulently by lumbermen in the same way that the Pre-emption Act was.

An expanding metropolitan economy, combined with the transport of goods to the marketplace, transformed the western landscape between 1850 and 1900. The demand for timber drew lumbermen to the Pacific Northwest who had already deforested eastern states. Combined with the ease of defrauding the government, immense profit margins provided strong incentive to eastern lumbermen to move west and log old growth redwood and Douglas-fir.

Between 1863 and 1882 redwood production and consumption grew over 100 percent. Many expressed optimism that the supply was limitless. For over a decade the mill at Big River cut into the redwoods and still the forest seemed endless and almost untouched. In his 1880–1882 biennial report, the State Mineralogist asserted that: "The continual lumber supply capacity of a redwood forest, under judicious care, is so prodigious as to be simply incalculable; none but a suicidal and utterly abandoned infanticidal policy, wantonly and untiringly practiced, can ever blot them out."

Wood wastage typified the timber operations in the late 1800s. Steep slopes were often harvested by rolling logs into a creek which was damned at its mouth. Winter runoff would carry the logs to the mill pond. However, occasionally the dam failed and the old-growth timber was washed out to sea. Tanbark oak was also over exploited. As soon as new roads were constructed within tanbark oak stands, the bark was cut out. Although a valuable commodity, the wood did not command as high a price in the marketplace as did the bark. As a result, trees were cut down, stripped of bark, and "the valuable wood was left to perish on the ground."

The wasteful practices and racketeering of the early Pacific Northwest timber industry did not go uncontested. Two burgeoning groups, forestry professionals and environmentalists, voiced concern and advanced potential solutions. The concern expressed evolved from the recognition of natural resistance already evident in California's Pacific Northwest forests. The rhetoric of wise use emerging from the country's first forestry school was expressed succinctly by one of Gifford Pinchot's earlier students, Aldo Leopold, who wrote in "The Maintenance of Forests" (1904) "how can this useless destruction be prevented? . . . The whole fault lies in careless and unnecessary methods in handling forest lands. The lumberman, in the mad race for his dollar is blind to the future." . . .

A new class of professional foresters asserted that a *laissez-faire* economy would facilitate deforestation. Professional foresters questioned whether unrestrained commodification of forest products advanced the common good. The solution to these threats to the welfare of the state, they believed, was the application of scientific methods to the management of natural resources. In response to deforestation, the California State Board of Forestry advocated establishment of restrictive timber harvesting policies which included requirements for reforestation. The board also advanced the need for fire protection and instigated the enforcement of existing laws. In 1886, the California State Board of Forestry estimated that the redwood would last for about one hundred and twenty years.

Early grassroots efforts to influence timber policy and acquire lands for preservation were made by the Sierra Club and the Sempervirens Fund. The first victory for a private conservation organization was realized in 1902, when the Sempervirens Fund successfully convinced the California legislature to purchase an old growth stand of coast redwood in the Santa Cruz Mountains, which subsequently became Big Basin Redwoods State Park.

Sinkyone people resisted the privatization of land and the commodification of nature because it limited their ability to hunt and gather. With Euro-American settlement, wildlife habitat was diminished and the traditional cycle of seasonal moving to allow depleted animal and plant populations to recover was disrupted. Logging debris and associated eroded sediments choked salmon runs leading to Indian starvation.

The Sinkyone were relegated to a diminishing land base of poorer environmental quality, forcing them to exploit the livestock of settlers for sustenance. In 1860, the California Senate recognized that "the march of civilization deprives the Indian of his hunting grounds and other means of subsistence. . . . He naturally looks at this as an encroachment on his rights, and, either from revenge, or what is more likely in California, from the . . . pressing demands of hunger, kills the stock of the settlers as a means of subsistence, and in consequence thereof, a war is waged against the Indian."

The spiritual transgressions on the Sinkyone people by settlers also stimulated resistance. Along with salmon, tanbark oak acorns formed the primary staple of Sinkyone people. As a result tanbark oak held an esteemed position within the Sinkyone belief system. If one dreamed of tanbark oak, this was a sign of good luck. Because souls inhabited all rocks, trees, and animals within the Sinkyone cosmology, to the Sinkyone stripping the bark off tanbark oak was a spiritual travesty. Jack Woodman, one of the last Sinkyones raised within that culture, recounted a spirit's story received by Woodman when he was "dancing and singing for a cure dance." . . . Naigaicho passed through Briceland. . . . He came over Elk Ridge and he saw where white men had peeled tanbark. He said to me, "It looks just like my people lying around . . . with all their skin cut off. Tanoak has big power. . . . He saw men breaking rocks and plowing up grass. . . . He felt worst about the tanbark."

In an effort to alleviate tensions between indigenous peoples and settlers, in 1853 Congress authorized the establishment of five reservations for the relocation of California Indians with the idea that the people would be converted to an agrarian lifestyle. The Mendocino reservation supported over 3,000 by 1857. But runaways from the reservation were common and over 8,000 native people remained "wild" at most only occasionally residing on the reservation. Indigenous resistance continued against adopting a settled, agrarian existence based on private land ownership. The dominant settlers imposed their world view violently. Estimates of indigenous peoples killed in the Coast Range during a four month period in 1860 exceeded the number of Indians killed during the century of Spanish and Mexican domination. Acts of genocide were commonplace in the region.

The natural resources exploited on Sinkyone land for the market economy were primarily three tree species: coast redwood (*Sequoia sempervirens*), Douglas-fir (*Pseudotsuga menziesii*), and tanbark oak (*Lithocarpus densiflorus*). All three constrained capitalist interests in the region due to growth patterns, physiological limits, reproductive characteristics, wood qualities, and ecological relationships. Abiotic conditions affecting access also obstructed efforts to extract timber products from Sinkyone land.

The large girth of the old-growth redwoods strained the capabilities of the early mills. Logs often had to be split before mill machinery could handle them. In response to this serious impediment to processing, the timber industry responded by constructing larger mills with heavier equipment.

Redwood reproductive capabilities limited the rate of reforestation of logged redwood lands. Although coast redwood has strong vegetative reproductive qualities, sexual production is often poor. And if multiple fires sweep through an area of stump sprouting redwoods, the stumps begin to lose vitality and the ability to respond. Given extreme volumes of waste wood left on the ground in logged areas, fires, if they spread through the area, often reduced site fecundity.

Attempts to convert logged redwood land to other profitable land uses were often thwarted. The extensive root systems of redwoods often persisted in the soil, making conversion of redwood lands to arable soil unprofitable. Attempts to convert logged over land to pasturage, often resulted in severe erosion due to steep slopes and the instability of the region's geologic formations.

Along with redwood, Douglas-fir filled almost all the building trade needs by the late 1800s. Given its superior sexual reproductive vigor, Douglas-fir was rapidly increasing while redwoods were not. Due to the relatively poor decay resistance properties of Douglas-fir wood, it was not as highly valued as redwood.

Exceptionally high quality tanbark stands grew on the summit and upper slopes of the ridges between Usal Creek and the Mattole River. Both the wood and the bark had economic value: "The wood furnishes one of the finest stove woods on the coast, and always fetches the highest price. . . . There is always a market for the bark." The bark, which is rich in tannin, was prized by tanners for the processing of leathers for the production of cargo chests and saddlery. Already by 1886, foresters and capitalists recognized the supply was rapidly being exhausted. Foresters in response advocated adopting a superior harvesting technique that did not girdle the trees and therefore allowed for future bark production by the same trees.

Because great sums of capital were invested in San Francisco tanneries, tanners expressed alarm at the diminishing supply and the inflating price of tanbark. Tanners became willing to experiment with other potential sources of tannin when they recognized that they could not make an adequate profit if required to import tannin sources from the East.

Clearcutting of redwood forests began in the 1890s; until World War II, clearcutting was restricted primarily to areas accessible from the ocean. The accessibility of timber and the cost to transport it to a market largely controlled the profitability of logging. Until an adequate road system was established, forest products were transported by schooner over the Pacific Ocean. Initially wood products were shipped to market via schooners which were loaded offshore using a cable and block system either from a wharf or the bluff. Positioning a schooner for loading required a great deal of seamanship to avoid being driven ashore or onto the rocks. Several short railroad lines were built to haul logs from the slopes to the ocean for loading. Local homesteaders as well as urban exporters supplied foodstuffs to the logging camps, which were occupied from spring until the winter rains began.

In 1888, a mill was established along with a small company town for the workforce. A wharf was built which extended out into the ocean from Usal beach. It was later destroyed by a tidalwave. By 1901, the

FIGURE 5. LOADING TIMBER ONTO SHIPS.
Before adequate roads existed, logs were hauled to the coast where schooners were loaded from bluffs using block and cable systems and apron shoots. Here logs are loaded at Union Landing, Mendocino County, California. Courtesy, The Bancroft Library, University of California, Berkeley.

woods were "logged out"; the mill was closed and subsequently destroyed by fire. The small town persisted including a store and a hotel in Hotel Gulch. The lagoon was filled with silt by the 1890s probably due to erosion from upstream logging. The result was a severe degradation of the fisheries habitat in the vicinity of the prehistoric Sinkyone village.

Between 1850 and 1950, the capitalistic community exploiting Sinkyone land was increasingly engaged in a dialectical relationship with nature and those who defended nature. Legal challenges began to temper their actions. In response, the timber industry adapted logging to conform to new restrictive policies. Transgressions could result in judicial repercussions. Conformance to social mores regarding appropriate land use began to be expected by the public. This emerging cultural value, which paralleled a pre-existing indigenous value, began to irritate the individualism of Euro-American settlers. The groundwork for later reformations, which placed greater restrictions on loggers, was laid down during this period in response to the wasteful practices of the industry. Reforestation, erosion control measures, and fire prevention all evolved as part of contemporary forest policy in part due to the advocacy efforts of the California Board of Forestry. Recognition of natural limits stimulated response. Fear of depleting tanbark oak and redwood motivated the planting of wattle and blue gum eucalyptus on Sinkyone land.

Between 1976 and 1987, the California Department of Parks and Recreation acquired the land as the Sinkyone Wilderness State Park. It is managed by Department of Parks and Recreation primarily for wilderness recreation (backpacking) and preservation. Native Americans also manage the park for ceremonial purposes as well as traditional gathering of plant materials under use permits.

Further Readings

Arvola, T. F. *Regulation of Logging in California, 1945–1975*. Sacramento: California Division of Forestry, 1976.

Axelrod, Daniel I. *History of the Coniferous Forests, California and Nevada*. Berkeley: University of California Press, 1976.

Bari, Judi. *Timber Wars*. Monroe, ME: Common Courage Press, 1994.

Brown, Alan K. *Sawpits in the Spanish Redwoods, 1787–1849*. San Mateo, CA: San Mateo County Historical Society, 1966.

Carranco, Lynwood F., and John T. Labbe. *Logging the Redwoods*. Caldwell, ID: Caxton Printers, 1975.

Chaney, Ralph W. *Redwoods of the Past*. Berkeley: Save-the-Redwoods League, 1964.

Chase, Doris Harter. *They Pushed Back the Forest*. Sacramento: Author, 1959.

Clar, C. Raymond. *Harvesting and Use of Lumber in Hispanic California*. Sacramento Corral of Westerners, 1971.

Cox, Thomas R. S. *Mills and Markets: A History of the Pacific Coast Lumber Industry to 1900*. Seattle: University of Washington Press, 1974.

Cox, Thomas, R. S. Maxwell, P. D. Thomas, and J. J. Malone. *This Well-Wooded Land: Americans and Their Forests from Colonial Times to the Present*. Lincoln: University of Nebraska Press, 1985.

Coy, Owen C. *The Humboldt Bay Region, 1850–1875: A Study in the American Colonization of California*. Los Angeles: California State Historical Association, 1929.

Eddy, J. M. *In the Redwood's Realm*. San Francisco: D. S. Stanley & Co. 1893.

Dana, Samuel Trask, and Sally Fairfax. *Forest and Range Policy: Its Development in the U.S.* New York: McGraw-Hill, 1980 [1956].

Fahl, Ronald J. *North American Forest and Conservation History: A Bibliography*. Santa Barbara: Clio, 1976.

Farquhar, Francis P. *History of the Sierra Nevada*. Berkeley: University of California Press, 1965.

Fritz, Emanuel. *The Development of Industrial Forestry in California*. Seattle: University of Washington College of Forestry, 1960.

Frome, Michael. *Whose Woods These Are: The Story of the National Forests*. Garden City, NY: Doubleday, 1962.

Hawley, A.T. *The Climate, Resources, and Advantages of Humboldt County, California*. Eureka, CA: J.E. Wyman and Son, 1879.

Hays, Samuel P. *Conservation and the Gospel of Efficiency: The Progressive Conservation Movement, 1890–1920*. Cambridge, MA: Harvard University Press, 1959.

Hutchinson, W. H. *California Heritage: A History of Northern California Lumbering*. Santa Cruz, CA: The Forest History Society, 1974.

Hyde, Phillip, and Francois Leydet. *The Last Redwoods*. San Francisco: Sierra Club, 1963.

Irvine, Leigh, *History of Humboldt County, California*. Los Angeles: Historic Record CO, 1915.

Johnston, Hank. *Rails to the Minarets: The Story of the Sugar Pine Lumber Company*. Corona del Mar, CA: Trans-Anglo Books, 1930.

Lewis, Oscar. *The Quest for Qual-a-wa-loo, [Humboldt Bay]*. San Francisco, 1943.

Lockmann, Ronald F. *Guarding the Forests of Southern California: Evolving*

Attitudes toward Conservation of Watershed, Woodlands, and Wilderness. Glendale, CA: A.H. Clark, 1981.

Merriam, John Campbell. *The Highest Uses of the Redwoods: Messages to the Council of the Save-the-Redwoods League, 1922–1941*. Berkeley, CA: Save-the-Redwoods League, 1941.

Monteagle, Frederick J. *A Yankee Trader in the California Redwoods*. Oakland, CA: East Bay Regional Park District, 1976.

Nash, Gerald, and R. W. Etulain, eds. *The Twentieth-Century West: Historical Interpretations*. Albuquerque: University of New Mexico Press, 1989.

Nash, Roderick. *Wilderness and the American Mind*. New Haven: Yale University Press, 1967.

Noyes, C. G. *Redwood and Lumbering in California Forests*. San Francisco: Edgar Cherry & Co., 1884.

Otter, Floyd L. *The Men of Mammoth Forest: A Hundred-Year History of a Sequoia Forest and Its People in Tulare County, California*. Ann Arbor: Edwards Brothers, 1963.

Palais, Hyman. "Pioneer Redwood Logging in Humboldt County," *Journal of Forest History* 17 (Jan. 1974).

Richards, Gilbert. *Crossroads: People and Events of the Redwoods of San Mateo County*. Woodside, CA: Gilbert Richards Publications, 1973.

Robbins, William G. *Lumberjacks and Legislators: A Political Economy of the U.S. Lumber Industry, 1890–1941*. College Station: Texas A&M University Press, 1982.

Robinson, Glen O. *The Forest Service: A Study in Land Management*. Baltimore: Johns Hopkins University Press, 1975.

Robinson, John. *The San Gabriels: Southern California Mountain Country*. San Marino, CA: Golden West Books, 1977.

Robinson, William W. *The Forest and the People: The Story of the Angeles National Forest*. Los Angeles: Title Insurance and Trust, 1946.

Runte, Alfred. *Public Lands, Public Heritage: The National Forest Idea*. Niwot, CO: R. Rinehart Publishers, 1991.

Schrepfer, Susan R. *The Fight to Save the Redwoods: A History of Environmental Reform, 1917–1978*. Madison: The University of Wisconsin Press, 1983.

Stanger, Frank M. *Sawmills in the Redwoods: Logging on the San Francisco Peninsula, 1849–1967*. San Mateo, CA: San Mateo County Historical Society, 1967.

Steen, Harold, ed. *The Origins of the National Forests*. Durham, NC: Forest History Society, 1992.

Strong, Douglas H. *Tahoe: An Environmental History*. Lincoln: University of Nebraska Press, 1984.

Strong, Douglas H. *Trees—Or Timber? The Story of Sequoia and Kings Canyon National Parks*. Three Rivers, CA: Sequoia Natural History Association in cooperation with the National Park Service, n.d.

Williams, Michael. *Americans and Their Forests: A Historical Geography*. Cambridge: Cambridge University Press, 1989.

Chapter 6

Rangelands Exploited

Documents

Richard Henry Dana on the Hide and Tallow Trade, 1840

California extends along nearly the whole of the western coast of Mexico, between the gulf of California in the south and the bay of Sir Francis Drake on the north. . . . Upper California has the seat of its government at Monterey, where is also the customhouse, the only one on the coast, and at which every vessel intending to trade on the coast must enter its cargo before it can commence its traffic. We were to trade upon this coast exclusively, and therefore expected to go to Monterey at first; but the captain's orders from home were to put in at Santa Barbara, which is the central port of the coast, and wait there for the agent who lives there, and transacts all the business for the firm to which our vessel belonged. . . .

As it was January [1835] when we arrived, and the middle of the south-easter season, we accordingly came to anchor at the distance of three miles from the shore, in eleven fathoms water, and bent a slip-rope and buoys to our cables, cast off the yard-arm gaskets from the sails, and stopped them all with rope-yarns. After we had done this, the boat went

From Richard Henry Dana, *Two Years before the Mast*, ed. Thomas Philbrick (New York: Harper Brothers, 1840), pp. 67–68, 71–73, 107–08, 119–21, 150–51, 173–74.

ashore with the captain, and returned with orders to the mate to send a boat ashore for him at sundown. . . .

Just before sun-down the mate ordered a boat's crew ashore, and I went as one of the number. We passed under the stern of the English brig, and had a long pull ashore. I shall never forget the impression which our first landing on the beach of California made upon me. The sun had just gone down; it was getting dusky; the damp night wind was beginning to blow, and the heavy swell of the Pacific was setting in, and breaking in loud and high "combers" upon the beach. We lay on our oars in the swell, just outside of the surf, waiting for a good chance to run in, when a boat, which had put off from the Ayacucho just after us, came alongside of us, with a crew of dusky Sandwich Islanders, talking and hallooing in their outlandish tongue. . . .

They ran the boat into the water so far that every large sea might float her, and two of them, with their trowsers rolled up, stood by the bows, one on each side, keeping her in her right position. This was hard work; for beside the force they had to use upon the boat, the large seas nearly took them off their legs. The others were running from the boat to the bank, upon which, out of the reach of the water, was a pile of dry bullocks' hides, doubled lengthwise in the middle, and nearly as stiff as boards. These they took upon their heads, one or two at a time, and carried down to the boat, where one of their number stowed them away. They were obliged to carry them on their heads, to keep them out of the water, and we observed that they had on thick woolen caps. . . .

After they had got through with the hides, they laid hold of the bags of tallow (the bags are made of hide, and are about the size of a common meal bag) and lifting each upon the shoulders of two men, one at each end, walked off with them to the boat, and prepared to go aboard. . . .

After a few days, finding the trade beginning to slacken, we hove our anchor up, set our topsails, ran the stars and stripes up to the peak, fired a gun, which was returned from the presidio, and left the little town astern, running out of the bay, and bearing down the coast again, for Santa Barbara. . . . As we drew near the islands off Santa Barbara, [the wind] died away a little, but we came-to at our old anchoring-ground in less than thirty hours from the time of leaving Monterey. . . .

We lay here about a fortnight, employed in landing goods and taking off hides, occasionally, when the surf was not high; but there did not appear to be one half the business doing here that there was in Monterey. In fact, so far as we were concerned, the town might almost as well have been in the middle of the Cordilleras. We lay at a distance of three miles from the beach, and the town was nearly a mile farther, so that we saw little or nothing of it. Occasionally we landed a few goods, which were taken away by Indians in large, clumsy oxcarts, with the yoke on the ox's neck instead of under it, and with small solid wheels. A few hides were brought down, which we carried off in the California style. This we had now got pretty well accustomed to; and hardened to also; for it does require a little hardening even to the toughest.

The hides are always brought down dry, or they would not be received. When they are taken from the animal, they have holes cut in the ends, and are staked out, and thus dried in the sun without shrinking. They are then doubled once, lengthwise, with the hair side usually in, and sent down upon mules or in carts, and piled above high-water mark; and then we take them upon our heads, one at a time, or two, if they are small, and wade out with them and throw them into the boat, which, as there are no wharves, we usually kept anchored by a small kedge, or keeleg, just outside of the surf. We all provided ourselves with thick Scotch caps, which would be soft to the head, and at the same time protect it; for we soon found that however it might look or feel at first, the "head-work" was the only system for California. For besides that the seas, breaking high, often obliged us to carry the hides so, in order to keep them dry, we found that, as they were very large and heavy, and nearly as stiff as boards, it was the only way that we could carry them with any convenience to ourselves. . . .

San Diego was a small, snug place, having very little trade, but decidedly the best harbor on the coast, being completely land-locked, and the water as smooth as a duck-pond. This was the depot for all the vessels engaged in the trade; each one having a large house there, built of rough boards, in which they stowed their hides, as fast as they collected them in their trips up and down the coast, and when they had procured a full cargo, spent a few weeks there taking it in, smoking ship, supplying wood and water, and making other preparations for the voyage home. . . .

I also learned, to my surprise, that the desolate-looking place we were in was the best place on the whole coast for hides. It was the only port for a distance of eighty miles, and about thirty miles in the interior was a fine plane country, filled with herds of cattle, in the centre of which was the Pueblo de los Angelos—the largest town in California—and several of the wealthiest missions; to all of which San Pedro was the sea-port. . . .

The next day we pulled the agent ashore, and he went up to visit the Pueblo and the neighboring missions; and in a few days, as the result of his labors, large ox-carts, and droves of mules, loaded with hides, were seen coming over the flat country. We loaded our long-boat with goods of all kinds, light and heavy, and pulled ashore. After landing and rolling them over the stones upon the beach, we stopped, waiting for the carts to come down the hill and take them; but the captain soon settled the matter by ordering us to carry them all up to the top, saying that, that was "California fashion." So what the oxen would not do, we were obliged to do. . . . The greatest trouble was with the large boxes of sugar. These, we had to place upon oars, and lifting them up, rest the oars upon our shoulders, and creep slowly up the hill with the gait of a funeral procession. After an hour or two of hard work, we got them all up, and found the carts standing full of hides, which we had to unload, and also to load again with our own goods; the lazy Indians, who came down with them, squatting down on their hams, looking on, doing nothing, and when we asked them to help us, only shaking their heads, or drawling out "no quiero."

Having loaded the carts, we started up the Indians, who went off, one on each side of the oxen, with long sticks, sharpened at the end, to punch them with. This is one of the means of saving labor in California—two Indians to two oxen. Now, the hides were to be got down; and for this purpose, we brought the boat round to a place where the hill was steeper, and threw them down, letting them slide over the slope. Many of them lodged, and we had to let ourselves down and set them agoing again; and in this way got covered with dust, and our clothes torn. After we had got them all down, we were obliged to take them on our heads, and walk over the stones, and through the water, to the boat. The water and the stones together would wear out a pair of shoes a day, and as shoes were very scarce and very dear, we were compelled to go barefooted. At night, we went on board, having had the hardest and most disagreeable day's work that we had yet experienced. For several days, we were employed in this manner, until we had landed forty or fifty tons of goods, and brought on board about two thousand hides; when the trade began to slacken, and we were kept at work, on board, during the latter part of the week, either in the hold or upon the rigging. . . .

For landing and taking on board hides, San Diego is decidedly the best place in California. The harbor is small and land-locked; there is no surf; the vessels lie within a cable's length of the beach; and the beach itself is smooth, hard sand, without rocks or stones. . . . We took possession of one of the hide-houses, which belonged to our firm, and had been used by the California. It was built to hold forty thousand hides, and we had the pleasing prospect of filling it before we could leave the coast; and toward this, our thirty-five hundred, which were brought down with us, would do but little. . . .

The hides, as they come rough and uncured from the vessels, are piled up outside of the houses, whence they are taken and carried through a regular process of pickling, drying, cleaning, etc., and stowed away in the house, ready to be put on board. This process is necessary in order that they may keep, during a long voyage and in warm latitudes. For the purpose of curing and taking care of these hides, an officer and a part of the crew of each vessel are usually left ashore. . . .

We pulled aboard, and found the long-boat hoisted out, and nearly laden with goods; and after dinner, we all went on shore in the quarter-boat, with the long-boat in tow. . . . Here the country stretched out for miles, as far as the eye could reach, on a level, table surface; and the only habitation in sight was the small white mission of San Juan Campestrano, with a few Indian huts about it, standing in a small hollow, about a mile from where we were. Reaching the brow of the hill where the cart stood, we found several piles of hides, and Indians sitting round them. One or two other carts were coming slowly on from the mission, and the captain told us to begin and throw the hides down. This, then, was the way they were to be got down: thrown down, one at a time, a distance of four hundred feet! . . .

Down this height we pitched the hides, throwing them as far out into the air as we could; and as they were all large, stiff, and doubled, like the cover of a book, the wind took them, and they swayed and eddied about, plunging and rising in the air, like a kite when it has broken its string. As it was now low tide, there was no danger of their falling into the water, and as fast as they came to ground, the men below picked them up, and taking them on their heads, walked off with them to the boat. It was really a picturesque sight: the great height; the scaling of the hides; and the continual walking to and fro of the men, who looked like mites, on the beach! This was the romance of hide-droghing!

Some of the hides lodged in cavities which were under the bank and out of our sight, being directly under us; but by sending others down in the same direction, we succeeded in dislodging them. Had they remained there, the captain said he should have sent on board for a couple of pair of long halyards, and got some one to have gone down for them. It was said that one of the crew of an English brig went down in the same way, a few years before. We looked over, and thought it would not be a welcome task, especially for a few paltry hides; but no one knows what he can do until he is called upon; for, six months afterwards, I went down the same place by a pair of top-gallant studding-sail halyards, to save a half a dozen hides which had lodged there.

Having thrown them all down, we took our way back again, and found the boat loaded and ready to start. We pulled off; took the hides all aboard; hoisted in the boats; hove up our anchor; made sail; and before sundown, were on our way to San Diego.

Guadalupe Vallejo Recalls the Rancheros, 1890

It seems to me that there never was a more peaceful or happy people on the face of the earth than the Spanish, Mexican, and Indian population of Alta California before the American conquest. We were the pioneers of the Pacific coast, building towns and Missions while General Washington was carrying on the war of the Revolution, and we often talk together of the days when a few hundred large Spanish ranches and Mission tracts occupied the whole country from the Pacific to the San Joaquin. No class of American citizens is more loyal than the Spanish Californians, but we shall always be especially proud of the traditions and memories of the long pastoral age before 1840. . . .

My mother has told me much, and I am still more indebted to my illustrious uncle, General Vallejo, of Sonoma, many of whose recollections are incorporated in this article.

When I was a child there were fewer than fifty Spanish families in the region about the bay of San Francisco, and these were closely connected by ties of blood or intermarriage. My father and his brother, the late General Vallejo saw, and were a part of, the most important events in the history of Spanish California the revolution and the conquest. My grandfather, Don Ygnacio Vallejo, was equally prominent in his day, in the exploration and settlement of the province. The traditions and records of the family thus cover the entire period of the annals of early California, from San Diego to Sonoma.

What I wish to do is to tell, as plainly and carefully as possible, how the Spanish settlers lived, and what they did in the old days. The story will be partly about the Missions, and partly about the great ranches. . . .

The principal sources of revenue which the Missions enjoyed were the sales of hides and tallow, fresh beef, fruits, wheat, and other things to ships, and in occasional sales of horses to trappers or traders. The Russians at Fort Ross, north of San Francisco, on the Bodega Bay, bought a good deal from the Missions.

The embarcadero, or "landing," for the Mission San Jose was at the mouth of a salt-water creek four or five miles away. When a ship sailed

From Guadalupe Vallejo, "Ranch and Mission Days in Alta California," *The Century Magazine*, 41, no. 2 (1890): 183–92.

into San Francisco Bay, and the captain sent a large boat up this creek and arranged to buy hides, they were usually hauled there on an ox cart with solid wooden wheels, called a carreta. But often in winter, there being no roads across the valley, each separate hide was doubled across the middle and placed on the head of an Indian. Long files of Indians, each carrying a hide in this manner, could be seen trotting over the unfenced level land through the wild mustard to the embarcadero, and in a few weeks the whole cargo would thus be delivered. For such work the Indians always received additional gifts for themselves and families.

A very important feature was the wheat harvest. Wheat was grown more or less at all the Missions. If those Americans who came to California in 1849 and said that wheat would not grow here had only visited the Missions they would have seen beautiful large wheatfields. . . . At the Mission San Jose a tract about a mile square came to be used for wheat. It was fenced in with a ditch, dug by the Indians with sharp sticks and with their hands in the rainy season, and it was so deep and wide that cattle and horses never crossed it. In other places stone or adobe walls, or hedges of the prickly pear cactus, were used about the wheatfields. Timber was never considered available for fences because there were no sawmills and no roads to the forests, so that it was only at great expense and with extreme difficulty that we procured the logs that were necessary in building, and chopped them slowly, with poor tools, to the size we wanted. Sometimes low adobe walls were made high and safe by a row of the skulls of Spanish cattle, with the long curving horns attached. These came from the matanzas or slaughtercorrals, where there were thousands of them lying in piles and they could be so used to make one of the strongest and most effective of barriers against man or beast. Set close and deep, at various angles, about the gateways and corral walls, these cattle horns helped to protect the inclosures from horse-thieves. . . .

Early in the century flour-mills by water were built at Santa Cruz, San Luis Obispo, San Jose, and San Gabriel. The ruins of some of these now remain; the one at Santa Cruz is very picturesque. Horse-power mills were in use at many places. About the time that the Americans began to arrive in numbers the Spanish people were just commencing to project larger mill enterprises and irrigation ditches for their own needs. The difficulties with land titles put an end to most of these plans, and some of them were afterward carried out by Americans when the ranches were broken up.

One of the greatest of the early irrigation projects was that of my grandfather, Don Ygnacio Vallejo, who spent much labor and money in supplying San Luis Obispo Mission with water. This was begun in 1776, and completed the following year. He also planned to cart the water of the Carmel River to Monterey; this has since been done by the Southern Pacific Railway Company. My father, Don J. J. Vallejo, about fifty years ago made a stone aqueduct and several irrigation and mill ditches from the Alameda Creek, on which stream he built an adobe flour-mill, whose millstones were brought from Spain. . . .

Hemp and flax were grown to some extent. A fine large cane, a native of Mexico, was planted, and the joints found useful as spools in the blanket factory, and for many domestic purposes. The young shoots of this cane were sometimes cooked for food. . . .

In 1806 there were so many horses in the valleys about San Jose that seven or eight thousand were killed. Nearly as many were driven into the sea at Santa Barbara in 1807, and the same thing was done at Monterey in 1810. Horses were given to the runaway sailors, and to trappers and hunters who came over the mountains for common horses were very plenty, but fast and beautiful horses were never more prized in any country than in California, and each young man had his favorites. A kind of mustang, that is now seldom or never seen on the Pacific coast, was a peculiar light cream adored horse, with silver-white mane and tail. . . .

A number of trappers and hunters came into Southern California and settled down in various towns. There was a party of Kentuckians, beaver-trappers, who went along the Gila and Colorado rivers about 1827, and then south into Baja California to the Mission of Santa Catalina. Then they came to San Diego, where the whole country was much excited over their hunter clothes, their rifles, their traps, and the strange stories they told of the deserts, and fierce Indians, and things that no one in California had ever seen. . . . After 1828 a good many other Americans came in and settled down quietly to cultivate the soil, and some of them became very rich. They had grants from the governor, just the same as the Spanish people. It is necessary, for the truth of the account, to mention the evil behavior of many Americans before, as well as after, the conquest. . . .

Perhaps the most exasperating feature of the coming-in of the Americans was owing to the mines, which drew away most of the servants, so that our cattle were stolen by thousands. Men who are now prosperous farmers and merchants were guilty of shooting and selling Spanish beef "without looking at the brand," as the phrase went. My father had about ten thousand head of cattle, and some he was able to send back into the hills until there were better laws and officers, but he lost the larger part. . . .

Every Mission and ranch in old times had its calaveras, its "place of skulls," its slaughter corral, where cattle and sheep were killed by the Indian butchers. Every Saturday morning the fattest animals were chosen and driven there, and by night the hides were all stretched on the hill-side to dry. At one time a hundred cattle and two hundred sheep were killed weekly at the Mission San Jose, and the meat was distributed to all, "without money and without price." The grizzly bears, which were very abundant in the country—for no one ever poisoned them, as the American stock raisers did after 1849—used to come by night to the ravines near the slaughter-corral where the refuse was thrown by the butchers. The young Spanish gentlemen often rode out on moonlight nights to lasso these bears, and then they would drag them through the village street and past the houses of their friends. . . . Innumerable stories about grizzlies are traditional in the old Spanish families, not only in the Santa Clara Valley, but also through the Coast Range from San Diego to Sonoma and Santa Rosa. Some of the bravest of the young men would go out alone to kill grizzlies. . . .

There was a group of warm springs a few miles distant from the old adobe house in which we lived. . . . Sometimes we heard the howl of coyotes, and the noise of other wild animals in the dim dawn, and then none of the children were allowed to leave the carreta. A great dark mountain rose behind the hot spring, and the broad, beautiful valley, unfenced, and dotted with browsing herds, sloped down to the bay as we climbed the cañon to where columns of white steam rose among the oaks, and the precious waters, which were strong with sulfur, were seen flowing over the crusted basin, and falling down a worn rock channel to the brook. Now on these mountain slopes for miles are the vineyards of Josiah Stanford, the brother of Senator Leland Stanford, and the valley below is filled with towns and orchards.

A Rancher Saves the Tule Elk, 1915

The complete extermination of any species of animal or plant in any part of its habitat is always a matter of regret. Even if the species be a noxious one, we nevertheless dislike to see it entirely wiped out in any locality in

From Barton Warren Evermann, "An Attempt to Save California Elk," *California Fish and Game*, 1, no. 3 (1915): 85–95, excerpt.

which it was naturally found. If it be a useful species, well known to the laity as well as to naturalists, its extermination is deplored; and when the species becomes extinct, when not a single individual is left anywhere upon the face of the earth, it is regarded as most regrettable. The world will never cease to regret the practical extermination of the buffalo. It will never cease to deplore the actual extinction of the great auk and the passenger pigeon. . . .

Among the important species of California animals now threatened with extinction is the California valley elk (*Cerrus nannodes*). This elk originally roamed in great numbers over the great interior valley of California. It was doubtless most abundant in the San Joaquin portion of the Sacramento-San Joaquin Valley, but its range probably included the entire valley and the adjacent foothills. It was certainly abundant as late as 1854. The early records contain many references to its abundance. One of the earliest records is to be found in the manuscript report of the Viscaino explorations made in 1602. Speaking of the animals in the vicinity of Monterey the statement is made: "Among the animals there are large, fierce bears, and other animals called elks, from which they make elk leather jackets." . . .

Whether the Marin County elk were of the same species as the San Joaquin Valley elk is not certainly known. It may be that the elk of the heavily forested, humid region along the coast from Marin County northward is a distinct species. The facts can be determined only by comparison of material from the two regions. But whatever may be the facts as regards this matter, it is clear that elk were very abundant in the San Joaquin Valley and adjacent foothills, certainly as late as 1850 to 1854. From that time they decreased rapidly. In the early seventies it is said the herd had been reduced to a few individuals—one report says to a single pair—and they were on the Kern County ranch of Messrs. Miller and Lux. It is said that the imminent extinction of the species came to the attention of Mr. Henry Miller of the Miller & Lux Company, and he immediately gave strict orders to all the employees of the company that the elk must not be disturbed under any circumstances, and that everything possible for their protection should be done.

That has been the policy of Messrs. Miller and Lux to this day. The animals were protected. The herd increased. In 1914 it was estimated to contain about four hundred animals. The state game law makes the killing of any elk a felony, punishable by imprisonment for a term not exceeding two years. Although the elk roam at will over the Miller & Lux ranch, doing—the company estimates—from $5,000 to $10,000 worth of damage every year to the alfalfa and Egyptian corn fields and to the fences, they have not been disturbed. That the species was not exterminated is due, without doubt, to the intelligent interest taken in its preservation by Mr. Henry Miller. It must be admitted, however, that Messrs. Miller and Lux are willing, in view of the very considerable loss the elk are causing them each year, to have the herd reduced somewhat by moving some of the animals to suitable reservations in other parts of the state. . . .

In 1905, a few elk were taken from this herd and placed in the Sequoia National Park, where they have done fairly well. These and the original Kern County herd are the only elk of this species in existence.

The development of the oil fields and the expansion of agricultural operations in the Kern County region have brought many dangers to the elk herd in that locality. To aid in saving the species from possible extermination it was proposed to place a few elk in each of the various reservations and parks in the state in the hope that they might thrive there and become the nuclei of new herds.

This was the hope of Messrs. Miller and Lux. One of the conditions of their offer was that the animals should be put only in places affording a favorable environment and where they would probably breed.

The offer of Miller & Lux was accepted by the Academy, and plans were made for carrying out the undertaking. . . . After giving the matter careful consideration it was decided to undertake the catching and shipping of the elk in October. Early in that month Messrs. Miller and Lux constructed a corral one fourth of a mile long and one eighth of a mile wide in an alfalfa field into which the elk were observed to come every night to feed. A wing one fourth of a mile long was run out from each corner of the end toward the foothills. Woven fence wire was put upon the wings at once, but only the posts for the corral proper were placed at that time. After the elk had come down into the field several nights and gotten used to the posts, heavy woven fence wire was placed on the two sides and the rear end of the corral, and the following night about 150 elk came into the corral; then the wire was placed on the posts at the entrance and the animals were trapped.

The wire fence was very strong and at least eight feet high; nevertheless, some broke through or jumped over it. A good many people came out in automobiles and otherwise to see the elk, and so frightened them that about 100 broke out the first afternoon. Those that remained became quite tame in a few days. Various, diverse and unexpected difficulties came up every day and it was not possible to predict what success would be attained in the undertaking. . . . However, it is gratifying to know that, in spite of all difficulties and uncertainties, Messrs. Miller and

Lux succeeded in capturing and placing in the cars for shipment a total of fifty-four elk. These were disposed of as follows:

> 1. To a thousand acre private reservation of Mr. J. M. Danziger, Los Angeles, six elk.
>
> This reservation is in the Santa Monica Mountains, near Los Angeles. The environment, it is believed, will prove very favorable.
>
> 2. To a six hundred acre private reservation of Mr. E. L. Doheny, Los Angeles, ten elk.
>
> This reservation also is in the Santa Monica Mountains, only a short distance from the Danziger ranch, and is under elk-proof fence.
>
> 3. To a seven hundred acre park of Mr. S. C. Evans, Riverside, four elk.
>
> This park adjoins the city limits of Riverside and furnishes ideal conditions.
>
> 4. To the San Diego City Park, twelve elk.
>
> The conditions here are not entirely as favorable as one would desire, but it is believed the elk will do well. This park was regarded as a favorable location in which to try the experiment of keeping the elk in relatively small enclosures.
>
> 5. To the Modesto City Park, two elk.
>
> 6. To the California Redwood Park Association, ten elk.
>
> This association is the governing body for the Big Basin reservation, which comprises some 55,000 acres. It is believed the conditions obtaining there will prove favorable.
>
> 7. To the Del Monte Park, ten elk.

These elk were turned loose in the large reservation of the Pacific Improvement Company near Monterey. The environment, it is believed, will prove favorable.

Recent reports received from the various parties to whom the elk were sent state that the animals are doing well in all cases.

A Woman Describes the Decline in Range Forage, 1945

During the period before the introduction of cattle into California, native grasses had perfected their ecological adaptation to the environment, and the land was covered with verdant grasses when the first Spanish settlers arrived. That the grasses then present were of exceptional pasturage quality is evidenced in the diaries and accounts of the newcomers. Such remarks as that of Father Francisco Palou in 1769 that there were places of, "Pure earth well covered with grass . . . and a very beautiful valley which when we saw it seemed to us to be nothing less than a cultivated cornfield or farm, on account of the mass of verdure," were common to most observers. . . .

In fact, so plentiful and nutritious was the herbage of the California range during the early years of occupation that little thought was given to the possibility of range depletion. Spanish and Mexican rancheros, and indeed the early American range-cattle owners, were not in the least disturbed by any impending damage to range forage. The range was grazed for decade after decade, and if the flora changed at all during the time, little attention was paid to it. Droughts occurred with their temporary grass shortages, but the cattle herds grew steadily larger in numbers, and with the lax and easy methods prevailing in the industry, range deterioration mattered little.

During the entire period, however, a change in the native flora had been taking place. Native bunch grasses were gradually being pushed out and new, less palatable plants were taking their places. Several factors operated to effect this change.

Probably the greatest single factor in range deterioration was the damage done to native plants by ranging livestock stock. Bunch grasses, with their rounded bases of many slender stems, which in some cases stand above the surface of the ground, are held in place only by a spreading root system of fine, thread-like extensions. They do not withstand trampling and close grazing. Cattle are prone to pull a loosely held bunch of grass

From Hazel Adele Pulling, "Range Forage and the California Range Cattle Industry," *Historian*, 7 (Spring 1945): 113–29.

entirely free from the ground in feeding. Cattle hooves, and especially the small pointed hooves of sheep, act as sharp spades in prying the tufted bunches out of the ground. This is particularly true when the grass is grazed following a rain when the soil is loose. Since bunch grasses are perennials which propagate largely by new growths from the roots, the tearing out of the entire plant removes all chance of renewal of the growth. Thus in the areas widely grazed during early decades, bunch grasses began to disappear.

Another important factor in the early change of range forage was the advent of plants hitherto foreign to California ranges, many of which it is believed were brought into the area by the Spanish missionaries. The fact that many of the new plants were native to Europe lends force to the belief that they were brought into California by the newcomers. . . .

At the same time, plants of no forage value were invading the range to usurp the places left vacant by over-grazed and trampled bunch grasses. Some of these plants were lamb's quarter (*Chenopodium album*), Napa thistle (*Centaurea melitensis*), nettle-leaf goosefoot (*Chenopodium murale*), rough pigweed (*Amaranthus retroflexus*), creeping buttercup (*Ranunculus reptans*), cheeseweed (*Malva parviflora*), wild carrot (*Daucus carota*), and Chilean tarweed (*Madia sativa*).

Of all the new plants which began their progress across the hills and valleys of California, four had become so plentiful by 1849 as to warrant the attention of travellers and settlers who followed in the wake of the Gold Rush. Wild oats by that time occupied considerable portions of the great central valley, the south central coastal area, and the coastal sections of southern California. Alfileria had occupied many of the semiarid slopes of the Sierra Nevada and the south central coastal areas. Bur clover had become plentiful on the hills and in the valleys of the western highlands. Black mustard had begun to drive the nutritious forage plants out of the lower valleys in the southern coastal areas to form thickets dense with brush. Wherever grazing herds penetrated, there the flora of the California range was undergoing change. . . .

The new plants were hardy and adapted to a wide range of soil and moisture conditions. None was as palatable as the native bunch grass, or the latter would not have been grazed in preference to the alien plants. Natural protective features were possessed by the invaders, too, which aided them in their struggle for existence. . . .

Of them all, wild oats attracted most attention of observers because of its prevalence throughout the state. Abundantly foliaged, highly palatable, and nutritious, the plant is eagerly consumed by cattle during its green state and after it has cured on the ground. It forms a year-long forage, beginning to grow as soon as the first autumn rain falls, remaining green during most of the winter and spring, and curing well during the summer.

Alfileria was particularly valued by settlers and rancheros during the early American period because of the excellent forage which the herb supplied early in the winter. . . .

Bur clover is described by W. L. Jepson as "a rare instance of an aggressive immigrant herb having a high economic value." It grows well in all types of soil and under a wide range of environmental conditions. . . . Its nutritive qualities approach those of alfalfa more than any other wild range grass, and, like alfalfa, the herb will leach if allowed to become wet after the plant had dried. . . . Part of the bur clover plant is valuable throughout the year. . . .

The California range, however, was not to remain in the favorable condition which prevailed during the early 1850s. The discovery of gold brought many people into the state who demanded more and better meat than the California herds yielded. A market such as then prevailed had never before existed in California, and stock owners made the most of it. Herds doubled and trebled by natural increase and by huge importations from other states. The great increase in numbers of cattle soon resulted in an overstocking of the range, particularly that of southern California, the south central coastal region, and the San Francisco Bay area.

The severe drought of 1862–1865 intensified the progress of grass deterioration. Successive years of rain failure resulted in the dying of whole plants, a lack of seed formation by those that survived, and in the inability of the seeds which did form to germinate. Grazing of plants of greater palatability and nutrition superseded the grazing of less valuable plants to a degree greater than ever before, so that their continuance was more precarious than it had formerly been. The destruction of each tuft of native bunch grass and each plant of the more valuable annuals before seeds had matured meant increased opportunity of other, less valuable, plants to take root. . . .

Besides over-stocking and drought, another factor was fast taking its toll of range grasses, and even of the range itself. Farmers were entering California in ever-increasing numbers, and range lands were disappearing before the plow. Damage to grasses was everywhere apparent. "Cultivation has destroyed the oats on the plains and the grasses on the river bottoms," stated the president of the State Board of Agriculture in 1868, "and the long dry seasons recurring every season kill out all ordinary varieties of grasses and clovers."

Soon it was apparent that damage to grasses was affecting the range-cattle industry. The *Los Angeles Evening Express* on April 5, 1873, remarked: "It is a fact to be regretted that the grass ranges are not what they formerly were. The herds keep them eaten off close to the ground, and we no longer see the old-time stretches of tall and waving pasturage. . . . We have not taken sufficient care of our stock and our grazing lands." . . .

The extent and degree of change in forage conditions varied in different areas at different times. The mountain ranges, especially the higher elevations of the Sierra Nevada and the north coastal ranges, retained longest their purely bunch-grass character. There were invasions in these areas of wild oats and alfileria, but annuals of lesser value did not find foothold to the extent that they did in the lower levels which were more grazed. An example may be seen in the county of Sonoma where, in 1879, bunch grass had become practically non-existent in the foothills, but where, higher in the hills, stands of bunch grass existed, along with wild oats, alfileria, white and bur clover, sufficient to support one head of cattle on every ten to fifteen acres of land. On lower levels of land even the latter types had begun to disappear. In their place had come plants, some of them entirely worthless as range forage, and all of less value than those which had formerly occupied the land.

While the exact date and nature of the later plant invasions cannot be stated, contemporary cattlemen have left evidence of the change and of what it meant to them. Thus from the journal of Richard Gird, cattleman of San Bernardino county, we find complaints of the rapid invasion of undesirable plants. . . . Gird, like other cattlemen of the state, was constantly beset with the presence of weeds. In July of [1881], he confided to his journal that "weeds . . . of different kinds seem determined to take the place." Two years later, after a season during which the annuals had not seeded well, Gird wrote in discouragement: "The grass has gone through its season and gone to seed before the summer sun came on, and weeds grew rank. Wide spread cockle burs on the hills, sunflowers most everywhere, and rag weeds on the bottoms." . . .

As time passed it was clear that, just as wild oats, alfileria, and bur clover had earlier crowded out much of the native bunch grass vegetation, their places were in turn being taken over by other invaders. Among the plants which made their appearance at the later dates, which spread rapidly, and which were particularly obnoxious to cattlemen, were several varieties of annual brome grasses.

Soft or downy chess, one of the annual bromes, is prone to invade areas which are overgrazed, where the soil has become closely packed by trampling. . . . Of even less value among the new plants was cheat grass, an unpalatable and unnutritious weed that has become especially prevalent in the foothills of the northern part of the state; red brome grass, sometimes called "broncho grass"; and needle brome grass, known as "ripgut" and "devil" grass, a plant which has invaded most of the lowlands of the state, often crowding all other plants out entirely through its numerous and self-planting seeds. Needle grass is particularly disliked by stockmen because its many long-awned, sharp, and pointed seeds lodge in the alimentary canal, eyes, and skin of grazing animals causing great mechanical injury.

Thus it is evident that the California cattleman faced, in the rather general deterioration of range grasses, a problem no less important to the success of his undertaking than were the more obvious ones of lessening range area, no-fence laws, drought, disease, and farmer competition. When one considers that in 1936 only thirty-six per cent of the grasses and herbs on the range were palatable and nutritious, as compared with a like condition in ninety-five per cent of the native grasses, as tested on a small ungrazed area near Sonora, Tuolumne County, it is apparent that range deterioration was a factor increasingly disturbing to the cattleman. . . . In answer to a questionnaire sent out by the Public Lands Commission of 1904 to 104 cattlemen, fifty-eight reported that the grass on their ranges had grown steadily worse with the years.

Under these conditions, it is not surprising that the character of the grass and herbage of the various sections of the state became increasingly one of the determining factors in the conduct of the industry. The transition of the "range-cattle industry" to that of "stock farming," a process which saw its inception during the 1890s, was in no small degree due to the changing character of the forage of the range.

Essays

The Rangelands

Raymond F. Dasmann

Of the six thousand species of vascular plants in California, nearly a thousand have been introduced from other parts of the world. Of the remaining five thousand, nearly thirty percent are endemics—a high percentage for any area, even for islands. Endemics are species that occur in one place and nowhere else except when deliberately moved by man. Why there are so many in California is a question without a single answer. The great variety of habitat isolation provided by the deserts and high mountains contribute to it. California has undoubtedly provided a refuge for species that were once more widespread, but have been reduced by severe climatic changes elsewhere—the coastal redwoods and Sierra big trees, for example.

The relative isolation of California has contributed to the richness of the flora, but has also given it a vulnerability which one usually associates with island species. In relative isolation from competitors or predators, many California species lacked the ability to compete successfully when more aggressive species were brought into the state. . . . This was true of those species that made up the original grasslands of California. If the buffalo or American bison had reached California, the grasslands over the years would have developed adaptations to the presence of a large grazing animal that traveled in great herds. The grasslands would have been a different kind from that developed in the presence only of deer, elk, and pronghorn antelope, for these species of large animals did not prefer the tall perennials that made up the original California prairie.

Of all the families of flowering plants, grasses and grasslike plants (the

From Raymond F. Dasmann, *California's Changing Environment* (San Francisco: Boyd & Fraser, 1981), pp. 22–8. Reprinted by permission of Materials for Today's Learning, Reno, Nevada.

sedges and rushes) attract the least interest. Everybody knows what grass is. Very few learn to identify the different kinds. . . . Livestock owners are an exception. Grasses are important to them, and they know how to differentiate the species. . . .

All of the grasses of California, unlike those of the Midwest and East, are adapted to summer drought. Their growing season is relatively short. The hills green up and initial growth starts after the first rains of the autumn. As the cooler temperatures of winter set in, grasses become semidormant. When temperatures warm in the spring, usually March or April in the lowlands, rapid growth begins. Within a period of two or three months all growth is complete and seed heads are formed. Annuals then die back. Perennials enter a period of summer dormancy.

Perennial grasses dominated in the original California flora. They were of two principal forms: bunch grasses which grow upright in dense tufts of stems from a perennial root crown, and sod-forming grasses which spread out horizontally from underground or aboveground rhizomes or stolons. These latter are the preferred lawn grasses since they form a continuous cover over the soil—Kentucky blue grass and Bermuda grass are among the more common species. Bunch grasses were represented in the original California flora by the purple needle grass (*Stipa pulchra*), squirrel-tail grass (*Sitanion*), and the wheat-grasses (*Agropyron*) along with many others. The sod-formers included the salt-grasses (*Distichlis*) which grow in alkaline or saline areas, or the galleta grass (*Hilaria*) of the desert regions. Perennial grasses may grow for many years, the older portions dying back, but new shoots growing up. Perennial grasses may live to be very old. Annuals, by contrast, pop up each year from seed, go through their whole life cycle in a few months, form new seed, then die. There were many native annual grasses in California, but they were subordinate to the perennials, occupying disturbed areas or areas of thin soil for the most part or desert areas where the annual rainfall was insufficient to support an established grass cover.

The Spanish who first came to California brought livestock with them—horses and burros, cattle, sheep, goats, and swine. These animals found conditions to their liking. They increased and multiplied. Some were kept near the missions, or later on the ranchos, but others escaped and ran wild. Horses, cattle, burros, and goats adapted to life in the wild. By the time the American settlers came to California, herds of wild cattle and horses were even more common than elk and antelope in the Central Valley. The wild burros did particularly well in the deserts and sagebrush lands. Goats thrived especially on the Channel Islands where there were no predators. Domestic sheep did not adapt well to a feral life.

Coyotes in California helped see to that. Initially the livestock business in California was based on cattle more than other species. A trade developed in hides and tallow, and this commerce became increasingly important until California's conquest by the United States.

The original grasslands were productive and the grasses evidently were of great nutritional value. Edwin Bryant, who came here in 1846, had the following comments:

> The horned cattle of California which I have thus far seen, are the largest and the handsomest in shape which I ever saw. There is certainly no breed in the United States equaling them in size. They, as well as the horses, subsist entirely on the indigenous grasses, at all the seasons of the year; and such are the nutritious qualities of the herbage, that the former are always in condition for slaughtering, and the latter have as much flesh upon them as is desirable. . . .

These comments were written at a ranch near Mount Diablo. As Bryant traveled into the San Francisco Bay area he noted: "From this plain we entered a hilly country [near Livermore], covered to the summits of the elevations with wild oats and tufts or bunches of a species of grass, which remains green through the whole season." In Santa Clara County he traveled "over a level and highly fertile plain, producing a variety of indigenous grasses, among which I noticed several species of clover and mustard, large tracts of which we rode through, the stalks varying from six to ten feet in height." Already, by Bryant's time, the changes had begun. The native grasses were still in place and "green through the whole season," but invaders, wild oats and mustard, were taking over.

The Spanish brought livestock, but they also brought food for their animals. In the hay, no doubt, were the seeds of many of the grasses and broad-leaved herbs (forbs) from the Mediterranean region of Europe. These were hardy plants, adapted already to the summer-dry, Mediterranean climate of California, but adapted also to thousands of years of heavy pressure from grazing livestock. They were the Eurasian survivors of ten thousand years of pastoralism. In California, wherever there was bare ground, they took hold. The invaders did not arrive all at once, or did not spread at the same time. . . .

By the time that the first plant ecologists explored California, the grasslands had changed beyond recognition. Only by careful detective work, looking over areas that had long been protected from grazing—

railroad right-of-ways, graveyards, or areas that for one reason or another had been fenced and protected—could the original flora be found. One worker, George Hendry, took the trouble to examine the straw from which the adobe bricks of the Spanish missions had been made. By working from the first missions to those last constructed, he was able to determine the sequence in which the introduced species became established and spread.

It was the heavy grazing and trampling that contributed most to the change. Aggravating this situation were the periodic droughts that have always been a factor in the California environment. . . . The drought of the 1860s was of unusual severity. William Brewer was traveling through California at that time and noted massive livestock losses and commented on the complete lack of vegetation in areas that would normally be grass-covered. Faced with total collapse, livestock owners took their animals into the mountains and even into the Mojave, where there was some forage to be found. George Nidever, who was then living on San Miguel Island, reported that in 1862 he had 6,000 sheep, 200 cattle, 100 hogs, and 32 horses on the island. In 1863 to 1864, he lost 5,000 sheep, 180 cattle, a few hogs, and 30 horses. In 1870 he sold out. As a result of such experiences, San Miguel lost soil and vegetation and became for a time virtually a desert island. . . . The drought of the 1970s, which frightened most Californians into conserving water, seemed unusually severe because of the stresses now being placed upon the state's water resources, resulting from increased population, increased demand, and wasteful water use. Had it lasted as long as the dry years from 1897 to 1930, California's modern and wasteful economy could have faced drastic restructuring.

The 1860s put an end to the cattle boom in California and caused livestock owners to shift to sheep, which require less water and are more easily moved to available water and forage supplies. This set in motion a pattern of transhumance, or migratory sheep grazing, which virtually devastated the southern Sierra Nevada up to its highest meadows.

The first great trail herds of sheep began their long treks through the Sierra. For a time the unexploited mountain grasslands and meadows supported these animals, but as the ranges became overgrazed, carrying capacity decreased rapidly. According to Roswell Welch of Porterville, a party traveling from Kernville to Mount Whitney in the 1890s found the land completely devastated by overgrazing by sheep. Most of the southern Sierra was in the path of the trail herds that used to leave southern Kern County and move over Breckenridge Mountain or around it, through the Walker or Olancha Pass area in Inyo County, thence up the east side of the Sierra, across Sonora Pass, and back via the west slope of the mountains to Kern County.

The establishment of the national forests and national parks, starting in 1890 to 1891, marked the beginning of the end of excessive exploitation of the high mountain ranges in California, but may also have had the effect of forcing greater pressure on the lowlands. The next drought period, beginning in 1890 in southern California and in 1897 farther north, had further effects. There was a gradual decrease of sheep at this time (from a high of six million in 1876) and a shifting of herds from California to the intermountain states. Nevertheless, in 1920 there were still nearly three million sheep in California and the numbers have not declined since.

In a document published in 1936, the Forest Service stated the results of its surveys and evaluation of rangeland conditions in the western United States. Some of the conclusions were shocking:

> Forage depletion for the entire range area averages more than half; the result of a few decades of livestock grazing.
>
> Three-fourths of the entire range area has declined during the last 30 years, and only 16 percent has improved.
>
> Probably not much over 5 percent of the entire range area is in a thoroughly satisfactory condition.
>
> About seven-tenths, or 523 million acres, of the range area is still subject to practically unrestricted grazing.

L. T. Burcham's studies of California in the 1950s essentially confirmed the Forest Service's findings. Not only had the carrying capacity declined by half so that it now took twice as many acres to support a cow or a sheep, but it was virtually impossible to maintain cattle on the range throughout the year without drastic decline in their condition and great danger of loss. The practice of livestock grazing has changed, with much greater reliance on irrigated pasture and feed lot, and much less pressure on the rangelands.

The degree of damage done to the rangelands differs from one part of the state to another. Recovery has been better in the higher rainfall areas and virtually nonexistent in the driest areas. The vegetation of the valley and foothill ranges has been changed permanently. Even with complete protection from grazing, or any system of grazing management short of replanting, the grasslands do not recover to the perennial grass cover of the past, but remain in annual grass cover. The invading grasses from Europe, like their human counterparts, are here to stay. Or so it seems at this time.

California's Grazed Ecosystems

Paul F. Starrs

Lands grazed by domesticated livestock undergo great changes through the reaches of time. While the long-term physical effects of grazing are well-documented in most lands, chronicling changes wrought in human lifeways is a more speculative undertaking, although the transformations are hardly less substantial. No land not an island, however, has been so swiftly altered by introduced livestock, especially with such far-reaching social and cultural effects, as California. In scarcely more than 200 years of occupation by European, Asian, and American people arriving from every point of the compass, California has grown from an empty land to a full one. This narrow window of years, utterly insignificant in comparison to Old World chronologies, offered time enough for the lifeways of California's inhabitants to change from subsistence to a world-wide and trend-setting role as a dominant force in the Pacific Rim economy. Through all this, the place of grazing animals is central: California pastoralism—ranching—provided a foundation, and base, and incentive for subsequent settlement.

While California today produces highly diversified goods, a basic Spanish-era economy of land speculation and livestock grazing has continued significance. Beef, mutton, wool, mohair, tallow, hides, and horses are only a few of the historic products of California livestock raising. As is often the case in a swiftly developing region, California's pastoralism was important first for the livestock products circulated into the world economy, more so for massive changes wrought in the physical landscape, but it was compelling most of all for what pastoralism made

From Paul F. Starrs, "Changing Landscapes of California Pastoralism: 200 Years of Change," in *Landscape Ecology: Study of Mediterranean Grazed Ecosystems*, edited by W. James Clawson, XVI International Congress, Nice, France, October 7–8, 1989, Man and the Biosphere Symposium, pp. 49–61. Reprinted by permission of the author.

possible: a land-owning based, yet readily-changed society with a solid, adaptable, and multi-ethnic cultural core. . . . It is still rangelands . . . that provide much of the undeveloped greenbelt land that surrounds the "golden state's" expanding cities. Grazing cattle and sheep are the first visual link between most residents and the state's agriculture, which produces twenty billion dollars a year in California. Grazing lands are habitat for wildlife, provide watershed protection, and safeguard acreage unsuited to crops or housing. . . .

That European travelers looking for new lands to settle and occupy should find California attractive is hardly surprising. Early English explorers came and went along the California shore, leaving only a scattered record of landfalls. Unlike the forays of English, French, or Russian envoys into the New World, which required radical reorientation of goals and techniques before permanent settlements could be established, the Spanish moved into lands they knew and understood. . . . Aridity, sparse vegetation, forests limited to montane regions, and even the wildlife and native human populations of the American Southwest and Pacific Slope were familiar or at least recognizable. Spanish (and later Mexican) social, political, and cultural institutions were readily adaptable to the new landscapes; that was part and parcel of the Spanish success. The question was how the empire was to be unfurled: like an octopus, the colonial Spanish realm had within its grasp a great range of territory, people, and potential riches, but it took time for the tentacles to take hold of every place within reach. And Alta California—including all of present-day California—lay at the most distant and awkward remove of the Spanish empire.

In an era when the Manila galleons controlled the Pacific, Spanish navigators not surprisingly visited California's islands early. Juan Cabrillo's expedition reached Santa Cruz Island, off the southern California coast, in 1542, but did not stay. Assessment reliably preceded occupation and settlement. Livestock were rarely brought on these exploratory forays, although the Channel Islands probably had populations of goats and sheep in the early 1700s, long before they were placed on the California mainland. Instead, maritime explorers examined the coast, its bays and inlets, charting the waters. Settle, though, they did not; despite a naval and mercantile dominance of the seas, ocean waters were too insecure a pathway for Spanish assertion of absolute territorial control. Instead, conquistadors and clerics habitually colonized by overland movement, not by sea. . . .

The *gente de razon*, or "people of reason" who formed the landholding elite of the Spanish frontier were first and foremost livestock ranchers.

They might practice dryland farming, raise some irrigated crops, and pay proper homage to the mother church, but these folk specialized in grazing livestock—sheep, goats, horses, and especially cattle—on extensive pastures acquired as mercedes, or grants, from colonial governments. Hispanic ranching was effective, and, in being dominated by elaborate traditional use practices, it was also secure. . . .

Through the seventeenth and eighteenth centuries, Baja California and Alta—or present-day—California lay at the northern edge of the Spanish empire. These Californias could be reached by sea, or less conveniently, overland across the Mojave and Sonora deserts. Yet colonial powers in Mexico City by the mid-eighteenth century recognized the importance of acquiring a foothold that might forestall, or at least greatly complicate, land acquisition in the Californias by rival colonial powers. By the middle 1700s, livestock raising in California was begun. . . .

In its brief and tempestuous block of years (1769–1821), the mission era in California established important facts. Livestock did well in California; better than well, they thrived on their own, with virtually no continuing care. Only the increase needed to be gathered, and then only when there was demand—beyond the missions and the few substantial towns in Alta California, markets for livestock would not open up until the mission era was virtually over. Nonetheless, the die was cast, and the influences of early livestock raising on plants, landforms, economies, and California life were *faits accomplis*.

With time, the formal Spanish missions faded with the growth of secular and international interest in Alta California. Ranching passed from the padres to private hands and eventually to a larger public. In California of the early 1800s, great changes in the politics of geography were taking place. Lands once reserved to the church and the governor in Monterey were initially passed by grant into private hands in 1786, marking the first of numerous rancho grants awarded during the Spanish and Mexican period. As Spain's hold on its empire weakened, control of the lands of New Spain—Mexico—was ceded to authorities in Mexico City in 1821. And Mexico was far quicker than Spain to issue patents to California land, especially under the 1824 Laws of Colonization, designed to establish a stable, loyal, and well-armed populace in sparsely settled areas. Replacing the church era was what is often described as the *californio* period, taking its name from the generic description of the Spanish, Mexican, and Anglo settlers in Alta California after mission secularization.

These Californios are still in some respects a mystery group. Their lives, society, work, and play are sufficiently mythologized to have cast doubt on what the life of a typical ranchero might have been like. Colonial grants of property—more than 800 were made during the Mexican period, from 1821 to 1846—fed a hunger for land that prompted a simultaneous and equal speculation in property and cattle raising. A fifth of all ranchos went to foreigners, who flocked to California in a land boom that foreshadowed the later Gold Rush. The fevered search for "real estate" remains a salient part of Californian life through today; some might say it has defined the very character of California. The Coast Ranges were dotted with ranches; property disputes arose, herds mixed, adjudication procedures were formalized. A body of traditional law encouraged the formation of livestock associations that policed the behavior of their members, a practice that spread throughout North America. Land and livestock defined new forms of California; they also provided the only real livelihood and employment available to average citizens until new economies developed after cession to the United States in 1848. . . .

As a rule, ranching work, society, and profits were as threadbare as today; while some rancheros were affluent, far more grant holders and cattle owners were not. Land, measured in square Spanish leagues, was their wealth—but title, while well-established, was rarely well-documented. Without land there would be no cattle or sheep, and without these, there was only retreat to city life. While some rancheros lived in town and left their properties in the hands of a managing mayordomo, all maintained strong ties to ranching life. Then, as now, few ranchers willingly gave up rural roots to enter the metropolis.

Livestock found a readier market as a trading commodity in the nineteenth century than it had for the padres in the eighteenth. In periodic *rodeos* (the term is a Spanish one), cattle were gathered, slaughtered in huge *matanzas*, or killings, and the hides stripped from the carcasses. While the meat was left in place, the fat, or tallow, was rendered in huge pots. Tallow jelled, as it cooled, finally forming an almost solid fat that was in great demand for candles and oil for lighting—especially though by no means exclusively—in the South American mines.

The landowning *californios* needed a work force, and found them in the vaqueros, or ranch hands. The *vaqueros* who participated in these round-ups were among the most picturesque figures in Mexican California history. . . . Vaqueros were, often by choice, workers, not owners, a distinction that would have continued importance in American-era ranching. If the *californios*, with their large tracts of granted rangeland proved predecessors to Anglo-period ranchers, then perhaps the vaquero and the Californio are indeed literal antecedents of the contemporary cowhand and rancher.

What is incontestable is the blossoming relationship between the hide and tallow industry, the rise of California ports and cities from which these commodities were shipped, and the role of ranching society in bringing newcomers to California in the 1830s and 1840s. Before the Gold Rush was the livestock boom—not in beef, which was of almost no significance, but in products and offal that today have paltry value (*charqui*, or beef jerky, was the only way to preserve large quantities of meat, and most ranchers hardly bothered). Land speculation, however, continued and built on its ranching foundation.

English, American, Mexican, Spanish, and French boats sailed up the Pacific Coast, ferrying wool, hides, and tallow from the California entrepots to foreign destinations. By the 1830s, California was part of a broader European and even Asian economy, caught up in a swirl of mercantile exchange. . . .

The borders, opening during the Mexican period, allowed a flow of population in all directions. California rancheros were involved in importations of livestock to Australia, to parts of the Great Basin, including Nevada and Utah, and even to Hawaii, where agents of King Kamehamea III brought a delegation of vaqueros and some of their cattle to expand herds on the islands.

In a matter of scant decades, California changed from a rural outpost—remote, impoverished, marginally successful—to part of a thriving Pacific trade economy. Nothing, though, failed like success, and by the 1840s, efforts were actively underway in the eastern United States to wrest California—as the plum among the northwestern-most properties of Mexico City—away from its rightful owners. War brought that change in ownership, and Mexico relinquished its claim to California in 1848. The Anglo period had begun.

The *Californio*, or Mexican–California period, created novel California landscapes. The biological transformation of Alta California, begun when the first sheep, goats, cattle, and horses entered the summer-dry rangelands that dominate much of California, accelerated as livestock populations increased with the number of ranchos and newly opening markets for livestock products. Immigrants flocked to California like moths to fire, seeking land, wealth, livestock, or, at a minimum, a bit of the action. As livestock numbers increased, wildlife disappeared, although through the 1860s and 1870s, accounts describe mixed herds of cattle, tule elk, deer, and grizzly bears; a formidable combination. . . . Livestock herds at first decreased in the 1830s as the harvest of hides and tallow took its toll, but numbers rose again in the early 1840s as more land grants were made. . . .

The Gold Rush might have proved the undoing of California ranching. Instead, it was the kiln that hardened the mold. . . . Livestock products were in constant demand in western mining communities, which extended far beyond California—cattle, especially, were trailed over distances as great as a thousand miles to market. Animals worth only a dollar or two in the Coast Ranges might increase in value tenfold by the time they arrived at a gold or silver boom town.

The 1850s and 1860s were giddy times for California livestock raisers. A herd might begin with modest numbers of animals and a small landholding. Yet demand for meat was so large that for a time during the gold boom on the 1850s, beef and mutton were unavailable in restaurants; wild game was harvested and put on the table instead. Flood years followed by drought in the 1860s savaged rangelands and herds, but increased profits for those who could stay the course. When private land was not always readily available, enterprising graziers found that moving animals some distance to feed was usually possible. Unclaimed public domain lands were considered free range, and were thoroughly exploited. The ecological effects of these moving herds, their tenders on a continuing search for fodder, meadows, and bedding grounds, are both easily imagined and seen today: the despoliation of western landscapes by overgrazing was more pronounced in the late 1800s and early 1900s than ever before or since. . . .

The Hispanic ranching period of California remained important, both in symbolism and in fact. Rarely has history been so successfully blended into promotional literature directed at prospective immigrants. The missions, padres, California Indians, the californio period, and especially the ranchos, vaqueros, and livestock industry appeared in stories, advertising broadsides, novels, and even on orange crate labels, advertising the Golden State. Using the past to promote development that would destroy the very traditions used to draw settlers is typical of California's singular and ironic vision of progress. Only very recently have tradition and historic preservation been allowed to interfere with commerce.

"Progress," however defined, is very much a part of California life. Southern California, popularly known as the "Southland," is emblematic of the march of California's growth. In three decades, from the 1890s to 1920s, Los Angeles changed from a quiet Spanish-heritage town to an endless series of orange groves. In another three decades, oranges disappeared as housing subdivisions filled lands that shortly before were grazing land. There is hardly another modern saga equal to this tale of physical modification of the earth. . . .

Livestock ranching is the oldest culture of California, aside from the long-gone landscapes of native Americans. Yet many of the state's landscapes—Southern California orange groves, farming around Sacramento, the Mother Lode gold country, logging towns along the north coast, are disappearing or gone. Change can, though, be directed. Our obligation is to create landscapes of interest, richness, and depth; pastoralism in California has long provided a historic, well-appreciated, and still-evolving landscape.

Cattle on a Thousand Hills

Robert G. Cleland

The spectacular cattle boom of 1849 and the early fifties was the natural outgrowth of the Gold Rush. Prior to 1848 California cattle were commercially valuable only for their hides and tallow; and the average price of full-grown steers seldom rose above four dollars a head. But the Gold Rush created an enormous and ever-expanding demand for beef, raised the price of cattle to levels never before dreamed of in California, destroyed the simple scale of values to which the ranchers had long been accustomed, and transformed herds of black, slim bodied cattle into far richer bonanzas than the gold fields of the Sierra yielded to a vast majority of the Argonauts.

Because of the urgent demand for livestock in the mining regions and in such newly created cities as San Francisco, Stockton, and Sacramento, the long-established custom of slaughtering cattle for hides and tallow rapidly gave place to the much more profitable business of selling the animals for beef. Prices began to feel the stimulus of the new martinet as soon as the first wave of gold seekers reached California. In the spring of 1849, Hugo Reid, the picturesque and capricious Scotsman of San Gabriel, wrote from Monterey to Abel Stearns that dried beef was con-

From Robert G. Cleland, "The Rise and Collapse of the Cattle Boom," in *The Cattle on a Thousand Hills: Southern California, 1850–1880*, 2nd ed. (San Marino, CA: Huntington Library, 1951), pp. 102–5, 106, 107-9, 110, 111, 116.

sidered cheap at 20 cents a pound, that tallow was selling for 48 cents a pound in San Francisco, and that there was not a candle for sale in all Monterey. Reid added that "Stinking mould yankee Candles, so soft you cannot handle them" were bringing 30 cents a pound, wholesale, in "the Yerba Buena."

Southern rancheros adopted the . . . plan of driving their cattle on the hoof to the northern markets. The practice, begun in the early days of the Gold Rush, quickly developed into one of the essential economic operations of the state, and continued with little interruption until the completion of the Southern Pacific Railroad to Los Angeles more than a quarter of a century later.

During that time tens of thousands of cattle were driven, either along the coast or through the sparsely inhabited San Joaquin Valley, from the southern ranges to northern California. In economic significance and picturesque detail, the traffic was comparable to the great cattle drives over the Bozeman Trail of Montana or the Abilene Trail of Kansas. . . .

The average herd included from 700 to 1,000 animals, though sometimes the number ran as high as 2,000 or 2,500. A trail boss (perhaps a *mayordomo*) and three or more vaqueros went with each herd. The cost of "carrying" a drove from Los Angeles to "the upper country" averaged, roughly, from two to four dollars a head. Wholesale stampedes, severe storms, forays by wild Indian tribes, or the depredations of cattle thieves occasionally took heavy and even disastrous toll, but under normal conditions the herd suffered little loss.

Some southern rancheros leased grazing rights in the vicinity of San Jose, Sacramento, or San Francisco Bay, where the stock fattened after the long drive; other owners sold direct to cattle buyers who came down in large numbers from the north either to purchase on their own account or to serve as agents for butcher shops and meat dealers in the larger cities. . . .

Inefficient vaqueros, great fields of mustard (in which some of the stock were lost), heavy fogs in the mountains, a poisonous weed that killed six heifers at one time, lack of grass and intense heat kept the trail boss of a band of 800 cattle belonging to Abel Stearns decidedly on tenterhooks during the early summer of 1862.

"I have had no stampeeds for the past eight nights," he wrote from a camping place above Santa Barbara. ". . . I had an awful time passing the beach at San Buenaventura; it was a matter of impossibility to keep my Stock from running, I had half of my Baqueros ahead with me and I know everything was done to keep them from running but we could not keep them back, the consequences were some ten or twelve got very lame and will not be able to travel." . . .

The rise in cattle prices, which began early in 1849, continued with only temporary setbacks for nearly seven years. During that time, beef cattle were quoted as high as $75 in San Francisco, and small calves brought from $20 to $25 each." Even on the distant southern ranges, the price of full-grown steers, representing an eight- or ten-fold increase over that of the hide-and-tallow days, rose to $30 and $40 a head.

But the seven fat years of inflated prices and treacherous prosperity proved, in the end, the curse of the native landowner. . . . Horace Bell, who came to California in the extravagant days of the cattle boom and saw . . . the prodigality and ostentation of the *paisanos*, left the following picture of the Los Angeles of that day:

> The streets were thronged throughout the entire day with splendidly mounted and richly dressed caballeros, most of whom wore suits of clothes that cost all the way from $500 to $1,000 with saddle and horse trappings that cost even more. . . . Everybody in Los Angeles seemed rich, everybody was rich, and money was more plentiful at that time, than in any other place of like size, I venture to say, in the world.

The improvidence and luxury of which Bell . . . wrote consumed the income of the Californians so completely that little or nothing was added to their capital, set aside as a prudent reserve against forthcoming years of drought and other disaster, or employed to restore the seriously depleted herds.

In failing to preserve their breeding stock, rancheros were especially shortsighted. Abel Stearns estimated that, during the boom period, from 25,000 to 30,000 cattle annually were sold out of Los Angeles County alone, and that the southern ranges suffered a yearly loss, over and above the natural increase, of at least 15,000 head.

Scarcely able at first to comprehend the fictitious prices created by the mining boom, southern rancheros soon came to accept them as a normal and permanent feature of the industry. But their delusion was short-lived. Due chiefly to large imports of sheep (especially from New Mexico) the development of farms devoted to the breeding and fattening of livestock on an extensive scale, and the introduction of cattle in immense numbers from the Mississippi and Missouri valleys, the demand for southern California range stock began to decline as early as 1855.

During the Gold Rush thousands of head of livestock were brought into California from the Middle West, by way of the overland trails; and early in the fifties the cattle trade between the Missouri frontier and the Sacramento Valley had developed into a definite, well-established business. According to an estimate published in the *Los Angeles Star*, over 20,000 head of cattle and nearly 25,000 sheep en route to California passed Fort Kearny during the spring and early summer of 1852. The following year Governor Bigler reported that over 20,000 cattle reached California by way of Beckwith's Pass; 9,000 by way of the Gila route; 15,000 by way of Sonora Pass; and some 13,000 by way of the Carson River Trail—making a total of nearly 62,000 which entered the state over the main emigrant roads alone. Thousands of head of Texas longhorns were also driven to California, and the trade became an important feature of the Texas cattle business during the middle fifties. . . .

By 1860 the trade in cattle between the Middle West and California had perceptibly declined; but the decrease was more than offset by large imports of sheep from New Mexico. According to one estimate, 100,000 sheep crossed the Colorado River, en route to California, during the fall and early winter of 1858–59. In February of that season, the *Los Angeles Star* reported the arrival of Joaquin Perca, Jesils Luna, and Vincente Otero with 35,000 sheep from New Mexico. The following fall it was said that 46,000 head from the Rio Grande Valley were passing through Arizona, bound for California; and between December and January 21 some 80,000 head arrived at the ferry on the Colorado.

The growth of the sheep industry in California, supplementing other factors already mentioned, led to a decline in cattle prices before the close of 1855. A severe drought, the next season, destroyed at least 10,000 cattle in Los Angeles County alone, and compelled many ranchers to market their herds at sacrifice prices. . . .

The year 1856 may therefore be looked upon as marking the end of the golden age of the cattle business in California. From that time on, the industry suffered one vicissitude after another until the great drought of the mid-sixties subordinated it to other forms of agriculture. By the summer of 1857 cattle dealers found the Los Angeles market "completely gutted"; "in 1858 drovers purchased only a third as many cattle as they had been accustomed to take in former years. During 1860 the standard price of breeding cows fell to ten dollars a head, and, even at that figure, there was no demand. . . .

In the midst of this acute depression most of the old *paisanos* discovered that they were heavily involved in debt and saddled with interest charges of staggering proportions. . . .

Failure to contend successfully with debts contracted at the prevailing interest, or to adapt themselves to other conditions imposed by the new economic order, compelled the old *paisanos*, one by one, to surrender their vast estates to alien hands and pass, almost unnoticed and forgotten, into the dim twilight of their once-romantic day.

Further Readings

Brown, William S., and Stuart B. Show. *California Rural Land Use and Management: A History of the Use and Occupancy of Rural Lands in California*. 2 vols. San Francisco: United States Forest Service, California Region, 1944.

Burcham, L. T. *California Range Land*. University of California, Davis: Center for Archaeological Research, 1982.

Burcham, L. T. *California Range Land, an Historic-Ecological Study of the Range Resource of California*. Sacramento: Division of Forestry, Department of Natural Resources, 1957.

Clawson, Marion. *The Western Range Livestock Industry*. New York: McGraw-Hill, 1950.

Clawson, Marion, and Wendell Calhoun. *Longterm Outlook for Western Agriculture: General Trends in Agricultural Land Use, Production, and Demand*. Berkeley: n.p., 1946.

Cleland, Robert Glass. *The Cattle on a Thousand Hills*. 2nd ed. San Marino, CA: Huntington Library, 1951.

Cleland, Robert Glass. *The Place Called Sespe: A History of a California Ranch*. Alhambra, CA: private printing, 1940.

Dana, Samuel, and Sally Fairfax. *Forest and Range Policy: Its Development in the U.S.* New York: McGraw Hill, 1980.

Dana, Samuel Trask, and Myron Edward Krueger. *California Lands: Ownership, Use and Management*. Washington, DC: American Forestry Association, 1958.

Dary, David. *Cowboy Culture: A Saga of Five Centuries*. New York: Knopf, 1981.

Dasmann, Raymond F. *California's Changing Environment*. San Francisco: Boyd & Fraser, 1981.

Demke, Siegfried G. *The Cattle Drives of Early California*. San Gabriel, CA: Prosperity Press, 1985.

Fairfax, Sally, and Carolyn Yale. *The Federal Lands*. Washington, DC: Island Press, 1987.

Frink, Mauricet. *When Grass Was King*. Boulder: University of Colorado Press, 1956.

Gates, Paul W., ed. *California Ranchos and Farms, 1842–1862*. Madison: The State Historical Society of Wisconsin, 1967.

Goetzmann, William H. *Army Exploration in the American West, 1803–1863*. New Haven: Yale University Press, 1959.

Goetzmann, William H. *New Lands, New Men: America and the Second Great Age of Discovery*. New York: Viking, 1986.

Gressley, Gene. *Bankers and Cattlemen*. Lincoln: University of Nebraska Press, 1996.

Hafen, LeRoy R., ed. *Mountain Men and Fur Traders of the Far West*. Glendale, CA: A.H. Clark Co., 1965.

Hays, Samuel P. *Conservation and the Gospel of Efficiency: The Progressive Conservation Movement, 1890–1920*. Cambridge: Harvard University Press, 1959.

Jordan, Terry G. *North American Cattle-Ranching Frontiers: Origins, Diffusion, and Differentiation*. Albuquerque: University of New Mexico Press, 1993.

Malin, James C. *The Grassland of North America: Prolegomena to Its History*. Lawrence, KS: n.p., 1947.

Mauldin, Harry K. *History of Cattle Industry in Lake County*. Lake County, CA: Lake County Historical Society, 1965.

Moynihan, Ruth B., et al. *So Much to Be Done: Women Settlers on the Mining and Ranching Frontier*. Lincoln: University of Nebraska Press, 1990.

Myres, Sandra L. *Westering Women and the Frontier Experience, 1800–1915*. Albuquerque: University of New Mexico Press, 1982.

Paul, Rodman. *The Far West and the Great Plains in Transition, 1859–1900*. New York: Harper and Row, 1988.

Pisani, Donald J. *Water, Land, and Law in the West: The Limits of Public Policy, 1850–1920*. Lawrence: University Press of Kansas, 1997.

Powers, Bob. *Cowboy Country*. Glendale, CA: A.H. Clark Co., 1988.

Preston, William. *Vanishing Landscapes: Land and Life in the Tulare Lake Basin*. Berkeley: University of California Press, 1981.

Pyne, Stephen J. *Fire in America: A Cultural History of Wildland and Rural Fire*. Seattle: University of Washington Press, 1997.

Robinson, W. W. *Ranchos Become Cities*. Pasadena, CA: San Pasqual Press, 1939.

Rowley, William D. *U.S. Forest Service Grazing and Rangelands: A History*. College Station, TX: A&M University Press, 1985.

Runte, Alfred. *Public Lands, Public Heritage: The National Forest Idea*. Niwot, CO: R. Rinehart Publishers, 1991.

Russell, Carl P. *Firearms, Traps, and Tools of the Mountain Men*. New York: Knopf, 1967.

Russell, J. H. *Cattle on the Conejo*. Los Angeles: Thomas Litho & Print. Co., 1959.

Trafzer, Clifford E. *American Indians as Cowboys*. Newcastle, CA: Sierra Oaks Publishing, 1992.

Wishart, David. *The Fur Trade of the American West, 1807–1840: A Geographical Synthesis*. Lincoln: University of Nebraska Press, 1979.

Wunder, John R., ed. *Working the Range: Essays on the History of Western Land Management and the Environment, Contributions in Economics and Economic History*. Westport, CT: Greenwood, 1985.

Chapter 7

Building the Hydraulic Empire

Documents

An Advocate Encounters the Irrigators, 1887

There is no part of the Pacific Coast so well adapted to irrigation works on a grand scale as the extensive plains lying contiguous to the Sierras, whose abundant water supply, to a great extent, goes to waste at present. . . . Kern, Tulare, Fresno, Merced, Stanislaus, San Joaquin, and northward to old Shasta, are all counties whose natural fertility is being greatly assisted by the use of water from the snow-mountains.

The raisin-grower who depends wholly upon water ditches differs so greatly from the wheat-farmer, or the cattle-owner, that he almost needs a different classification among men. He lives on broad, level plains, and yet within a few hours' ride of the mountains, a six hours' journey of the Pacific Ocean. He can stimulate a vegetable growth over the deep alluvial soil, that no other region known to civilized man is able to surpass. The winters and springs are exquisitely beautiful and temperate; the summers are hot but healthy. Soil and climate combine to make his realm a

From George Freeman, "Among the Irrigators of Fresno," *Overland Monthly*, 9 (1887): 621–27.

grape land, for wine and for raisins; a fruit land; a land for cotton, tobacco, rice, and sugar. . . .

One who visits the irrigators, therefore, will find that they are not isolated communities; socially, they are well organized, close-knit, and homogeneous, far beyond the ordinary farming communities. And this, which is much to their credit, is one of the beneficent results of the colony idea. The California irrigator is the typical figure of the great change at present passing over the industries of the State. His statue should stand beside that of the pioneer of '49, in the Capitol; the ditch-digger of '77–'87 beside the flume-builder of the early fifties. One-third of the arable land of the State is naturally a desert, the average rainfall being wholly inadequate to ensure crops, and build up stable communities. Without irrigation, this extent of country is only utilized for scanty pasturage, and the value of the lands is merely nominal, the soil remaining dry thirty or forty feet deep. Now, many of the most valuable and attractive fruit lands of the State are situated within the irrigated district; the largest interior cities of California seem likely to grow up in regions developed by great irrigation systems, and the close of the age of wheat farming for other than home consumption, is near at hand. This is the work, not of the capitalists, but of men of small means, trying to increase the value of their separate holdings. And they want State control, but not State management, of water.

Owing to the great abundance of water, irrigation in the San Joaquin has hitherto been a simple matter. The system based upon the methods of Egypt, Palestine, India—open ditches, water distributed over the fields by sub-channels—is proving sufficient for all the needs of the community. The water comes from the fountainheads of streams, and is taken out near the mountains, and distributed through large ditches, which aggregate in the single county of Fresno, more than nine hundred miles in length. Some farmers level their lands, and flood the soil; others plough furrows, and conduct small streams over the surface until it is sufficiently moist. The owners of the land usually own stock in the ditch companies, and so control the supply.

The conversion of a desert into a garden is not a difficult task if the proper conditions are present. The water supply must be adequate. . . . If streams are the source of supply, they must be unfailing, or else the water must be stored in reservoirs. . . .

The water ditches as at present constructed, are not only useful, but even picturesque. In places their banks are lined with tall trees—poplars, walnuts, and evergreens. The walls that retain the water are low and of an easy grade, so that vines and trees can well be planted on them. When lands become very valuable, and every inch is utilized, the prettiest parts of the valley will be along the canals. The water, as has been explained, is taken from the river near the foothills—from rivers, in many cases, whose contents would otherwise waste themselves in sloughs and marshes—and the skill and labor of the irrigator has only to turn a portion of this water from the river beds, and distribute it where it is so greatly needed, to make a well watered garden out of the former desert, and to put the richest of fields of grain, vegetables, or other products in the place of half-starved flocks and herds. Within the past five years, this result has been accomplished many times over; bearing and highly profitable vineyards are today established on soil thinly covered with sparse weeds only that time ago. It even seems to the residents of the irrigated belt of the State as if some enchantment had helped them. . . .

The social and moral side of colony life has been dwelt upon at length by many writers. There does not seem to be a dissenting voice to the statement that intensive, horticulture, carried on upon twenty-acre tracts, rose-bowered, tree-shaded, and beautiful, means a very hopeful condition of affairs. It means suburban rather than rustic life. It gives every educational and religious surrounding: books, pleasant companions, recreation, and study are brought within the reach of the entire community. No one is poor and no one is exasperatingly rich. The figures just given show how very comfortable honest, hard-working colonists may make themselves. . . .

It would be a pleasant thing to follow the trickling stream from under the Whitney glacier, under pine, cedar, and oak, from ravine to ravine, and range to range, into the rivers below, and across the well cultivated fields of Kern, Fresno, or Tulare. At last the welcome water has done its work, has moistened and enriched the lands once desert, and converted them into prosperous and beautiful homes. . . .

A Writer Describes Irrigation Districts, 1897

Considering the extent that it is broken by mountains, California may be said to be an arid State. In the winter season the great cyclonic belt which enters this coast upon the breast of the Pacific current pushes down from its northern residence and visits the upper portions of the

From John Bennett, "The District Irrigation Movement in California," *Overland Monthly*, 29 (1897): 252–57.

State with some twenty-six inches of rainfall, then withdraws and leaves brown levels under blue skies for eight months. All that area east of the forty-first meridian of longitude and south of the thirty-sixth parallel of latitude, excepting the rim of coast, is desert—about one third of the entire State. It is a region of baked plains and blazed mountains, of strange flora and infernal fauna, and of furnace heats in the summer season which dance upon the arid air above vast unbroken stretches which in the winter time are cold and dreary.

In this desert district the annual rainfall is about three inches. On the coast line of its degrees of latitude it is thirteen inches, and as you bow northward with the contour of the State you progress in the ascending scale of precipitation.

But bordering on all those plains, up and down through the northern and central portions and across the southern breadth of the State, there are great mountain ranges. Their peaks receive the winter rains in the form of snows which lie banked upon them through the periods when the valleys have the greatest moisture, and when summer has parched the broad levels and sent their heat up the slopes the stored snows are released and gradually move down upon the areas to which it appears it was the scheme of Providence that they should be administered in the season of need.

Such are the natural conditions of land and water in the State of California. It may be said in a rough way that there is enough water at all times within the State to saturate for the purposes of agriculture all its arable land. But when the mountain reservoirs yield their accumulated deposits, they do not distribute them over the thirsty areas, but pour them forth through channels traversing the famished plains, leaving to man the application of the waters to the arid soil.

Irrigation, therefore, is a question which inheres to the very character of California. . . . Through irrigation principally, can its surface be subdued to the uses of man. It has therefore been the policy of the law to promote irrigation. . . . The first phase of the law affecting this question was to permit on part of individuals the appropriation of water. The water, instead of remaining appurtenant to the land, becomes alien to it; it is a thing apart which one can own, though he does not possess the soil—though without it the soil is useless. Under this arrangement, therefore, it became possible for speculators by simply posting a notice of appropriation to grab whole rivers, and by relocating from time to time in a ball toss between themselves and friends as the notices expired they could maintain their hold until the natural element thus sequestered grew through demand into a property.

The speculator in California, therefore, has not, as in most Western States, grabbed the land, he has gobbled the water. Unless he is a railroad or has become successor to the title of some old Mexican grant, the government of the United States has never allowed him to appropriate over one square mile of the land. . . . The monopolist of California is . . . the water company, an organization which appropriates and dams the stream and sells the water like so much merchandise. . . .

Whenever fifty or a majority of the holders of title, or the evidence of title, to lands susceptible of one mode of irrigation from a common source and by the same system of works, desire to provide for the irrigation of the same, they may propose the organization of a district under the provisions of the [1887 Wright] act, and when so organized such district shall have the powers conferred, or that may be thereafter conferred, by law upon such irrigation districts.

A petition shall first be presented to the board of supervisors of the county in which the greater portion of the lands are situated. . . . The supervisors shall then give notice of an election to be held in such proposed district for the purpose of determining whether or not the same shall be organized under the provisions of the act. . . . Such notice shall require the electors to cast ballots which shall contain the words: "Irrigation District—Yes," or "Irrigation District—No." . . . If upon a canvass of the votes so cast it appears that at least two thirds of the votes are cast for "Irrigation District—Yes," the board of supervisors shall enter an order declaring such territory duly organized as an irrigation district. . . .

It can scarcely be imagined that a law fairer in its provisions could be devised. Throughout its whole aim is to make it possible for communities living on arid lands to irrigate their lands, and to make their mortgages given for that purpose so secure that they might be salable at the lowest rates of interest. Few attempts, however, were made at organization under the law before it was discovered that parcels of the areas to be shaped into districts were related to certain diverse elements who would fight the law to a final knockout.

These were the existing water companies and the land speculator. It is natural to suppose that water companies would antagonize such a scheme, which might bring about competitors who would do the irrigating at cost. . . . But the more aggressive, because the more diffused, element of disturbance was the land speculator. If the persons owning lands within the boundaries of districts had all been resident farmers, mechanics, merchants, or other industrials, there would perhaps never have been a lawsuit contesting the act; but if those producers and disseminators of wealth within the districts did not realize it prior to the passage of the act, they have a reason to do so now that a landowner as landowner, is a singular anomaly. While the value of his land increases only through contiguous improvement for which he is to be taxed, he is an enemy to all improvement. His attitude is not that of a user of land and a producer of wealth, but he is a consumer and absorber of the benefit of others' industry. He does not hold land to make it pleasurable and profitable to communities, to allow its inherent bounties to be converted by labor into benefits, but he holds it in order that the desirability of its accessibility and its incident value may be increased by the industry, thrift, and morality, of dwellers in contiguous parcels.

It is the gloomy misfortune of California also, that by far the greater part of her land area is owned by persons who for the most part, do not even reside within the State, and except where they live in her metropolis, and the lands are in San Francisco county, few indeed of them abide in the counties where the lands are located. It was expected by the framer of the act that this class would make trouble. Of no fact was he more certain than that of all men, they were entitled to the very slightest consideration. Since they generally owned vast areas, sometimes reaching into entire valleys and the half of whole counties, if the act should provide that the vote to determine whether or not a district to be formed should be an acreage vote, like a share vote in a stock company, there would likely never have been a single irrigation district formed in the State, for one vacant landowner possessing more land than all the farmers, shopkeepers, and others, in the whole proposed district together, would vote his acreage against the proposition, and that would end it. . . .

[The State Supreme Court] held that districts, when organized, have all the elements of corporations formed to accomplish a public use. The results of a drainage law and one which has for its purpose the irrigating of immense bodies of arid land must necessarily be the same as respects the public good; the one is intended to bring into cultivation and make productive a large area of land which would otherwise remain uncultivated and unproductive of any advantage to the State, being useless and incapable of yielding any revenue of importance toward the general purposes of the State government by reason of too much water flowing over or standing upon it or percolating through it. The other has for its main object the utilizing and improving of vast tracts of arid and unfruitful soil, desert-like in character, much of it, which if water in sufficient quantity can be conducted upon and applied to it may be made to produce the same results as flow from the drainage of large bodies of swamp and overflow lands. Such a general scheme by which irrigation may be stimulated, the taxable property of the State increased, the relative burdens of taxation upon the whole people decreased, and the comfort and advantage of many thriving communities subserved, would seem to redound to the advantage of all the people of the State to a greater or less extent. It is true that incidentally private persons and private property may be benefited, but the main plan of the legislature, viz., the general welfare of the whole people inseparably bound up with the interests of those living in sections which are dry and unproductive without irrigation, is plain to be seen pervading the whole act in question. This is not a law passed to accomplish exclusively and selfish private gain; it is an extensive and far-reaching plan by which the general public may be vastly benefited; and the legislature acted in good judgment in enacting it. . . .

The temper of the people of the State is certainly that it is much to be regretted that the principle of the Wright law does not obtain in more of our laws. If it did, we should have less people swarming in sloth and poverty in the cities and less areas of brown bare lands in the country; the active agency of human labor and the passive element of production would be brought together and the result would be a bounteous production of wealth which would redound to the benefit of everyone within the State. If there was ever a law worthy of defense it is surely this one, for it deals with the very vitality of the State. . . . [These areas] will be the garden spots of the United States, a delight to the State of California, and peopled by the happiest and most prosperous communities on earth.

A Novelist Extolls the Watered Land, 1911

Dismounting, Mr. Worth tied his horse to a scraggly, half-buried mesquite and, taking his canteen from the saddle, climbed laboriously up the steep, sandy slope. He would look over the country from that point and then make straight for camp, for it was getting well on in the afternoon. From the top of the hill he could see the wide reaches of the King's Basin Desert sweeping away on every side. At his feet the bare sand hills themselves lay like huge, rolling, wind-piled drifts of tawny snow glistening in the sunlight with a blinding glare. Beyond these were the gray and green of salt-bush, mesquite, and greasewood, with the dun earth showing here and there in ragged patches. Still farther away the detail of

From Harold Bell Wright, *The Winning of Barbara Worth* (Chicago, IL: The Book Supply Co., 1911), pp. 152–54.

hill and hummock and bush and patch was lost in the immensity of the scene, while the dull tones of gray and green and brown were over-laid with the ever-changing tints of the distance, until, to the eyes, the nearer plain became an island surrounded on every side by a mighty, many-colored sea that broke only at the foot of the purple mountain wall.

The work of the expedition was nearly finished. The banker knew now from the results of the survey and from his own careful observations and estimates that the Seer's dream was not only possible from an engineering point of view, but from the careful capitalist's standpoint, would justify a large investment. Lying within the lines of the ancient beach and thus below the level of the great river, were hundreds of thousands of acres equal in richness of the soil to the famous delta lands of the Nile. The bringing of the water from the river and its distribution through a system of canals and ditches, while a work of great magnitude requiring the expenditure of large sums of money, was, as an engineering problem, comparatively simple.

As Jefferson Worth gazed at the wonderful scene, a vision of the changes that were to come to that land passed before him. He saw first, following the nearly finished work of the engineers, an army of men beginning at the driver and pushing out into the desert with their canals, bringing with them the life-giving water. Soon, with the coming of the water, would begin the coming of the settlers. Hummocks would be leveled, washes and arroyos filled, ditches would be made to the company canals, and in place of the thin growth of gray-green desert vegetation with the ragged patches of dun earth would come great fields of luxuriant alfalfa, billowing acres of grain, with miles upon miles of orchards, vineyards and groves. The fierce desert life would give way to the herds and docks and the home life of the farmer. The railroad would stretch its steel strength into this new world; towns and cities would come to be where now was only solitude and desolation; and out from this world-old treasure house vast wealth would pour to enrich the peoples of the earth. The wealth of an empire lay in that land under the banker's eye, and Capital held the key.

But while the work of the engineers was simple, it would be a great work; and it was the magnitude of the enterprise and the consequent requirement of large sums of money that gave Capital its opportunity. Without water the desert was worthless. With water the productive possibilities of that great territory were enormous. Without Capital the water could not be had. Therefore Capital was master of the situation and, by controlling the water, could exact royal tribute from the wealth of the land.

Critics Evaluate the California Water Plan, 1971

California began its development as an agricultural state. Yet most of the agricultural lands of California receive less than 15 inches of rainfall per year. Therefore, from Spanish mission days, the development of irrigation has been one of the principal concerns of the state. . . .

The California Water Plan is a master plan for the control, conservation, protection, and distribution of the waters of California, to meet present and future needs for all beneficial uses and purposes in all areas of the State to the maximum feasible extent. It is a comprehensive plan which would reach from border to border both in its constructed works and in its effects. . . .

At the very outset, it is necessary to understand one cardinal point: Water resources planning in California has been formulated in terms of the "surplus" of water in the northern part of the state and the "deficiency" of water in the south. Consequently, the problem of water planning in California has reduced to the question of how to transport water from the north to the south. . . .

The United States Bureau of Reclamation's Central Valley Project was designed as an undertaking of the State of California in its initial State Water Plan in 1930. Primarily, it was intended to develop the agricultural potential of the Central Valley. . . . The initial units of the overall plan for water development within the basin were first defined in a report issued by the State in 1930 and entitled "State Water Plan." This plan envisaged the transfer of an overabundant water supply from the northern part of the Sacramento Valley to the southern part of the San Joaquin Valley, where a deficiency existed. The units proposed as the first phase toward the solution of this problem included Shasta Dam, Delta Cross Channel, Delta-Mendota Canal, Friant Dam, Madera Canal, Friant-Kern Canal, Contra Costa Canal, and a power transmission system.

The initial units proposed in the State report are essentially those authorized for construction under Reclamation law by the Congress in 1937 and since built by the Bureau of Reclamation. In addition to the

From Garmuch Gill, Edward Gray, and David Seckler, "The California Water Plan and Its Critics," in David Seckler, ed., *California Water* (Berkeley, CA: University of California Press, 1971), 3–8, 12–14, 17–27. Reprinted by permission.

Map A

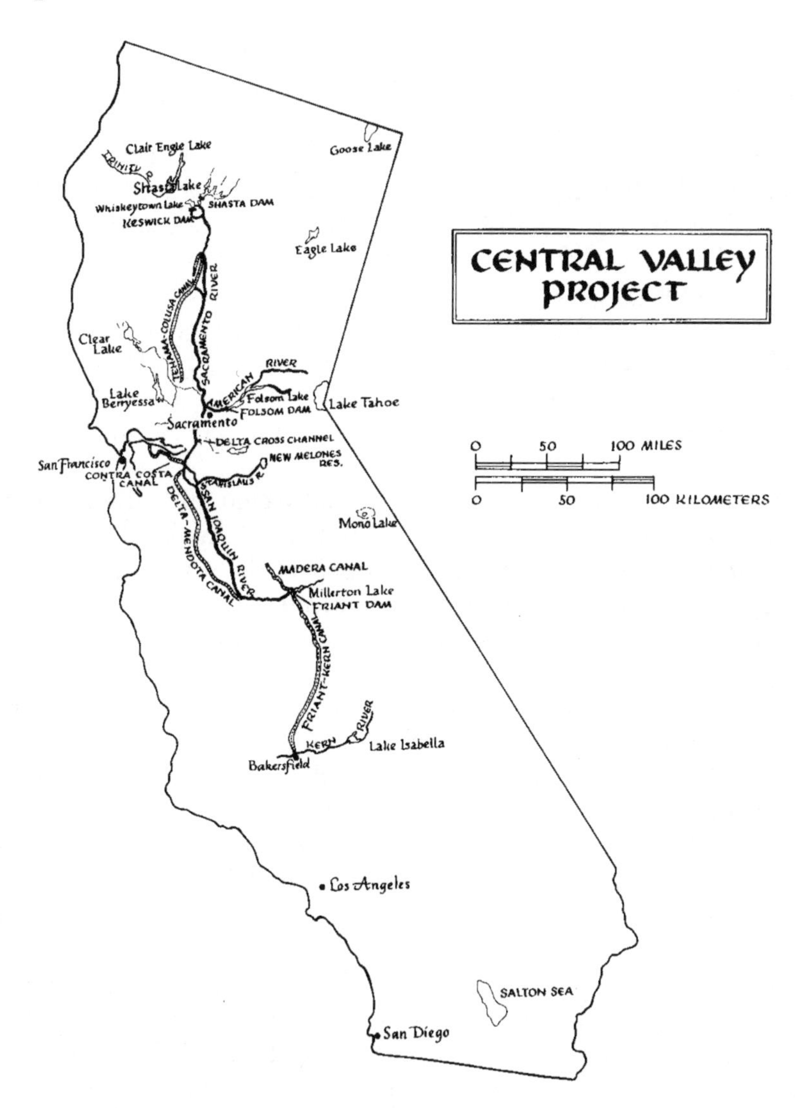

MAP 8. THE FEDERAL GOVERNMENT'S CENTRAL VALLEY PROJECT.
The CVP supplies water and hydroelectricity for California's farms, as well as fresh water for urban, industrial, and agricultural needs in the San Joaquin valley. Work began in 1937, and the system delivered its first power in 1944. Map by Madge Kelley from Norris Hundley, The Great Thirst: Californians and Water, *1770s–1990s. Berkeley: University of California Press, 1992, p. 253. Reprinted by permission of Madge Kelley.*

Map B

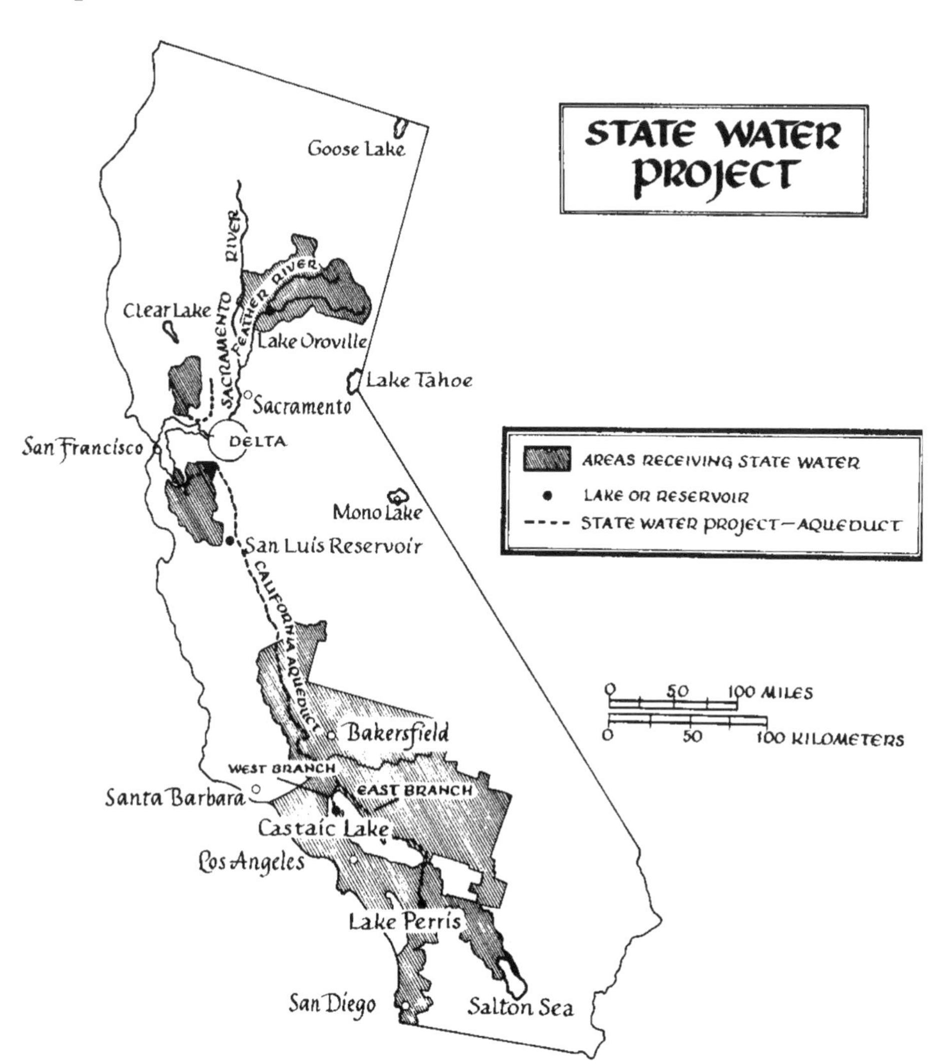

MAP 9. THE CALIFORNIA STATE WATER PROJECT.
The primary goal of the SWP was to store water from snowmelt and rivers in northern California and deliver it to southern California, supplementing the federal project, but without the government's 160-acre limitation. Construction, which began in 1962 with Oroville Dam on the Feather River, the key element in the project, was completed in 1967. Map by Madge Kelley from Norris Hundley, The Great Thirst: Californians and Water, 1770s–1990s. *Berkeley: University of California Press, 1992, p. 288. Reprinted by permission of Madge Kelley.*

units mentioned above, Keswick Dam was authorized and built to regulate the fluctuating power releases from Shasta Dam and add to the power capacity of the project. In 1949 Congress authorized the construction and operation of the American River Division, including Folsom Dam and Power plant, Nimbus Dam and powerplant, and the Sly Park Unit as part of the Central Valley Project. In 1950, the Sacramento Valley Canals were authorized, including the Corning Canal, Red Bluff Diversion Dam, and the Tehama-Colusa Canal. . . . The Trinity River Division, authorized in 1955, was scheduled for completion in 1963. The latest authorized addition (1960) is the San Luis Unit. This unit, a joint Federal-State venture, consists of the San Luis Dam and Forebay Dam, 123 miles of canal, a pumping-generating plant, and three pumping plants. . . . All features of the project have been integrated into a unified operations system. . . .

The activities of the United States Bureau of Reclamation in California were not universally applauded. . . . In May 1957, the [state's] Department [of Water Resources] published "The California Water Plan." . . . By the close of the 1959 session, the Legislature had authorized, by the Burns-Porter Act, the issue of $1.75 billion in general obligation bonds to provide the main financial base of the [State Water] Project.

In the words of the Department of Water Resources:

> The State Water Project is an amazing venture. It is the first statewide water resources development in the United States, and is the largest single water development in the world to be financed at one time. It is the first water project which will construct four reservoirs having recreation as their primary purpose. The Project contains the highest dam in the United States. Oroville Dam will tower 770 feet above its foundation when completed late this year. The Project will have the largest underground powerplant in the nation—the Oroville Powerplant—which, in conjunction with its companion powerplant downstream at Thermalito, will produce power enough for a city of one million people. The Project's largest pumping plant will lift more water higher than any other in existence, raising 110 million gallons an hour in a single lift up the Tehachapi Mountains. . . .

Additional to the main objective of supplying irrigation and municipal water, State Water Project facilities provide flood control and recreation benefits, and generate power. These facilities are also important in regulating stream flows down the Lower Feather and Sacramento rivers into the Delta. . . .

The California Water Plan has encountered a heavy barrage of criticism from its beginning in the 1930's to the present day. . . . The criticisms fall into two broad and often overlapping categories: (1) the State Water Project is economically and financially unsound, and (2) it will severely damage the environment of California.

The economic-financial arguments may he conveniently grouped into four basic categories:

Supply and demand. The idea of southern "deficiencies" and northern "surpluses" has been exaggerated. . . .

Evaluation. Many critics questioned the neglect of cost increases that were almost certain to occur either because of underestimation or inflation. . . .

Alternatives. It is said that both the present and future alternatives to the State Water Project have been ignored. . . . [Joe S.] Bain et al. said: "A possibly feasible alternative would have been postponement of some version of the present [Feather River Project] by two or three decades. . . . It would also be more prudent, because several decades from now changes in technologies and developments in population patterns not presently anticipated may indicate the desirability of different solutions to any problem of water deficiency in the southern part of California. The unseemly haste which has marked the early undertaking of the Feather River Project can probably be attributed more to the desire of Southern California interests to secure rights to Northern California water while they are still available than to any misapprehension on their part concerning the fact that the Project is being undertaken prematurely from an economic standpoint."

Capital gains. Many critics have voiced concern over the capital gains that will accrue to large landowners because of the State Water Project. These objections are based on two considerations: first, a public project that subsidizes the wealthy is bad in itself; and second, the U.S. Bureau of Reclamation, it is contended, would have undertaken at least the agricultural segment of the State Water Project had the state not done so. Thus, California has given up valuable federal subsidies by choosing to go it alone in order, they say, to escape the 160-acre limitation clause. . . . [Erwin] Cooper regards the motivation behind the State Water Project as:

> The idea of moving the Feather River out of its own watershed began as the brainchild of the big San Joaquin Valley

> land powers: the corporate farms, land management firms, railroads with huge acreages. . . . Their future lay in irrigated farming. The men who controlled them, along with much else in California, were unanimously dedicated to the proposition that the only way to beat the Bureau of Reclamation's 160-acre limitation on water for farmers was to have the state, rather than the federal government, operate a water distribution system. In this way their lands would escape fragmentation by the acreage auctioneer.

The environmental critics of the State Water Project include, prominent among others, the Sierra Club. The Sierra Club has been joined by many other organizations—the Planning and Conservation League, for example, the Committee of Two Million (a fisherman's organization)—and by a number of individuals. . . .

> (1) The feeling exists that Los Angeles has already grown beyond supportable dimensions; and . . . further growth should not be encouraged by State Water Project water. . . .
>
> (2) The impact of the Project on the Delta and San Francisco Bay is not known, but it is most likely dangerous. It could lead to loss of the fishing resource and the wildlife marshes in the Delta, and to chemical changes of the waters of the Bay-Delta leading to disturbance of the biological balance (with consequent growth of algae and oxygen reversal).
>
> (3) With continued withdrawals of water from the Delta, additional sources of water will have to be found. This means that the dam builders will turn their attention to the rivers of the north coastal area, specifically the Eel, the Klamath, and the Trinity. These rivers and their valleys constitute one of the last refuges of nature in California. In addition to destroying the region's natural amenities and its economic base in the recreational use of its resources, these dams will destroy the periodic flushing action of floods through the river systems. The deep pools in the rivers will thus silt full, destroying breeding grounds and therefore the valuable fishery resources of the area. Weeds and brush will encroach onto the banks of the rivers, supplying a base for further silt accumulation and destroying the accessibility and beauty of the rivers. Because the silt will accumulate in the river beds and reservoirs, the ocean beaches

> will lose their source of replenishment and decay back into the sea. These effects are already observable from existing water "development" projects in the region; further development will generate an ecological disaster.

The future of the California Water Project is unclear. [But] developing concern over the environment will almost certainly influence decisions about water development to a far greater extent than in the past. [Moreover], the accelerating pace of technological change has provided a spectrum of alternatives hardly imagined at the time the basic features of the Project were taking shape in the 1940's and 1950's.

Among the more outstanding technological horizons in the future are waste-water reclamation, nuclear desalination, the exploitation of newly discovered thermal water resources in the south, and groundwater utilization. Both the state's Department of Water Resources and the federal Bureau of Reclamation have played active roles in developing these technologies. . . . The Department of Water Resources has operated an experimental desalting plant at Point Lobos for several years. . . . The U.S. Bureau of Reclamation is designing development projects based on irrigation by sprinkler rather than by furrow, which is expected to improve substantially the efficiency of water use in agriculture. . . .

Thus, the California Water Project at the beginning of the 1970's is in a rather mixed situation. Though there is a firm grasp on the traditional commitment to north–south diversion, exploratory moves are being made to incorporate the new perspectives of environmental quality and recent technology. The strong push of environmentalists may be sufficient to direct California into a new and altogether different approach to water planning in the years ahead.

An Environmentalist Laments the Loss of Wild Rivers, 1982

The debate over construction lasted thirty-six years, and here is the outcome. I stand at the bottom of New Melones Dam, looking up a scarp of

From Tim Palmer, "The Dam," in *Stanislaus: The Struggle for a River* (Berkeley: University of California Press, 1982), pp. 46-60. Reprinted by permission.

limestone rubble, shell of the new mountain that blocks the river. The top of the dam is my horizon, 625 feet up, sixty-two stories, about as high as San Francisco's Transamerica Building without its tower, and a lot more permanent looking. Sixteen million cubic yards of dirt and rock, enough to fill 48 million wheelbarrows, or 2 million dump trucks, one of the larger piles of earth ever assembled by man. . . .

People have said that California is above all one great plumbing system, and it holds true for the Stanislaus. A hundred years' evidence of water tradesmen is patent; they have done a job on this river, the New Melones part being just the latest in a century of change. . . .

New Melones is a part of the Central Valley Project (CVP) of the federal Bureau of Reclamation. . . . In the state's studies of the 1920s, New Melones Dam was first mentioned as a secondary sit-in fact, barely discussed. Water planners found no need for the dam's conservation yield (irrigation water). The long-range State Water Plan later proposed thirteen reservoirs in the San Joaquin basin, New Melones being the largest and second most costly at $26.2 million. . . .

Established in 1795 or 1802, depending on the teller, the Army Corps of Engineers is the oldest public works agency and the largest engineering operation in the world. About 300 Army personnel and 30,000 civilians operate the civil works division, which builds dams. At any given time, the corps is working on about 300 projects: New Melones is only one of these. . . . At first roads were the main thing the corps built, but not for long. In 1852 Congress gave it responsibility for navigation-maintenance of harbors and rivers. This is how ocean freighters reach the Central Valley cities of Stockton and Sacramento. With the generosity and sympathy that followed floods in the thirties, the flood Control Act of 1936 gave a popular job to the corps. They would try to stop the floods.

The corps's role in California was unique because of the Gold Rush. . . . To correct the damage [done by hydraulic miners], state officials turned to the corps, which responded with channelization and levees. . . .

The other federal water agency is the Bureau of Reclamation, founded under the Newlands Reclamation Act of 1902 as a colonizing or settlement agency under Western enthusiast Teddy Roosevelt. The aim was to bring families to unsettled regions of the West (the bureau operates in eighteen states), not to subsidize large farms or ranches. Safeguarding this point, the law stated that no federally supplied water would go to individuals owning more than 160 acres. But the law was often overlooked. In the Westlands water district near Fresno, the average farm exceeds 2,000 acres, agribusiness corporations being subsidized at about $1,500 per acre per year. . . . Once could say that [Congressman John] McFall is the father of the dam. He grew up near the river. Savvy in local politics, the young Democrat became mayor of the farming town of Manteca, later serving in the California Assembly. At thirty-eight, he was elected to Congress and flew off to Washington . . . sitting on the Public Works Committee, the greenhouse of corps projects. . . .

On October 3, 1962, McFall's bill passed in the House of Representatives, and subsequently in the Senate, guided by California Senators Clair Engle and Thomas Kuchel. On October 13, President Kennedy signed the 1962 Flood Control Act, providing for New Melones and scores of other projects. . . . Along with its size, the purposes of the dam were expanded to include irrigation, hydroelectric power, water quality, fish and wildlife enhancement, and recreation. McFall molded a broader base of support.

On Christmas Day, 1964, leaden rains pounded the Valley and melted the . . . Sierra snow. . . . The Stanislaus overflowed its banks, swamping farms and development on the flood plain. Damages totaled $1.6 million, mostly below Oakdale. . . .

It was this issue—flood control—that inspired local support, that stifled local opposition, that led to the dam's construction. Different reservoir sizes or flood-control alternatives were not considered. The strategy was straightforward—get New Melones built. Identical resolutions asking for the dam were adopted by the association, flood-prone communities, and the California Association of Soil Conservation Districts. Alert in times of opportunity, the corps requested $1 million to start the project. This was not good enough for McFall, who elaborated on recent flood damage and urged Congress to better the corps with $1.5 million in 1965. . . .

In 1970 Central Valley conservationists, primarily members of the Regional Group of the Sierra Club, saw more opportunities. Through Stanislaus River Chairman Roger Gohring, they pushed for environmental improvements and a canoe "trail," with nine access areas and lower-river parks to be bought by the corps. John McFall would call this the "string of pearls." . . .

The corps wasted no time in kicking off the project in 1965. With the first appropriation of $1,500,000, a ground-breaking ceremony was held. . . . Construction finally began in 1966. Flood control was the main reason for support of the dam, and the major issues in the long debate had been localism versus federalism and the Army Corps of Engineers versus the Bureau of Reclamation. New Melones supporters had encountered no opposition because of the dam itself.

The environmental movement had not yet arrived; the wilderness canyon was a secret. Like the Colorado's fabled Glen Canyon, dammed in the 1960s, the upper Stanislaus was a place that few people knew about. The first known paddlers were a group of Sierra Club members in 1960; the first commercial outfitter appeared in 1962.

During the early New Melones debate, the United States had no Environmental Policy Act requiring environmental impact statements. The nation had no wild and scenic rivers system, no wilderness system, no coherent water policy. When plans were completed for the demise of the canyon, the green-and-white dynamo of a river, and the marble-cobbled beaches below 800-foot walls of white limestone, almost no one cared about the place, and nobody questioned the logic and validity of the corps's calculations. In all this debate over New Melones Dam, no one mentioned the place to be flooded, the land to be buried beneath hundreds of feet of slackwater.

Reporters Assess the 1982 Reclamation Reform Act, 1988

Prior to passage of the Reclamation Reform Act (RRA), acreage limitation was conceived of entirely in terms of land ownership and residency. A single farmer could hold title to no more than 160 acres of federally irrigated land and had to live on the farm. In the absence of official rules to clarify the law's meaning, interpretations over the decades held that a husband and wife could receive federal water on a joint holding of 320 acres. Additionally, the Bureau arbitrarily decided that ownership limitations applied only within a given water district. Multi-district ownership of 160 acres per district was allowed.

The concepts underlying these interpretations included the notion that reclamation policy was designed to settle the West and that resident

From Don Villarejo and Judith Redmond, "The Reclamation Reform Act of 1982," in *Missed Opportunities, Squandered Resources* (Davis: California Institute for Rural Studies, 1988), pp. 42–46. Reprinted by permission.

family farmers would own the land they farmed. California, however, was developing along a different model of landholding. Absentee landowners leased substantial acreages to farm businesses who operated the land. Today, about 49% of California cropland is leased from a non-operator landlord. This practice differs sharply from that of other regions where owner-operators are the norm.

Without published rules to guide enforcement, vast farming operations could be built upon leasing practices: dozens of 160-acre parcels, each under different title, could be farmed by a single leaseholder. . . . The average size of these farms was 7,733 acres per farm, roughly 12 square miles. The enormous size of these operations as compared to the 160-acre limit led even impartial bodies to suggest that the law was being imperfectly enforced. . . . "This is not the 160-acre farm to which reclamation law intended spreading the benefits." . . .

One provision of reclamation law permitted owners of land in excess of 160 acres to receive federally subsidized water if they entered into a recordable contract to sell their excess holdings within 10 years of the first water delivery. Thus, even very large holdings by absentee landlords could be farmed by lessees if the properties were covered by such a contract. . . . (As it turned out, many . . . landowners planned merely to redistribute the title to their land rather than selling it to new farmers.)

Finally, some excess landholders reaped the benefit of the reduction of the groundwater overdraft. . . . This meant that excess landowners had an additional option. Depending upon the pumping depth on a particular land parcel, they could choose to irrigate by groundwater pumping alone and avoid selling (or redistributing ownership) of that parcel. . . .

On August 13, 1976, the Secretary of Interior imposed a moratorium on sales of excess land pending rules to enforce acreage limitation. In an independent action, the Westlands [Water District of California] blocked the proposed rules by arguing that an Environmental Impact Statement to determine the effect of acreage limitation was necessary. Faced with the prospect of genuine enforcement of the 160-acre limit, Westlands landowners went to the Congress to get the law changed. Eventually, the 1982 amendments [to the 1902 Reclamation Act—the Reclamation Reform Act (RRA)] were passed. . . .

[The Reclamation Reform Act] set an absolute limit of 960 acres of owned land for an individual owner or entity (property owned by husband and wife and minor children is considered a single unit). This was made somewhat stronger by also insisting that the limit applied to combined holdings receiving project water throughout the 17 western states. . . . Under prior law [the Department of the] Interior had limited holdings to 160 acres per district. In California, with its dozens of water districts, setting a limit on multi-district holdings could be quite important. Finally, corporations with more than 25 stockholders were limited to ownership of 640 acres westwide.

One of the most significant aspects of RRA is that, with respect to water pricing, the notion of "landholding" has replaced "land ownership." The former term refers to all of the operations on which federal water is being used by the farm operator. This change more accurately reflects existing patterns of land use in areas where the leasing of cropland is widespread. Thus, water pricing is now to be primarily applied to farm operators based on measures of landholdings. . . .

In the period immediately following implementation of RRA, there have been three important changes in the pattern of land ownership. First, the amount of land held by large (more than 960 acres) owners has declined, although most changes in this size range occur as a result of recordable contracts under prior law. Second, the amount of land held by small (160 acres or less) owners has [also] declined. Third, medium-size ownership units have enjoyed a considerable expansion of their share of privately owned land. Thus, to the extent that Congress intended to shift ownership of federally irrigated land from the largest to the smallest owners the policy has not worked, but it has shifted title in owned land into the medium-sized holdings.

ESSAYS

Hydraulic Society in California

Donald Worster

No region on earth has had more to do with shaping the twentieth century than California. That is as true of agricultural history as it is of mass culture, sexuality, urbanization, atomic bombs, and the shift from bourbon to wine. Put a historian down anywhere in the state, and he or she will find something profound to say about modernity. My own choice is to be deposited by the side of a concrete-lined irrigation canal in Kern County, by a stream that is not a stream, where no willows are allowed to grow or herons or blackbirds nest. That intensely managed piece of nature tells us a great deal about contemporary rural life and land use, some of it profound, some of it disturbing, all of it indicative of a worldwide momentum. The factual history of that canal and the agriculture dependent on it is rather well known in particulars. . . .

In 1939 two books appeared on the emerging character of California agriculture. They each had a galvanic impact on public thinking, and are both still in print, still capable of jolting us out of moral torpor. The first was John Steinbeck's *The Grapes of Wrath*, the second, Carey McWilliam's *Factories in the Field.* Together they framed a compelling interpretation of that rural order which to date has not been persuasively challenged. A succession of able historians has filled out and updated the original accounts, but no one has yet changed the fundamental terms of discussion. Forty years covers a long span in the modern history of ideas, however, and we may be ready today to try another angle of vision.

The human consequences of industrialization and the making of a new rural proletariat were the themes that superseded all others for Steinbeck

From Donald Worster, "Hydraulic Society in California: An Ecological Interpretation," *Agricultural History*, 56, no. 3 (July 1982): 503–15. Reprinted by permission.

and McWilliams. Describing California, they echoed eighteenth- and nineteenth-century protests (from writers like Oliver Goldsmith and William Cobbett) against the enclosure acts in England, and they reminded readers too of the cities of Depression America, where millions were made helpless by the failing of industrial capitalism. Karl Marx was an obvious if remote influence, particularly on McWilliams, who tried hard to argue in 1930s Marxian socialist terms that California agriculture had evolved out of a feudal state shaped by Spanish land barons. Capitalist entrepreneurs had taken over from them to build an industrial empire, which rested on the backs of stoop-and-pick workers recruited from all over the world. The eventual outcome of that history, he predicted, would be an uprising of the exploited undergroup collectivizing the factory farms so that their productivity would become available for the common welfare. Class conflict, in other words, was what the California story in agriculture was all about.

Altogether missing from that interpretation—and again it is McWilliams I have in mind especially—was the peculiar ecological situation in California: the interaction of men and women there with the land, which is after all at the heart of any farming system. It will be my own argument that a new ecologically oriented inquiry will give us not only a fuller account of the state's agricultural evolution but also, and this may be the most surprising result, it will establish a more satisfactory framework than old-fashioned, generic Marxism for making sense of that history of class exploitation. It will do so by showing that nature's fate is humanity's as well.

At no point in his novel did Steinbeck so much as mention the fact that the vineyards of California, its orange groves and cottonfields and vast tomato patches, were supported by an elaborate irrigation regime. McWilliams noted the fact only in passing. The contemporary ecology based agricultural historian, in contrast, finds in irrigation a key formative element, an underlying infrastructure out of which social relations grew. What is exemplified in the state is not only "factory farming," as McWilliams and Steinbeck made us aware, but more specifically a modern hydraulic society—a social order founded on the intensive management of water. That regime did not evolve in isolation from the industrial system, of course, but all the same it was a distinctive emergent, reflecting the geography and arid climate of the state. We ought to be scrutinizing, therefore, not only historical analogies like Cobbett's England but also those desert landscapes of the world where large-scale irrigated agriculture has flourished as it has in California.

The phrase "hydraulic society" comes from two students of ancient cultures, Julian Steward and Karl Wittfogel. They maintained that in the great river valleys of Mesopotamia, Egypt, India, and China a striking cultural convergence took place during the 3,000 years before Christ. In each of those places it became necessary, under the pressure of population growth, to plug all the rivers with storage dams, diverting the water into elaborate networks of canals and ditches to irrigate the peasants' fields.

Construction and maintenance of those massive public works required the marshaling of vast corvees—faceless armies of laborers—for at least a part of every year. . . . The engineered waters of the Tigris, the Euphrates, and the Nile were then, according to the theory of hydraulic society, the environmental basis for the first authoritarian, complexly hierarchical civilizations. I take this to be the essence of the hydraulic thesis: the domination of nature is an ambition that first appears stark and unchecked in the archaic desert empires, and thereafter the ambition, wherever and whenever it recurs as a compelling cultural idea, is always associated with the domination of some people by other people. This is a grand, imposing argument, defying a narrow, reductive, "scientific" formulation (domination being necessarily a difficult concept to quantify), and undoubtedly open at numerous points to challenge and skepticism. . . .

In California and the West has emerged the most elaborate hydraulic system in world history, overshadowing even the grandiose works of the Sassanians and the Pharaohs. In 1976 the federal Bureau of Reclamation alone operated 320 water-storage reservoirs, 344 diversion dams, 14,400 miles of canals, 900 miles of pipelines, 205 miles of tunnels, 34,620 miles of laterals, 145 pumping plants, 50 powerplants, and 16,240 circuit miles of transmission lines. That technology has remade completely the western river landscape. . . .

The rivers of California, particularly those that flow out of the Sierra Nevada and down the Great Central Valley, are in season raging torrents of energy. For thousands of years they defied their own channels every spring, flooding the sloughs, the soddy meadows, and what John Muir called "the bee pastures of heaven." Aboriginal farmers accepted in most cases that tumultuous authority: when the floods relented, they rushed to plant corn and beans in the mud, hoping to harvest them before the soil dried out. Only in a few places in what would become California were diversion ditches—really shallow furrows—dug. The early Spanish and American invaders, on the other hand, made more and bigger furrows, and midstream on the smaller tributaries they threw up brush dams to force a steady stream into their gardens. That scale of intervention made little ecological disturbance, but it allowed a satisfactory living for a pioneer family wanting nothing more than self-sufficiency.

Few of the Americans who began farming in California, however, were interested in pursuing a self-sufficient life. They brought ideas about markets and profitability along with plow and seed to the western river valleys. Most of the rich agricultural soil, they soon discovered, lay on the larger floodplains, where simple dams of brush and rocks were feeble constraints, completely inadequate to tame a San Joaquin carrying melted snow from some of the highest peaks on the continent. Even the joining together of two or three settlers' efforts could not constitute enough human force to manage the major rivers. By the early 1870s California had reached its first diversion plateau. It could go no farther toward market development without inventing a larger, more potent harness—a yoke that would unite many people's energies. But who would hold the reins?

The marketplace . . . was the most powerful determinant of the course taken by the American irrigated agro-ecosystem. It soon produced one of the key institutional devices that Californians used to climb off their first plateau: the private corporation. The San Joaquin and King's River Canal and Irrigation Company was incorporated in 1871, with headquarters in San Francisco where most of its shareholders lived, to construct a large new irrigation system. When this system was completed two years later, it ran almost forty miles and watered 16,000 acres a year. "The farmers and settlers in the San Joaquin Valley," the company's promotion brochure read, "are too poor to carry out the necessary canals and ditches by themselves, and require the cooperation of capitalists." Water now became a commodity to be mass-produced and mass-consumed. . . .

Paradoxically, California irrigation, in its reorganization on the Great Valley floor, seemed at first to be tending toward a more decentralized, democratic rural economy. Immense wheat farms, some as large as 50,000 acres, were broken up for sale to individuals or to colonies of cultivators. A forty- or eighty-acre orchard under the ditch was about all a single man or woman could handle, and it afforded a substantial enough income for a while. The state that had once been described as "the rich man's paradise, and the poor man's hell" now began to resemble, in the eyes of some observers, "the land of the common people," a rural eden of "small estates, of small enterprises, of small fortunes." But here was the rub: title to land without title to water meant little in a dry country. So long as small farmers had to buy water from a distant corporation, they had obtained a new prosperity at the cost of a substantial portion of their independence.

In addition to corporate capitalism, there was a second ladder devised to climb from the lower to a higher level of diversion. It was called the irrigation district, and it was a unique product of the arid region. The 1887 Wright Act authorized California farmers to organize quasi-governmental entities that could build common irrigation works and tax local residents to pay for them. The district was, in other words, a public corporation brought into being by a majority vote of landowners and often coercing a recalcitrant minority to share the expense. It was a means of getting more income out of a river without surrendering to urban capitalists. By 1920 there were 71 irrigation districts in the state, most of them put together in the boom years of 1915 to 1920. The largest, covering over a half-million acres, was in Imperial Valley, land of broiling sun, rich green crops, and recurring violence between growers and their hired laborers. Imperial had been carved out of the desert by high-risk capitalists who had tapped the Colorado River bonanza, but even after its reorganization along nominally more democratic, locally managed lines, the district could not shake off its origins. Being competitive in marketplace America meant here, as in other districts, adopting to a large extent the private corporation's top-down pattern of internal authority. . . .

A long training in forming irrigation districts, in conforming to unified water plans, and in submitting to strong leaders helps greatly to explain why California eventually became the most powerful agricultural region in the United States. . . .

Irrigation farmers . . . had unrivaled access to credit, to the capital needed for maximizing their technological efficiency, and they gained political leverage to protect their position even in a highly urban state. Most important, they secured on their own terms a labor pool large enough to harvest their produce cheaply and, through collective strength, they kept those laborers firmly under control decade after decade: first the Chinese, then the Japanese, the Filipinos, the Blacks, the Okies, and the Mexicans—California's polyglot, wage-based version of the Egyptian corvees.

Thus flourished the second phase of the hydraulic society's life cycle lasting from the 1870s to the 1930s. Then once again a plateau was reached, threatening stagnation, or worse, sudden economic decline. There was not enough pure water to support a constantly expanding irrigation demand. Every drop that could be economically taken directly from the rivers was taken, and farmers began to mine their underground supplies at breakneck speed. . . .

California farmers were, naturally enough, not prepared to give up the markets they had won, the technology they had invested in, or the social order they had built on irrigation. But getting to the next, higher plateau necessarily involved an appeal to the state or federal government for aid. . . . In the late thirties federal engineers began planning the intricate Central Valley Project, the first stage of which was completed over the next decade, propping up the factories in the field with subsidies from the national treasury.

Since California farmers agree with, and are the only means to realize the ruling commercial values of modern hydraulic society, it is improbable that federal centralization will mean any upheaval in local structures of wealth. In fact the story of California irrigation in this latest, post-thirties phase has been the story of the establishment of concentrated private hegemony, at once economic and ideological, over publicly developed engineering works. But make no mistake about this: considerable autonomy has been delivered to remote, outside forces, to a collusion of bureaucratic and marketplace power centers. The individual farmer and the small community have become less than ever masters of their fate.

The most likely prospect for major historical change at this point comes not from government, not from a shrinking number of workers, not from California irrigators themselves, but directly and indirectly from nature. Most ancient hydraulic empires collapsed at the point where there were mounting ecological difficulties and a lag or breakdown in the managerial skill needed to meet them. . . . The state of California, some recent evidence indicates, may be approaching a similar alkalinity fate, as even the casual traveler can see along Interstate 5 in the southern San Joaquin Valley. All the environmental problems encountered in the 1930s have resurfaced, along with several unprecedented ones; meanwhile, the remedies are becoming more expensive and complicated to apply. For the moment these complex problems may still seem solvable, but add to them the newer threats of inflating energy costs and pests resistant to pesticides, and the future becomes even more insecure, more unpredictable. It seems unlikely, in any case, that a massive, intricate irrigated agriculture, especially one tied to an marketplace engine, can save itself forever from self-destruction, though it may be that the trap's closing will be evaded for a long while yet.

We have been telling, someone may say, a parable of modern life. . . . In the vast arid spaces of California and the West, I have been suggesting, the universal modern predicament appears with a stark, uncluttered honesty not always found in other landscapes. Here we are able to see etched in sharpest detail the interplay between humans and nature and to track the social consequences it has produced—to discover the process by which, in the remaking of nature, we remake ourselves.

When Is a River Not a River?

David Igler

Few legal cases in late nineteenth-century California sparked such bitter controversy as the pre-eminent water rights ruling of *Lux v. Haggin*. From the beginning of the litigation in 1879, *Lux v. Haggin* served as a public forum for a wide range of ongoing conflicts in California's arid San Joaquin Valley—"land monopolists" against "small farmers," San Francisco control of hinterland resources, and perhaps most notably, "riparian" landowners versus water "appropriators." While these highly politicized issues made *Lux v. Haggin* into a case with various meanings for different audiences, there nonetheless remained a fairly basic question for the court to decide: what is a watercourse? In order for Charles Lux and his fellow plaintiffs to establish their water rights as riparian landowners, they had to prove that the Buena Vista Slough which crossed their Kern County swampland was in fact a watercourse. If it was a watercourse, then they were riparian landowners. Conversely, if Buena Vista Slough was indistinguishable from Buena Vista swamplands—if no watercourse existed as Haggin and the other defendants contended, then no riparian rights existed either. The definition of a river therefore stood at the center of *Lux v. Haggin*. Over the course of the lengthy trial, however, this attempt to codify a river's existence developed into a much broader discussion about the very character of nature in California.

The text of *Lux v. Haggin* thus provides a unique glimpse at Californians' representations of nature, and how those representations functioned within the discourse of nineteenth century law. These representations, I

From David Igler, "When Is a River Not a River? Or, Reclaiming Nature's Disorder in *Lux v. Haggin*," *Environmental History*, 1, no. 2 (April 1996): 52–69. Reprinted by permission.

will argue, largely portrayed nature as orderly, consistent, and definable—a socially constructed nature both adaptable to the permanency of human law and amenable to the preservation of private property rights. Yet nature, and more specifically its hydrological processes, proved anything but orderly and consistent in the San Joaquin Valley of the late nineteenth century. Rather, this valley was a complex and dynamic landscape of natural extremes. Constant disturbance and natural change—even chaos—reflected more accurately the dominant natural forces of the region. But by conceptualizing and accepting a static model of nature, despite glaring evidence to the contrary, the contending interests in *Lux v. Haggin* simplified imaginatively the same natural landscape that their competing land reclamation projects simplified physically. The misguided notion of nature's unchanging "order" ultimately served to rationalize the litigants' environmental engineering, exacerbating tensions within human and natural communities.

In depicting nature as disorderly, I do not wish to discount certain balances of ecological systems within natural communities. Rather, I would suggest that forces of natural chaos and order interact dynamically, culminating in complex, historical patterns of natural change. A construction of nature as disorderly does not, contrary to the concerns of some environmental historians, legitimize human control or despoliation of the natural environment. Rather, a depiction of nature as disorderly fully embraces the caution expressed by Donald Worster—"What, after all, does the phrase 'environmental damage' mean in a world of so much natural chaos?"—by separating nature's autonomous and unpredictable action from human attempts to control and simplify natural landscapes. Locating and accepting nature's discord as well as its "discordant harmonies, to borrow Daniel Botkin's phrase, allows nature to assume a broader and more active role in our "stories" of the interaction between humans and the natural environment. As disorderly nature asserts its agency in human affairs, we create a variety of ideas, technological innovations, and institutions to negotiate its meaning and impact. But an acceptance of nature's disorder ultimately cautions us against presumptuous and destructive behavior—both as agents of change in the natural world and as historians of that change.

Legal and environmental historians of the Far West have largely treated *Lux v. Haggin* as the showdown between advocates of "riparian" water rights and supporters of "appropriation" water rights. Riparian landholders like Charles Lux owned property adjoining California's rivers, and their water rights derived from the riparian location of their land. "Appropriation" water rights, on the other hand, were supported by agriculturalists whose land did not border a river but had nonetheless filed for an appropriation of a river's flow. Throughout California, public support clearly sided with James Haggin and the doctrine of appropriation. Since most agriculturalists in the arid San Joaquin Valley did not own riparian land, irrigation by means of appropriation appeared to serve the "public interest" by spreading the Valley's water resources most equitably. In Kern County during the late 1870s, these two interests clashed when Haggin's large appropriations apparently diminished the flow of the Kern River to the detriment of downstream riparian landowners. Among those affected was the massive land and cattle enterprise Miller & Lux.

In *Lux v. Haggin*'s district court appearance, the court sided with Haggin and the doctrine of appropriation. Three years later, however, California's Supreme Court reversed the lower court ruling. The Supreme Court twice ruled in favor of Charles Lux and his assembled group of riparian landowners, and this support for riparianism subsequently had a widespread impact on water rights throughout the American West. In what came to be known as the "California Doctrine" of water rights, the court adopted the riparian principle as supreme but it also acknowledged some appropriation claims made prior to those of riparian landowners. . . .

Prior to the reclamation (and disappearance) of Tulare Lake during the late nineteenth century, the region surrounding Kern River and Buena Vista Slough was known as the Tulare Basin or Tulare Lake Basin. Separated from the San Joaquin Valley by the slight rise of the Kings River alluvial fan, the Tulare Basin constituted the southern end of the San Joaquin Valley, which together with the Sacramento Valley to the north comprised California's 460-mile long Central Valley. . . . Over the past 100 million years, sediment carried down from the Sierra Nevadas by the Kings, Kaweah, Tule, White, and Kern rivers continually reshaped the Tulare Basin floor, and since the Basin lacked an outlet to the Pacific Ocean, spring runoff from the Sierras created Tulare Lake, Buena Vista Lake, and Kern Lake. This was a landscape of hydrological extremes: during a dry year, the levels of Tulare, Buena Vista, and Kern lakes would drop drastically, while the next year the three lakes could form one continuous body of water stretching across the entire valley floor. In wet years—perhaps more accurately, flood years—the inundated rivers and streams often carved new courses across the landscape. As one observer noted in 1886, the location of the Basin's rivers was "nearly as fleeting as the clouds that sail over [the] land.". . .

The origins of *Lux v. Haggin* centered around a project supported by most every agriculturalist in California—land reclamation. Like the

Sacramento-San Joaquin Delta region in the middle of the Central Valley, the Tulare Basin was rich with land ready for reclamation. The Basin's west side held hundreds of thousands of acres of fluctuating swampland, and surrounding this "marshy quagmire" were even larger acreages of alkali desert. Reclamation, in short, attempted to replace these two extremes through hydrological engineering—diverting the water that fed swampland and spreading it over the dry lands. The purpose and means of reclaiming this land from nature was quite clear. As one local editor stated in 1878, "All that is desired is that these barren plains should be made to blossom as the rose. And all that is necessary to make them bloom is to give them away in chunks." This use of a natural metaphor to promote land reclamation accurately reflected a prevailing attitude among Californians that nature had left its work unfinished. In the name of Progress, Californians would complete nature's will through irrigation and agricultural husbandry. . . .

Land reclamation required either large-scale capital or farmer-cooperative enterprises. State land policies favored the former approach—giving the land away in "chunks." . . . Charles Lux and James Haggin were two of the biggest beneficiaries of state land giveaways throughout the decade prior to *Lux v. Haggin*. Haggin and his group of San Francisco capitalists (incorporated as the Kern County Land Company in 1875) spent the late 1870s acquiring desert lands in the Tulare Basin following their successful lobbying drive for the 1877 Desert Land Act. In a concerted attempt to control water appropriation rights to the Kern River, Haggin's group also bought up all the available stock in Kern County irrigation companies. . . . By the time that *Lux v. Haggin* went to court, Haggin had monopolized the largest appropriative rights to the Kern River.

Charles Lux, on the other hand, made few promises of public interest. As a partner in the massive land and cattle corporation Miller & Lux, his concern with land ownership centered around adding to a landed estate that encompassed over one million acres in California, Nevada, and Oregon. Long recognized as the state's most notorious land monopolist, Miller & Lux was more importantly a clear predecessor to California agribusiness with a vertically-integrated corporate portfolio of canal companies, slaughterhouses, banks and company stores. . . . During the lengthy dry season of 1876–77, Miller & Lux began constructing the Kern River Canal to facilitate their swampland reclamation along Buena Vista Slough, but the small flow that the Kern River held that year was intercepted by Haggin's upstream appropriations. In response, Miller & Lux organized a group of Kern County riparian landowners and they filed 84 suits against James Haggin and hundreds of other Kern County appropriators. *Lux v. Haggin*, filed in 1879, was the test case for control of the Kern River's flow, a flow made essential by both sides' reclamation projects.

In court, the legal status of Buena Vista Slough quickly became a key point of contention between these rival water lords. Whereas Haggin's team of attorneys argued that Miller and Lux were not riparian landowners because Buena Vista Slough was indistinguishable from the surrounding swampland—hence, no defined watercourse—Miller & Lux counsel Hall McAllister established the watercourse's existence as the backbone of their complaint against Haggin. Buena Vista Slough, McAllister claimed, "had run from time immemorial . . . watering and refreshing these lands and conferring upon them their entire vitality and value." Witness after witness, many of whom were employed by Miller & Lux, testified to the permanent character of the Slough's bed, banks, and flow of water. They all agreed that Buena Vista Slough traversed a vast swampland, but the swampland, McAllister argued, did not negate the independent status of the Slough as a watercourse. "The swamp district is not a quagmire," he contended, it was orderly and consistent, "traversable both by man and beast."

This theme of nature's order was a crucial argument for Miller & Lux because the very status of riparian land ownership hinged upon a river's unchanging location. If a river changed location, so changed the property rights associated with that watercourse. Nineteenth-century American water law, as forged in the humid eastern states, included few (if any) cases caused by a river naturally changing its course. The legal characteristics of a river indeed hinted at its very fixity, as defined by Joseph Kinnicut Angell's prominent Treatise on the Law of Watercourses: "A watercourse consists of bed, banks, and water." McAllister cited Angell's definition of a river, noting Angell's important qualifier that "the water need not flow continually; and there are many watercourses which are sometimes dry." Eastern water law also spoke to other natural characteristics of rivers in the arid west. . . . Eastern legal precedent therefore recognized the existence of a variety of water systems, from "boggy" places to gushing rivers with well-defined banks. But a "natural stream of water," nonetheless, remained fixed in an absolute place from "time immemorial" to the present, meaning a river did not change the location nature had bestowed upon it. . . .

If the physical evidence offered by Miller & Lux depicted a watercourse both orderly and consistent, their quasi-ecological arguments explained that nature as a whole was also guided by order. "Nature," McAllister argued, "appropriated these waters long before California was settled [and] she fixed the appropriation by the physical features of the country. . . . The whole question is this: Shall this river and this flow of water be left as Nature left it?" Nature, McAllister intimated, was no capricious mistress: "she" did not change her handiwork once completed. Yet reclamation—whether of swampland or desert land—by definition altered nature. In fact, it "reclaimed" land from an errant and chaotic nature, from nature gone awry.

The plaintiffs argued [that] while riparian reclamation complemented nature's order, appropriation fundamentally altered nature. [But] James Haggin's co-counsel John Garber sought to discredit this neo-preservationist attack. Haggin's appropriations did not destroy Kern River, Garber countered. The Kern would continue to flow, and Haggin's appropriations from the river supported the "industrious citizens who have thus built up the resources of the State, fertilized its desert wastes, and turned what was a wilderness into a garden spot." It was Miller & Lux's wasteful swampland, Garber argued, that "squander[ed]" the Kern's flow. . . . Without a "defined, tangible, [and] provable" river, [Haggin's attorneys] concluded, Miller & Lux had no riparian rights for the court to respect.

A series of questions about "nature" thus stood at the center of this highly-politicized water rights battle. What is a watercourse? At what point does a river, stream, or swampy slough cease to be a watercourse? Does nature, furthermore, confer certain rights upon a watercourse or landscape that humans must respect? These questions, which today we ponder from an epistemological position infused with environmental ethics, were to the litigants in *Lux v. Haggin* questions of private property and courtroom rhetoric. Neither Lux nor Haggin were suggesting that reclamation—this nineteenth-century crusade to engineer the landscape—should be impeded by notions of nature's autonomy. Nobody believed that a river actually possessed "riparian rights" of its own, as McAllister had rhetorically suggested. Least of all Judge Benjamin Brundage, who on November 31, 1881 ruled that "no continuous or defined channel" existed in Buena Vista swamp, and therefore the plaintiffs were not riparian landowners. . . . Haggin had won, the desert must be made to bloom. . . .

In 1884 the riparianists took their appeal to the California Supreme Court. Miller & Lux's lawyers again positioned their reclamation enterprise as sanctioned by nature. . . .

In the court's final decision, California Supreme Court Justice E. W. McKinstry affirmed Miller & Lux's riparian rights along Buena Vista Slough, but he also gave optimism to appropriators by acknowledging the "public use" of irrigation. Riparian rights, he conceded, could be condemned by irrigation companies or irrigation districts with just compensation. Engineering the landscape by whatever means most expedient, served society's best interest. . . .

If Henry Miller and Charles Lux were grateful for nature's beneficence, the year following the decision showed them equally ready to share nature's wealth with their former rivals. In a privately-negotiated agreement with James Haggin, Miller & Lux guaranteed themselves a consistent supply of water from the Kern, gave the remaining flow to the upstream appropriators, and the two sides shared the cost of building a holding reservoir. . . .

Lux v. Haggin . . . reveals the determined effort to engineer a natural landscape for the sole purpose of resource extraction. To this purpose, the litigants sought justification in nature's "laws" because they viewed reclamation as ultimately fulfilling nature's unfinished business. "Barren" deserts, unproductive swampland, and wild rivers had no place in nature's "garden." That nature may have intended a different landscape—one of complex extremes, one ruled by natural disturbance and disorder, one that suggested all the "obstinate" natural conditions in the American West—remained unimaginable. By thinking about the Tulare Basin landscape primarily as a place to change and not as a landscape engaged in a continual process of changing itself, appropriators and riparians alike precluded the more troubling issue of nature's autonomy. "Reimagining" nature can begin with a recognition of its autonomy, complexity, and disorder.

Hydraulic Society Triumphant

Norris Hundley

Californians from the days of the Gold Rush had dreamed of transforming the Central Valley into an agricultural paradise. The hydraulic miners had challenged that vision with their environmentally destructive

From Norris Hundley, "Hydraulic Society Triumphant: The Great Projects," in *The Great Thirst: Californians and Water, 1770s to 1990s* (Berkeley: University of California Press, 1992), pp. 232–34, 239, 240–43, 252–55, 271–72. Reprinted by permission of the Regents of the University of California and the University California Press.

hoses that washed away entire mountainsides, but even before their more abusive practices were outlawed, geologists and other experts had concluded that sustained productivity in the valley would require major manipulation of the state's waterscape. To them this judgment seemed reaffirmed by the lessons learned by 1900 from the collapse of the cattle and wheat dry-farming ventures, as well as from the generally poor showing of most mutual water companies and irrigation districts.

The Central Valley appeared to be a geographical paradox. A phenomenally rich and broad alluvial plain, some 450 miles long and 40 to 70 miles wide, it was watered by streams draining the Sierra Nevada on the east and the coastal ranges on the west. As speculators, developers, farmers, and would-be farmers saw it, however, nature had short-changed the area. Precipitation was light and came mostly after the growing season had ended. . . .

The search for a comprehensive water plan for the Central Valley went back at least to 1856 when the California surveyor general called for a "system of reclamation . . . for the whole State where required." Thereafter, others echoed that theme and both the state and federal governments launched investigations. No one advanced a substantive proposal, however, until 1919 when increased population and heightened difficulties in the valley and delta prompted Robert B. Marshall to publicize a dramatic plan. A highly respected former member of the U.S. Geological Survey, he had spent years surveying the topography of California. He drew on that knowledge to design a bold project that promised something for every major part of the state. His "Marshall Plan" called for a large dam on the upper Sacramento River and two aqueducts for varying distances on either side of the Central Valley to reclaim vast sections of the Sacramento and especially San Joaquin valleys from their current waterless or low-water conditions, to provide water to San Francisco Bay cities, to improve the navigability of the Sacramento River, and to prevent salt water intrusion into the delta. For southern California, the project would divert the Kern River to Los Angeles and the south coast by way of a tunnel through the Tehachapi Mountains. Some 12 million acres would be reclaimed in the Central Valley as well as in neighboring Santa Clara, Livermore, and Concord valleys, while Los Angeles would receive four times the volume of water then arriving through the Owens aqueduct. Revenue for the vast scheme would come from the sale of water and electricity generated at state power plants erected along the system. . . .

During the . . . [1930s] several developments combined to overcome the major obstacles to a comprehensive water plan. The most troublesome involved riparian law. California courts, it will be recalled, had broadened the doctrine to permit riparians (those who owned property adjacent to a river) to engage in irrigation so long as they recognized that all riparians on the same stream possessed a similar right. As a practical matter, this meant that riparians had reciprocal rights and an obligation to use water reasonably with respect to one another. The problem was that riparians insisted they had no obligation to be reasonable—that is, avoid wasteful practices—in their dealings with appropriators on the same stream. (Appropriators . . . ordinarily lived some distance from the rivers whose waters they used under the law of appropriation.) In 1926 the state supreme court in *Herminghaus v. Southern California Edison* agreed with them. Riparians, announced the court, could not be deprived of their rights to a river's flow, including flood water, even if they used the water wastefully. The decision effectively prevented appropriators from building dams on rivers claimed by riparians and trapping the flood waters absolutely necessary for new development at some distance from the streams, or for recharging distant and declining underground aquifers.

The ruling shocked the public into howls of protest which culminated in 1928 with a popularly voted initiative amending the state constitution and prohibiting any "waste or unreasonable use" of water. The "rocking chair" theory of riparian rights was gone; riparian owners could no longer just sit by an adjoining water course and, even though not using it, claim a right to its undiminished flow. To survive a challenge from an appropriator, a riparian had to put water to "reasonable beneficial use." Debate would continue on the meaning of "reasonable" and "beneficial," but the amendment removed a significant hurdle to statewide water planning and established a principle that legal experts agree remains "the central theme of modern California water rights law." . . .

In 1931 . . . State Engineer Edward Hyatt unveiled a plan that, while still impressive, was considerably scaled back in comparison to Marshall's proposal. . . . Stripped to its essentials, the Hyatt proposal, like Marshall's, called for: (1) a major reservoir on the Sacramento River (at the present-day site of Shasta Dam) to store flood waters; (2) regulated releases to improve navigability along the lower Sacramento and to prevent saltwater intrusion in the delta, while also providing additional fresh water for the growing cities, industries, and farms along the southern delta; (3) an interconnected canal and reservoir system for taking the Sacramento River water, which had been stored behind Shasta Dam and slowly released, out of the delta and down into the much drier San

Joaquin Valley; and (4) an aqueduct from the Colorado River to the southern California coastal plain. The entire undertaking, as in the Marshall Plan, was to be paid for by water and power sales. . . .

President Franklin D. Roosevelt, open to almost any proposal that would create jobs and soften the depression's harshness, proved sympathetic. In 1935 he released emergency relief funds so that construction could begin, stipulating that the allocation was made "in accordance with the reclamation laws." . . . What had started as a state undertaking now became a federal project, delighting the Reclamation Bureau, which was still heady from its gains in the Boulder Canyon Act. The change was to have profound consequences not only for the Central Valley but also far beyond California's borders. . . .

The federal takeover assured realization of the Central Valley Project, but its completion came slowly, piecemeal (some aspects still remain unfinished), and often with bitter clashes over policy that linger into the present. In its essentials, the project by the early 1950s consisted of several major dams to control floods, improve navigation on the lower Sacramento River, and provide irrigation water: Shasta and Keswick dams on the Sacramento and Folsom Dam on the American River, all of which would capture floodwaters and release them when needed into the Sacramento Valley's streams for irrigation, and New Melones Dam on the Stanislaus River and Friant Dam on the San Joaquin. (Folsom and New Melones were later additions to the original CVP, built by the Army Corps of Engineers and operated by the Reclamation Bureau.) In addition, it included four principal canal systems moving water in several directions for different purposes: the Tehama-Colusa Canal bringing water down into a section of the northern Sacramento Valley, on the west side; the Delta Cross Channel and Contra Costa Canal to impede saltwater intrusion into the delta and provide water for the farms, cities, and industries of the delta area; the Delta Cross Channel and Delta-Mendota Canal for moving water into and through the San Joaquin Valley; and the Friant-Kern and Madera canals serving irrigation needs in the San Joaquin Valley with runoff collected at Friant Dam from the Sierra Nevada. The combined dams and canals collected and transported annually more than 3 million acre-feet. . . .

Work on the Central Valley Project began in 1937 with the first power available for sale in 1944. Delivery of water to the San Joaquin Valley took even longer, not arriving until 1951 through the Delta-Mendota Canal, some fourteen years after the Reclamation Bureau had begun construction and had expended nearly a half billion federal dollars, more than two and a half times the $170 million authorized by the state in 1933 and a gigantic infusion of federal money into the state economy. . . . The delays were also due to bitter differences over two far-reaching policy issues: public versus private power and the 160-acre limitation provision. Both raised the question as to who should benefit from the cheap power and water that would be made available at taxpayers' expense. . . . By nearly all measures, California agribusiness and its allies seemed by the late 1940s and early 1950s to have transformed the Central Valley Project into a vehicle of great corporate aggrandizement. . . .

Further Readings

Cooper, Erwin. *Aqueduct Empire: A Guide to Water in California, Its Turbulent History and Its Management Today, Western Lands and Waters Series*. Glendale, CA: A. H. Clark, 1968.

de Roos, Robert. *The Thirsty Land: The Story of the Central Valley Project*. Stanford, CA: Stanford University Press, 1948.

Dunbar, Robert G. *Forging New Rights in Western Waters*. Lincoln: University of Nebraska Press, 1983.

Engelbert, E. A., and A. F. Scheuring, eds. *Competition for California Water*. Berkeley: University of California Press, 1982.

Fradkin, Philip. *A River No More: The Colorado River and the West*. Berkeley: University of California Press, 1996.

Gottlieb, Robert, and Margaret FitzSimmons. *Thirst for Growth: Water Agencies as Hidden Government in California*. Tucson: University of Arizona Press, 1991.

Hanson, Warren D. *Water and Power: A History of the Municipal Water Department and Hetch-Hetchy System*. San Francisco: City and County of San Francisco, 1994.

Hart, John. *Storm over Mono: The Mono Lake Battle and the California Water Future*. Berkeley: University of California Press, 1996.

Hays, Samuel P. *Conservation and the Gospel of Efficiency: The Progressive Conservation Movement, 1890–1920*. Cambridge, MA: Harvard University Press, 1959.

Hine, Robert V. *California's Utopian Colonies*. Berkeley: University of California Press, 1983.

Hoff, Jeff. *The Legal Battle over Mono Lake*. San Francisco: The State Bar of California, 1982.

Hoffman, Abraham. *Vision or Villainy: Origins of the Owens Valley–Los Angeles Water Controversy*. College Station, TX: Texas A&M University Press, 1981.

Hundley, Norris, Jr. *The Great Thirst: Californians and Water, 1770s–1990s*. Berkeley: University of California Press, 1992.

Hundley, Norris, Jr. *Water and the West: The Colorado River Compact and the Politics of Water in the American West*. Berkeley: University of California Press, 1975.

Jackson, Donald C. *Building the Ultimate Dam: John S. Eastwood and the Control of Water in the West*. Lawrence: University Press of Kansas, 1995.

Kahrl, William L. *Water and Power: The Conflict over Los Angeles' Water Supply in the Owens Valley*. Berkeley: University of California Press, 1982.

Kahrl, William L., ed. *The California Water Atlas*. Sacramento: Governor's Office of Planning and Research, 1979.

Kelley, Robert. Battling the Inland Sea; American Political Culture, Public Policy, and the Sacramento Valley, 1850–1986. Berkeley: University of California Press, 1989.

Lee, Lawrence B. *Reclaiming the American West: An Historiography and Guide*. Santa Barbara, CA: ABC-Clio, 1980.

Maass, Arthur, and Raymond L. Anderson. *. . . And the Desert Shall Rejoice: Conflict, Growth, and Justice in Arid Environments*. Cambridge, MA: MIT Press, 1978.

Martin, Robert Ben. *The Hetch-Hetchy Controversy: The Value of Nature in a Technological Society*, 1982.

McCool, Daniel. *Command of the Waters: Iron Triangles, Federal Water Development, and Indian Water*. Berkeley: University of California Press, 1987.

McDonald, Angus Henry. *One Hundred and Sixty Acres of Water; The Story of the Anti-Monopoly Law*. Washington, DC: Public Affairs Institute, 1958.

Miller, M. Catherine. *Flooding the Courtrooms: Law and Water in the Far West*. Lincoln: University of Nebraska Press, 1993.

Nadeau, Remi A. *The Water Seekers*. Garden City, NJ: Doubleday, 1950.

Palmer, Tim. *Stanislaus: The Struggle for a River*. Berkeley: University of California Press, 1982.

Pisani, Donald J. *From the Family Farm to Agribusiness: The Irrigation Crusade in California and the West, 1850–1931*. Berkeley: University of California Press, 1984.

Pisani, Donald J. *To Reclaim a Divided West: Water, Law, and Public Policy, 1848–1902*. Albuquerque: University of New Mexico Press, 1992.

Pisani, Donald J. *Water, Land, and Law in the West: The Limits of Public Policy, 1850–1920*. Lawrence: University Press of Kansas, 1997.

Powell, John Wesley. *Report on the Lands of the Arid Region of the United States*. Cambridge, MA: Harvard University Press, 1963 [1878].

Preston, William L. *Vanishing Landscapes: Land and Life in the Tulare Lake Basin*. Berkeley: University of California Press, 1981.

Reisner, Marc. *Cadillac Desert: The American West and Its Disappearing Water*. New York: Viking, 1986.

Robinson, Michael C. *Water for the West: The Bureau of Reclamation, 1902–1977*. Chicago: Public Works History Society, 1979.

Shallat, Todd A. *Fresno's Water Rivalry: Competition for a Scarce Resource, 1887–1970, Essays in Public Works History*. Chicago: Public Works Historical Society, 1979.

Stegner, Wallace. *Beyond the Hundredth Meridian: John Wesley Powell and the Second Opening of the West*. Boston: Houghton Mifflin, 1954.

Vilas, Martin S. *Water and Power for San Francisco from Hetch-Hetchy in Yosemite National Park*. San Francisco: Martin Samuel Vilas, 1915.

Villarejo, Don, and Judith Redmond. *Missed Opportunities, Squandered Resources*. Davis, CA: California Institute for Rural Studies, 1988.

Walker, Richard. *The California Water System: Another Round of Expansion*. Berkeley: Institute of Governmental Studies, University of California, 1979.

Walton, John. *Western Times and Water Wars: State, Culture, and Rebellion in California*. Berkeley: University of California, 1992.

Wilkinson, Charles. *Crossing the Next Meridian: Land, Water, and the Future of the West*. Washington, DC: Island Press, 1992.

Worster, Donald. *Rivers of Empire: Water, Aridity, and the Growth of the American West*. New York: Oxford University Press, 1985.

Chapter 8

FROM THE FAMILY FARM TO AGRIBUSINESS

DOCUMENTS

Frank Norris on Wooing the Land, 1901

The day was fine. Since the first rain of the season, there had been no other. Now the sky was without a cloud, pale blue, delicate, luminous, scintillating with morning. The great brown earth turned a huge flank to it, exhaling the moisture of the early dew. The atmosphere, washed clean of dust and mist, was translucent as crystal. Far off to the east, the hills on the other side of Broderson Creek stood out against the pallid saffron of the horizon as flat and as sharply outlined as if pasted on the sky. The campanile of the ancient Mission of San Juan seemed as fine as frost work. All about between the horizons, the carpet of the land unrolled itself to infinity. But now it was no longer parched with heat, cracked and warped by a merciless sun, powdered with dust. The rain had done its work; not a clod that was not swollen with fertility, not a fissure that did not exhale the sense of fecundity. One could not take a dozen steps upon the ranches without the brusque sensation that underfoot the land was alive; roused at last from its sleep, palpitating with the desire of repro-

From Frank Norris, *The Octopus: A Story of California* (New York: Doubleday, 1901), pp. 126–31.

duction. Deep down there in the recesses of the soil, the great heart throbbed once more, thrilling with passion, vibrating with desire, offering itself to the caress of the plough, insistent, eager, imperious. Dimly one felt the deep-seated trouble of the earth, the uneasy agitation of its members, the hidden tumult of its womb, demanding to be made fruitful, to reproduce, to disengage the eternal renascent germ of Life that stirred and struggled in its loins.

The ploughs, thirty-five in number, each drawn by its team of ten, stretched in an interminable line, nearly a quarter of a mile in length, behind and ahead of Vanamee. They were arranged, as it were, *en echelon*, not in file—not one directly behind the other, but each succeeding plough its own width farther in the field than the one in front of it. Each of these ploughs held five shears, so that when the entire company was in motion, one hundred and seventy-five furrows were made at the same instant. At a distance, the ploughs resembled a great column of field artillery. Each driver was in his place, his glance alternating between his horses and the foreman nearest at hand. Other foremen, in their buggies or buckboards, were at intervals along the line, like battery lieutenants, Annixter himself, on horseback, in boots and campaign hat, a cigar in his teeth, overlooked the scene.

The division superintendent, on the opposite side of the line, galloped past to a position at the head. For a long moment there was a silence. A sense of preparedness ran from end to end of the column: All things were ready, each man in his place. The day's work was about to begin.

Suddenly, from a distance at the head of the line came the shrill trilling of a whistle. At once the foreman nearest Vanamee repeated it, at the same time turning down the line, and waving one arm. The signal was repeated, whistle answering whistle, till the sounds lost themselves in the distance. At once the line of ploughs lost its immobility, moving forward, getting slowly under way, the horses straining in the traces. A prolonged movement rippled from team to team, disengaging in its passage a multitude of sounds—the click of buckles, the creak of straining leather, the subdued clash of machinery the cracking of whips, the deep breathing of nearly four hundred horses, the abrupt commands and cries of the drivers, arid, last of all, the prolonged, soothing murmur of the thick brown earth turning steadily from the multitude of advancing shears.

The ploughing thus commenced, continued. The sun rose higher. Steadily the hundred iron hands kneaded and furrowed and stroked the brown, humid earth, the hundred iron teeth bit deep into the Titan's flesh. Perched on his seat, the moist living reins slipping and tugging in his hands, Vanamee, in the midst of this steady confusion of constantly varying sensation, sight interrupted by sound, sound mingling with sight, on this swaying, vibrating seat, quivering with the prolonged thrill of the earth, lapsed to a sort of pleasing numbness, in a sense, hypnotised by the weaving maze of things in which he found himself involved. To keep his team at an even, regular gait, maintaining the precise interval, to run his furrows as closely as possible to those already made by the plough in front—this for the moment was the entire sum of his duties. . . .

The ploughing, now in full swing, enveloped him in a vague, slow-moving whirl of things. Underneath him was the jarring, jolting, trembling machine; not a clod was turned, not an obstacle encountered, that he did not receive the swift impression of it through all his body, the very friction of the damp soil, sliding incessantly from the shiny surface of the sheers, seemed to reproduce itself in his finger-tips and along the back of his head. . . . Everywhere there were visions of glossy brown backs, straining heaving, swollen with muscle; harness streaked with specks of froth, broad, cup-shaped hoofs, heavy with brown loam, men's faces red with tan, blue overalls spotted with axle-grease; muscled hands, the knuckles whitened in their grip on the reins, and through it all the ammoniacal smell of the horses, the bitter reek of perspiration of beasts and men, the aroma of warm leather, the scent of dead stubble—and stronger and more penetrating than everything else, the heavy, enervating odour of the upturned, living earth.

At intervals, from the tops of one of the rare, low swells of the land, Vanamee overlooked a wider horizon. On the other divisions of Quien Sabe the same work was in progress. Occasionally he could see another column of ploughs in the adjoining division—sometimes so close at hand that the subdued murmur of its movements reached his ear; sometimes so distant that it resolved itself into a long, brown streak upon the grey of the ground. . . . Everywhere throughout the great San Joaquin, unseen and unheard, a thousand ploughs upstirred the land, tens of thousands of shears clutched deep into the warm, moist soil.

It was the long stroking caress, vigorous, male, powerful, for which the Earth seemed panting. The heroic embrace of a multitude of iron hands, gripping deep into the brown, warm flesh of the land that quivered responsive and passionate under this rude advance, so robust as to be almost an assault, so violent as to be veritably brutal. There, under the sun and under the speckless sheen of the sky, the wooing of the Titan began, the vast primal passion, the two world-forces, the elemental Male and Female, locked in a colossal embrace, at grapples in the throes of an infinite desire, at once terrible and divine, knowing no law, untamed, savage, natural, sublime.

Susan Minor Depicts Sisterhood on the Farmlands, 1919

Two hand-cars dashed furiously and somewhat jerkily down the railroad track, each manned by a desperate looking crew of five or six. These hardened individuals were similarly attired in curiously cut trousers and coats of dark blue cotton stuff, faded, muddy, and torn, their heads bound about in red or blue bandanas. They carried themselves with gay abandon and though they pulled at the levers of the hand-cars with tense fury they laughed joyously.

The second car seemed to be in pursuit of the first for ever anon one or another of the leading crew glanced behind and then urged the rest to yet wilder pulling on the propelling handle.

Were they a bandit gang of the woolly days of California?

Were they staging a moving picture?

Look carefully around the edges of the bandanas. Look carefully at the hands gripping the metal bar. Long hair is tidily tucked away under the coarse, gay head wrappings. The hands are brown and hard but too small for men's hands.

The desperate bandits are really farm laborers out for a holiday, those weary women workers of the Woman's Land Army.

Weary in body we might be at times. I know because I was one of them, that we did get backaches and leg-aches from stooping to the ground for peaches and prunes, from reaching above our heads for peaches and prunes, from bending over grapevines. But weary in spirit—never!

The first California Land Army Camp where I labored was situated about fifteen miles south of Chico in the broad, level, hot and fertile Sacramento Valley, between the distant, dim Sierras and the equally dim Coast Range.

Here, close by hundreds of acres of peaches, prunes, figs, olives, and nuts, forty-five of us lived in eight little screened and electric lighted bungalows built for us on what had been, in the season, a wheat field and was now, except that part occupied by our camp, the drying field for fruit. . . .

We rattled away to the orchards every day except Sunday at 7 A.M.,

From Susan Minor, "Sisters All," *Overland Monthly*, 73 (May 1919): 391–95.

on the floor of the motor truck with our tin fruit-picking pails, singing, swaying our relaxed bodies with the motion of the machine, swinging our legs over the edge. And here we rattled back in the late afternoon after eight, nine or ten hours of work, dusty, perspiring, limp, contented, still singing.

After work, by way of recreation, we went swimming in the neighboring irrigation canal that cleared today of past fatigue and future fears or cantered about the plain on a horse that the lender soon had to remove from us, as we were riding him to his grave. Then came a show-bath, a change into some costume other than uniform, and, lo, Richard was herself again. In some paradoxical way these exertions had rested us.

Once a week those of us who cared to do so, struggled with our back-to-nature hair, girt a corset about our back-to-nature figures, powdered our sunburned noses, donned the best apparel at our command, and were motored to a dance in the nearest town, so getting a taste of the outside world that gave us food for thought and conversation for many a day as we climbed ladders and filled fruit boxes. . . .

These women who lived together in such easy good-fellowship, who rose while it was yet dark, and with many semi-humorous protests stumbled out in bath-robes, sweaters, kimonos, or unaugmented pajamas to go through setting-up exercises beneath moon and stars, who got sweaty and grimy and stuck up with rotten fruit, who combed their hair by hand-mirrors, and used shelves for dressing tables, and who not only tolerated such an existence as a patriotic need, but reveled keenly in the gay freedom of it, who paused a moment in their work to glory in the long rows of green peach trees against the cloudless sky, in a bunch of red grapes glittering with the clinging dew, in a line of feathery almond trees against a lavender horizon; these women varied in private life from factory girl to college student; ranged from eighteen to forty-five years, or thereabout, and were, almost without exception, from cities.

There you would see, sitting side by side on benches at the long, white-oilcloth covered tables, and eating with avidity from enamel plates, a waitress who used with freedom "ain't" and "his'en" and double negatives in good-natured argument, with a charming Swiss lady who read, or at least, had with her to read, Darwin and Nietzsche. There you might listen when in the evening or while at work, the Hawaiian girl, a Mills College student, sang the "Gypsie Trail," or one of her native songs, when the beautiful shadowy-eyed French girl sang "The Bluebird," or "The Marseillaise," or the Swiss woman yodle songs of the Alps. There you might stroke the amiable, homely kitten named "Measley," which had been adopted and fondly nourished by us when it strayed to us in a weak and emaciated condition. There you might possibly be permitted to blanket on one of the popular dormitories, two hay stacks, for which a number of girls nightly deserted their cots until a three days' rain soaked the stacks.

This easy good-fellowship that was regardless of age, education, and previous condition of servitude arose from a combination of causes. The cotton uniform, soon muddy and fruit stained, in which some girls looked picturesque, but no one overwhelmingly beautiful leveled all distinction except those of personality, manners and speech. As in the old Western frontier days our past had no direct bearing on our life in camp. The camp life was an isolated experience and we were taken for what we were worth at the moment. After a while we began to inquire into each other's previous occupation and habits, but from curiosity only and from no desire to establish any social standards.

The nature of the work also was a democratizer, work requiring speed in judging the color of fruit, speed of motion, and grit to maintain that speed. The possession of these qualities made two or three girls accustomed to piece-work in factories, two or three college students, a waitress, and an office clerk, the fastest workers. And when day after day in the orchard you have sprawled together for fifteen heavenly minutes of relaxation, morning and afternoon, on the ground beside the common water can, when any two of you have raced any other couple to see which could fill more boxes with the big, golden peaches, when the telephone girl beside you at the table offers you her tomatoes, and asks for nothing in return, when a college student loans you her cherished washboard to facilitate the laundering of your nightgown, a fellow-feeling grows. . . .

And we worked, oh, how hard we worked! We wanted to be really good laborers, not just good "for women." In spite of the fact that we were paid by the hour, and not by the amount accomplished, the rivalry for speed was keen. It was a matter of honor and pride to pick rapidly and yet pick only the best fruit, and the fastest workers were greatly revered by the slow ones.

The result has proved highly satisfactory to our employers, and almost every rancher and fruit company that employed women last year wants them back this year. We are said to have a better eye than men, for judging fruit in the right stage for picking and to be quicker of hand. We are slow in moving ladders and boxes of fruit, but this slowness is compensated for, in the opinion of our employers, by greater industry, intelligence, and reliability. . . .

The organization that makes possible this general introduction of women into agriculture is the Woman's Land Army of America. It came into existence in February, 1918, after the successful experiment of the preceding summer in placing women on the fields in New York State. It is managed by a national board of directors having headquarters at 19 West 44 street, New York City. This board is made up of women from various parts of the country and operates through state boards having offices in each of the forty states in which the Land Army has to this time been organized. In the fall of 1918 the Land Army was affiliated with the United States Department of Labor at the request of the Secretary of Labor in recognition of its excellent work.

John Steinbeck on The Grapes of Wrath, *1939*

The moving, questing people were migrants now. Those families which had lived on a little piece of land, who had lived and died on forty acres, had eaten or starved on the produce of forty acres, had now the whole West to rove in. And they scampered about, looking for work; and the highways were streams of people, and the ditch banks were lines of people. Behind them more were coming. The great highways streamed with moving people. There in the Middle- and Southwest had lived a simple agrarian folk who had not changed with industry, who had not formed with machines or known the power and danger of machines in private hands. They had not grown up in the paradoxes of industry. Their senses were still sharp to the ridiculousness of the industrial life. . . .

In the West there was panic when the migrants multiplied on the highways. Men of property were terrified for their property. Men who had never been hungry saw the eyes of the hungry. Men who had never wanted anything very much saw the flare of want in the eyes of the migrants. And the men of the towns and of the soft suburban country gathered to defend themselves; and they reassured themselves that they were good and the invaders bad, as a man must do before he fights. They said, These goddamned Okies are dirty and ignorant. They're degenerate, sexual maniacs. These goddamned Okies are thieves. They'll steal anything. They've got no sense of property rights.

And the latter was true, for how can a man without property know the

From John Steinbeck, *The Grapes of Wrath* (New York: Viking, 1939), pp. 293–95, 360–63. Reprinted by permission.

ache of ownership? And the defending people said, They bring disease, they're filthy. We can't have them in the schools. They're strangers. How'd you like to have your sister go out with one of 'em?

The local people whipped themselves into a mold of cruelty. Then they formed units, squads, and armed them—armed them with clubs, with gas, with guns. We own the country. We can't let these Okies get out of hand. And the men who were armed did not own the land, but they thought they did. And the clerks who drilled at night owned nothing, and the little storekeepers possessed only a drawerful of debts. But even a debt is something, even a job is something. The clerk thought, I get fifteen dollars a week. S'pose a goddamn Okie would work for twelve? And the little storekeeper thought, How could I compete with a debtless man?

And the migrants streamed in on the highways and their hunger was in their eyes, and their need was in their eyes. They had no argument, no system, nothing but their numbers and their needs. When there was work for a man, ten men fought for it— fought with a low wage. If that fella'll work for thirty cents, I'll work for twenty-five.

If he'll take twenty-five, I'll do it for twenty.

No, me, I'm hungry. I'll work for fifteen. I'll work for food. The kids. You ought to see them. Little boils, like, comin' out, an' they can't run aroun.' Give 'em some wind-fall fruit, an' they bloated up. Me, I'll work for a little piece of meat.

And this was good, for wages went down and prices stayed up. The great owners were glad and they sent out more handbills to bring more people in. And wages went down and prices stayed up. And pretty soon now we'll have serfs again.

And now the great owners and the companies invented a new method. A great owner bought a cannery. And when the peaches and the pears were ripe he cut the price of fruit below the cost of raising it. And as cannery owner he paid himself a low price for the fruit and kept the price of canned goods up and took his profit. And the little farmers who owned no canneries lost their farms, and they were taken by the great owners, the banks, and the companies who also owned the canneries. As time went on, there were fewer farms. The little farmers moved into town for a while and exhausted their credit, exhausted their friends, their relatives. And then they too went on the highways. And the roads were crowded with men ravenous for work, murderous for work.

And the companies, the banks worked at their own doom and they did not know it. . . . The great companies did not know that the line between hunger and anger is a thin line. And money that might have gone to wages went for gas, for guns, for agents and spies, for blacklists, for drilling. On the highways the people moved like ants and searched for work, for food. And the anger began to ferment.

The spring is beautiful in California. Valleys in which the fruit blossoms are fragrant pink and white waters in a shallow sea. Then the first tendrils of the grapes, swelling from the old gnarled vines, cascade down to cover the trunks. The full green hills are round and soft as breasts. And on the level vegetable lands are the mile-long rows of pale green lettuce and the spindly little cauliflowers, the gray-green unearthly artichoke plants.

And then the leaves break out on the trees, and the petals drop from the fruit trees and carpet the earth with pink and white. The centers of the blossoms swell and grow and color: cherries and apples, peaches and pears, figs which close the flower in the fruit. All California quickens with produce, and the fruit grows heavy, and the limbs bend gradually under the fruit so that little crutches must be placed under them to support the weight.

Behind the fruitfulness are men of understanding and knowledge and skill, men who experiment with seed, endlessly developing the techniques for greater crops of plants whose roots will resist the million enemies of the earth: the molds, the insects, the rusts, the blights. These men work carefully and endlessly to perfect the seed, the roots. And there are the men of chemistry who spray the trees against pests, who sulphur the grapes, who cut out disease and rots, mildews and sicknesses. Doctors of preventive medicine, men at the borders who look for fruit flies, for Japanese beetle, men who quarantine the sick trees and root them out and burn them, men of knowledge. The men who graft the young trees, the little vines, are the cleverest of all, for theirs is a surgeon's job, as tender and delicate; and these men must have surgeons' hands and surgeons' hearts to slit the bark to place the grafts, to bind the wounds and cover them from the air. These are great men.

Along the rows, the cultivators move, tearing the spring grass and turning it under to make a fertile earth, breaking the ground to hold the water up near the surface, ridging the ground in little pools for the irrigation, destroying the weed roots that may drink the water away from the trees.

And all the time the fruit swells and the flowers break out in long clusters on the vines. And in the growing year the warmth grows and the leaves turn dark green. The prunes lengthen like little green bird's eggs, and the limbs sag down against the crutches under the weight. And the hard little pears take shape, and the beginning of the fuzz comes out on the peaches. Grape blossoms shed their tiny petals and the hard little beads become green buttons, and the buttons grow heavy. The men who work in the fields, the owners of the little orchards, watch and calculate. The year is heavy with produce. And men are proud, for of their knowledge they can make the year heavy. They have transformed the world with their knowledge. The short, lean wheat has been made big and productive. Little sour apples have grown large and sweet, and that old grape that grew among the trees and fed the birds its tiny fruit has mothered a thousand varieties, red and black, green and pale pink, purple and yellow; and each variety with its own flavor. The men who work in the experimental farms have made new fruits: nectarines and forty kinds of plums, walnuts with paper shells. And always they work, selecting, grafting, changing, driving themselves, driving the earth to produce.

And first the cherries ripen. Cent and a half a pound. Hell, we can't pick 'em for that. Black cherries and red cherries, full and sweet, and the birds eat half of each cherry and the yellowjackets buzz into the holes the birds made. And on the ground the seeds drop and dry with black shreds hanging from them. . . . The purple prunes soften and sweeten. My God, we can't pick them. . . . We can't pay wages, no matter what wages. . . .

Then the grapes—we can't make good wine. People can't buy good wine. Rip the grapes from the vines, good grapes, rotten grapes, wasp-stung grapes. Press stems, press dirt and rot.

But there's mildew and formic acid in the vats.

Add sulphur and tannic acid.

The smell from the ferment is not the rich odor of wine, but the smell of decay and chemicals.

Oh, well. It has alcohol in it, anyway. They can get drunk.

The little farmers watched debt creep up on them like the tide. They sprayed the trees and sold no crop, they pruned and grafted: and could not pick the crop And the men of knowledge have worked, have considered, and the fruit is rotting on the ground, and the decaying mash in the wine vats is poisoning the air. And taste the wine—no grape flavor at all, just sulphur and tannic acid and alcohol.

This little orchard will be a part of a great holding next year, for the debt will have choked the owner.

This vineyard will belong to the bank. Only the great owners can survive, for they own the canneries too. And four pears peeled and cut in half, cooked and canned, still cost fifteen cents. And the canned pears do not spoil. They will last for years.

The decay spreads over the State, and the sweet smell is a great sorrow on the land. Men who can graft the trees and make the seed fertile and big can find no way to let the hungry people eat their produce. Men who have created new fruits in the world cannot create a system whereby their fruits may be eaten. And the failure hangs over the State like a great sorrow.

The works of the roots of the vines, of the trees, must be destroyed to keep up the price, and this is the saddest, bitterest thing of all. Carloads of oranges dumped on the ground. The people came for miles to take the fruit, but this could not be. How would they buy oranges at twenty cents a dozen if they could drive out and pick them up? And men with hoses squirt kerosene on the oranges, and they are angry at the crime, angry at the people who have come to take the fruit. A million people hungry, needing the fruit—and kerosene sprayed over the golden mountains.

And the smell of rot fills the country. . . .

There is a crime here that goes beyond denunciation. There is a sorrow here that weeping cannot symbolize. There is a failure here that topples all our success. The fertile earth, the straight tree rows, the sturdy trunks, and the ripe fruit. And children dying of pellagra must die because a profit cannot be taken from an orange. And coroners must fill in the certificates—died of malnutrition—because the food must rot, must be forced to rot.

The people come with nets to fish for potatoes in the river, and the guards hold them back; they come in rattling cars to get the dumped oranges, but the kerosene is sprayed. And they stand still and watch the potatoes float by, listen to the screaming pigs being killed in a ditch and covered with quicklime, watch the mountains of oranges slop down to a putrefying ooze; and in the eyes of the people there is the failure; and in the eyes of the hungry there is a growing wrath. In the souls of the people the grapes of wrath are filling and growing heavy, growing heavy for the vintage.

Cesar Chavez Warns of Pesticide Risks, 1993

There is nothing we care more about than the lives and safety of our families. There is nothing we share more deeply in common with the people of North America than the safety of the food we all rely upon.

The chemical companies that manufacture the pesticides and the growers who use them want us to believe that they are the health-givers, that because of pesticides people are not dying of malaria and starvation. They have convinced the politicians and the government regulators that pesticides are the cure-all, the key to an abundance of food. So they don't ban the worst of these poisons because some farm workers give birth to children who contract cancer or to babies who are born with deformities. They don't imperil millions of dollars in profits today because, some day, some consumers might get cancer. They allow all of us who place our faith in the safety of the nation's food supply to consume grapes and other produce that contain residues from pesticides that cause cancer and birth defects. We accept decades of environmental damage these poisons have brought upon the land. The growers and the chemical companies, the politicians and the bureaucrats, all say that these are acceptable levels of exposure. Acceptable to whom? . . .

Acceptable to all the other farm workers and their children who have known tragedy from pesticides? Acceptable to 300,000 farm workers who are poisoned each year in the United States, according to a 1985 study by the World Resources Institute? Or the eight hundred to one thousand who die each year from exposure to pesticides, according to a study by the U.S. Food and Drug Administration?

There is no acceptable level of exposure to any chemical that causes cancer. We cannot tolerate any toxic substance that causes miscarriages, stillbirths, and deformed infants. . . .

Compared with other jobs, farm workers are one of the least-protected groups in the nation. They are specifically excluded, either totally or partially, from health and safety standards under the federal Occupation-

From Cesar Chavez, "Farm Workers at Risk," in Richard Hofrichter, ed., *Toxic Struggles: The Theory and Practice of Environmental Justice* (Philadelphia: New Society Publishers, 1993), pp. 163–70, pp. 164–67, 169. Reprinted by permission of New Society Publishers, New Haven, Connecticut.

al Safety and Health Act as well as the Fair Labor Standards Act. Many are excluded from state worker compensation and unemployment insurance laws. Thus we lack effective legal remedies. Growers are not even required to tell workers the specific chemicals being used or to provide protective clothing. For one of the most hazardous occupations in the nation, three and one half million farm workers live under a double standard. . . .

In 1965, our infant farm workers' union led a major strike against Delano-area grape producers. That walkout was also headed for the same fate as its predecessors, until we tried something different: We asked consumers across North America to boycott California table grapes.

In 1992, the United Farm Workers marked its 30th anniversary. It has been almost twenty-seven years since our union cause first touched the hearts and consciences of people across America and around the world by letting them know about the abuses suffered by farm workers and their families.

In 1967, the farm workers dramatically transformed the simple act of refusing to buy fresh table grapes into a powerful statement against unfairness and injustice. The grape boycott was a hallmark of the 1960s and '70s. It rallied millions of Americans to the cause of migrant farm workers. And it worked. The first real collective-bargaining relationship was established between an agricultural employer and farm workers when we signed our first union contract with a Delano-area grape grower. . . .

Although UFW contracts in 1967 provided protection from dangerous pesticides, most farm workers still remained unprotected in 1982, when corporate growers gave the gubernatorial election campaign of Republican George Deukmejian more than $1 million. . . .

Many of these same growers also spent millions of dollars to help kill Proposition 128, California's 1990 Big Green initiative supported by environmental groups and the UFW, which would have protected California's last stands of privately held redwoods and banned cancer-causing pesticides.

Growers and other opponents of Proposition 128 would be mistaken to view the election results as a rejection of the need for reforms, such as the protection of old-growth forests and restrictions on the use of pesticides that produce birth defects. . . .

All my life, I have been driven by one dream, one goal, one vision: To overthrow a farm-labor system in this nation that treats farm workers as if we are not important human beings. Farm workers are not agricultural implements of beasts of burden to be used and discarded. . . .

How could we progress as a people, even if we lived in the cities, while the farm workers—men and women of our color—were condemned to a life without pride? How could we progress as a people while the farm workers, who symbolized our history in this were denied self-respect? How could our people believe that their children could become lawyers, doctors, judges, and professional people while permitting this shame and injustice to continue? . . .

Tens of thousands of the children and grandchildren of farm workers, and the children and grandchildren of poor Latinos, are moving out of the fields and the barrios. That movement cannot be overturned. Once social change begins, it cannot be reversed. You cannot uneducate the person who has learned to read. You cannot humiliate the person who feels pride. You cannot oppress the people who are not afraid anymore.

Boycott grapes!

Essays

Order on the Land

William Preston

Agricultural alteration of the Tulare Lake Basin increased dramatically in the 1850s, 1860s, and early 1870s. Unmanaged grazing not only altered patterns of natural vegetation but profoundly changed the productive capabilities of certain lands, for overgrazing led to severe soil compaction and accelerated erosion. The change from cattle to sheep was particularly detrimental to fragile basin communities: "They ate into the roots of the lush grass and left the quick rains to cut the soil. The wool in the hand was always worth the next season's feed to the sheep-herder." Through habitat alteration, domestic livestock also crowded native fauna off the basin plains. The disappearance of native animals accelerated as farmers colonized the basin, for wildlife was seen as a direct threat to crops and herds and was actively exterminated. . . .

As Tulare County settlers devoted more of their energies to cultivation, they induced further changes in regional patterns of vegetation. The exotic oats introduced by the Spanish were themselves succeeded by filaree as the dominant grass cover. Where grazing had already limited the rejuvenation of woodlands, farmers—in quest of fuel, building materials, and cleared land—rapidly reduced the extent of basin timber. Neither did the detrimental effects upon soil quality end with the transition to farming. Dryland grain cultivation, without crop rotation or fallowing, damaged soil quality to an extent that alarmed observers as early as 1868:

From William Preston, *Vanishing Landscapes: Land and Life in the Tulare Lake Basin* (Berkeley: University of California Press, 1981), pp. 117–18, 130–31, 134–36, 154, 157–59, 161, 162–63. Reprinted by permission of the Regents of the University of California and the University of California Press.

> We are now "running upon wheat" in California. Crop after crop of it is being raised, in opposition to the law which requires rotation, while in addition to this excessive drought upon the life of the land, we are doing nothing whatever to restore to the soil the element of which even judicious cultivation despoils it. Poverty as hopeless as that of the Sahara must inevitably overtake a county that is thus willfully given over to vandal cultivation. . . .

The most striking modification of the Tulare Lake Basin during the period between 1857 and 1871 was the reordering of its natural landscapes into cultural landscapes of geometrical forms. The smooth ecotones of nature were replaced by sharp edges and angles: fields, fences, farmsteads, town plats, and roads marked the land off into a new series of zones and territories. The elements of the life layer could not spread freely across this new geometrical fretwork, and they were actively replaced with a new, agricultural life layer. Native flora were cleared away and replaced with crops in rectangular fields; native animals were killed and replaced by tame livestock; standing water was drained, and flowing water was rechanneled into geometrically arranged canals. Human dominance was for the first time loudly proclaimed by the landscapes of the Tulare Lake Basin. . . .

The railroad and barbed wire fencing brought an end to the open range. Property values soared, and land uses were geared more and more directly to market conditions. The basin's landscapes began to lose their regional identity, becoming [as Frank Norris noted] "merely part of an enormous whole, a unit in the vast agglomeration of wheat land the whole world round, feeling the effects of causes thousands of miles distant—a drought in the prairies of Dakota, a rain in the plains of India, a frost on the Russian steppes, a hot wind on the llanos of the Argentine." The wheat boom began earlier in other parts of California. Experimental shipments of wheat from San Francisco to New York, Australia, and Great Britain in the late 1850s found an eager market for the hard, dry, white grain of the Central Valley, and by 1860 more than half of the state's wheat crop was unloaded in Great Britain alone. California was the premier wheat state in the Union by the mid-1870s, and it continued to rank first or second until the 1890s, when competition and local diversification changed the emphasis of California agriculture.

The Tulare Lake Basin proved very well suited to grain farming, and by the time it joined in the wheat boom, a market for its grain was well established. The railroad facilitated marketing and settlement, and new technologies and attitudes supported the development of huge bonanza wheat farms. . . . Minor inconveniences such as hog-wallow terrain were ignored, for the early machinery was light and maneuverable, and remaining obstacles were challenged by innovative techniques and ever-increasing mechanization. Local innovations in farm equipment and dry-land-cultivation techniques diffused widely through the American West and, indeed, the world.

Grain farming had begun to emerge in the basin even before the railroad era. It was neither capital intensive nor technologically complicated: settlers of limited means could take up wheat farming with a few horses or oxen, a plow, some seed, and a likely parcel of land. Suitable land was still plentiful, although land prices were rising fast. . . . Squatting was also an option, especially on railroad land, which was often opened to settlement with promises of preferential treatment before the company had secured full title. Settlers assumed that when their land came up for sale they would be charged low government prices, but land values rose quickly, in part because of their improvements. When the time came to sell, the Southern Pacific demanded payment at inflated prices. This, in combination with discriminatory rate structures and other injustices, brought substantial popular resentment to bear against the Southern Pacific in the 1870s. Conflicts culminated in May, 1880, in the bloody Mussel Slough incident, an outright battle between settlers and the railroad which underlined the role of monopolies in California and prompted Frank Norris to write *The Octopus* (1903), a muckraking expose of the Southern Pacific's illicit dealings in which Norris likened the railroad to an octopus with "tentacles of steel clutching into the soil."

No matter who owned the land or how large the farm was, success was measured in tons of wheat harvested. . . . The railroad had brought new capital, higher land taxes, and improved equipment. Huge mechanized grain operations were increasingly practical and increasingly necessary, and it became still easier for large landholders—settlers or speculators—to get rich. Mechanized farming was especially successful in the basin, and it benefited from the research activities of the Department of Agriculture. Gang plows, scrapers, combination seeder-harrows, and twine-knotters for binding grain were enthusiastically adopted, and by the late 1880s massive steam combines pulled by three dozen draft animals were plying the land. . . .

The aim was to play out the soil and then move on to something new. More attention was given to increasing the amount of land in cultivation than to perpetuating its fertility. Insufficient fertilization and fallowing caused rapid soil depletion in the grain districts: "When, at last, the land, worn out, would refuse to yield, they would invest their money in something else. By then, they would all have made fortunes. They did not care. 'After us, the deluge.'" On the drier lands west of the Kings River, such gambling ended almost as quickly as it had begun. In 1899 Orlando Barton lamented:

> The plains are given up to desolation. Eight or ten years ago large crops of wheat were raised on this land, which was all located, and farmhouses built on nearly every quarter section. . . . But not a spear of anything green grows on the place this year. . . . The houses of former inhabitants are empty, the doors swing open or shut with the wind. Drifting sand is piled to the top of many fences. The windmills, with their broken arms, swing idly in the breeze. Like a veritable city of the dead, vacant residences on every side greet the traveler by horse team as he pursues his weary way across these seemingly endless plains. . . .

As the world supply of wheat and barley rose, grain profits plummeted, reaching their low point in 1894. Basin farmers began to turn to intensive cultivation of other, more profitable crops. Fruits, vegetables, and dairy products found good markets, and improvements in agricultural and transportation technology made their production increasingly lucrative. . . .

Rapid agricultural settlement and the expansion of irrigation systems and diversified cropping produced dramatic changes in the basin's landscapes, and these changes, in turn, were evident in the perceptions and comments of residents and travelers. The basin no longer seemed to be a vast and isolated wilderness. The herds of wild animals and the broad expanses of native vegetation were gone, yet the cultural landscapes of most parts of the basin were not yet sufficiently impressive to offset the overriding images of flatness and monotony. Instead of offering "picturesque scenery like the prairies of the west, it is a vast undulating plain or dead level, with an occasional tree or park of oaks to diversify the general monotony." Still, the success of new farming ventures in certain parts of the basin commanded interest and attention. The valley trough and the regions west and south of the Tule River were least appealing in this respect. . . . For the first time, specific types of terrain (such as hog wallows) and soils (such as redlands) were described and analyzed without reference to their natural vegetation cover, although regional appraisals were still grounded in observations of soil fertility and climate as these related to the potential for agricultural settlement. . . .

The most significant change in environmental awareness after the construction of the railroad through the basin was a sudden recognition of people's role in changing the face of the land. Concern went beyond the recognition that a human landscape was emerging: cause-and-effect relationships between human practices and environmental degradation were recognized. Such observations helped people identify problems and formulate more conservative agricultural practices, and strengthened the conviction of settlers that the land could be altered still further. The natural vegetation had by this time been widely destroyed. Majestic oaks and riparian timber belts were the sole pre-agricultural remnants, and even they were disappearing: "The oaks are old and ragged, many are fast decaying, and when gone, the country will be nearly bare, as there are few young trees growing up to take their place." Farmers undertook to clear away the remaining trees from their properties: "Everyone regretted the necessity of having to cut the oak forest, but progress or economics or common sense would allow no other way out. Land values in the valley had increased and proportionately so had the taxes on the land . . . so of necessity the owners had to clear the land and make it suitable for farming." The need for retaining some kind of timber in the basin, "not only for firewood but for the general purposes of civilized life," was clear. Pampas grass, pitahaya, palm, maple, catalpa, locust, and pomegranate were introduced for shade or ornamentation, and several species of foreign and domestic trees were suggested as possible commercial timber species for the dry plains: eucalyptus, China tree, European oak, and local cottonwood were among these. Gradually introduced tree species began to change the appearance of the basin's urban and rural landscapes. . . .

The alterations of the environment that occurred during this period, however, were by no means entirely beneficial. Several adverse effects of the massive diversion and reapplication of water were already noticeable: the extent of Tulare Lake and its surrounding marshlands were radically altered, and the lake's salinity increased greatly. As late as 1882 the lake's waters had been "copious with fish," harvested commercially in large quantities, but by the winter of 1888–89 the lake fish had begun to die off. The lake was soon too saline for many species and it supported almost none by 1900. . . .

The patterns of land use that had begun to take hold by the end of the railroad era proved remarkably flexible and contributed to the success of further agricultural and urban settlement of the basin. Although diversification reflected closer awareness of local environmental variations, however, such variations were not accepted with resignation. Farmers undertook to eliminate environmental restrictions on land use by heavy applications of technology, supported by a pervading attitude of dominance over nature together with a pervasive optimism. Technological attitudes and developments complemented the older agricultural traditions of the Tulare Lake Basin: commercialism, large-scale landholding, rampant speculation, and reliance upon hired labor.

This Bittersweet Soil

Sucheng Chan

In the 1880s Chinese tenants began to enter the semiarid San Joaquin Valley, where they planted vineyards and orchards along rivers and dug miles of irrigation ditches—work for which they seldom received any compensation. They also became more numerous in the valleys along the northern, central, and southern California coast, where they practiced diversified farming. Although scattered narrative sources indicated that Chinese also worked in the citrus groves of Los Angeles, Orange, Riverside, and San Bernardino counties, they did so only as seasonal workers. County archival records provide little evidence that they leased any citrus groves. Chinese tenants in southern California were primarily vegetable and celery growers. In Santa Barbara they also produced seeds for sale. In general, Chinese entry into tenant farming in a particular region coincided with the date when that region was opened up for intensive cultivation, but it would be wrong to assume that they undertook only intensive agriculture. . . .

Chinese tenant farmers were most numerous and their historical significance greatest in the Sacramento-San Joaquin Delta. It would add a romantic touch to our story were it possible to claim that Chinese immigrants came to farm in the Delta because it reminded them of home—the Pearl River delta of Kwangtung Province, where most of them had lived. Such a case cannot be made, however, because though the Chinese no doubt saw the deltaic marshlands as they rode steamers up the Sacra-

From Sucheng Chan, *This Bittersweet Soil: The Chinese in California Agriculture, 1860–1910* (Berkeley: University of California Press, 1986), pp. 158–62, 165–68, 170.

mento River on their way to the mining regions, they did not enter the Delta to farm on their own initiative. They first went to work in the area when white landowners recruited them to do reclamation work. Only after they became aware of the extraordinary fertility of Delta soil did they lease land to grow crops.

What is this place called the Delta? . . . California's two great river systems, the Sacramento and the San Joaquin, and their respective tributaries, drain more than one-third of the state's area. Their floodplains make up the great Central Valley, the northern half of which is called the Sacramento Valley, the southern half, the San Joaquin Valley. . . . The Delta, twenty-four miles across at its widest point and forty-eight miles along its longest axis, is . . . shaped like a rough parallelogram, with its northern, eastern, southern, and western apices marked by the cities of Sacramento, Stockton, Tracy, and Antioch, respectively.

Tall bulrushes, known as tules, covered the marshy swamp. Before people tamed this land, the tules grew and died each year, decaying to become a rich organic soil known as peat. In the islands of the central Delta, the peat layer is as thick as forty feet. Many of the peat islands are shaped like shallow bowls with central depressions—known as backswamps—rimmed by natural and man-made levees, the former formed by sediment carried by the waters of the meandering sloughs, the latter built to protect the land from periodic floods.

Farming in the Delta has been both profitable and risky. Aside from raging torrents that have sometimes washed out entire islands, the Delta suffers from soil subsidence: over time, the centers of the islands have sunk lower and lower so that today almost three-fifths of the Delta's area is at or below sea level. Modern roads have been built on top of the levees, and many a tourist has had the strange experience of driving along roads that are on the same level as the tops of fruit trees growing in the orchards on both sides. The problem of subsidence has been compounded by the rise in the height of riverbeds as more and more sediment settles in the channels. . . . Ironically reclamation itself has created increased risks: as levees were built to enclose more and more islands and tracts, whenever there was a large runoff the water in the rivers—being confined to their constricted channels—had made it harder to drain the areas that were wet. Powerful pumps have to be used, adding to the cost of farming. In more recent years, salinity has also become a cause for worry as more and more of the water from the Sacramento River and its tributaries is transported through man-made channels to the southern part of the state, leaving less to flow through the Delta to flush out the salt brought in by seawater flowing inland through San Francisco Bay during high tides.

Delta soil conditions have affected various crops differently. Potatoes are a case in point: virgin peat soil provides extraordinarily large yields, but at the same time, peat soil with its high organic content develops a fungus after it has been planted to potatoes for several years. So farmers who wish to grow potatoes from one year to the next must move from field to field. Another important Delta crop, asparagus, depletes certain nutrients in the soil, and after twenty years or so the yields drop drastically. But unlike land planted to other crops, fertilizers do not improve the yield of old asparagus fields on peat land. Yet, it is uneconomical to dig the aging asparagus plants out because their rhizomes—the source of food for the asparagus shoots—grow very large, spreading at least eight feet across in the ground. Digging up old asparagus fields is a backbreaking job, and fields are generally not dug up until yields have greatly declined. Various fruit trees, too, have had to be planted selectively. The Delta's high water table is detrimental to peach, plum apricot, and cherry trees, which, accordingly, grow well only on the high natural levees. The Bartlett pear, on the other hand, grows well despite the high water table and has been the most important tree fruit produced in the area.

The Delta has also had a rich social history. Perhaps nowhere else in California has an area's social structure so closely mirrored its biological and physical ecological balance. People of many different ethnic backgrounds have worked together closely for 130 years to make the Delta into one of the state's most productive agricultural areas. The Delta is one of the few places left in rural California where third-, fourth-, and even fifth-generation descendants of some of the early settlers continue to farm the land in a complex pattern of interaction between man and nature, and between one ethnic group and another. In this historical development, Asian immigrants have played a central role. More than any other place in rural California, the Delta has provided the conditions conducive to the establishment of relatively stable and settled Asian communities. Chinese have been involved in all stages of farm-making in the Delta: reclaiming the swamps, clearing the land, breaking up the sod for cultivation, leasing part of the land to grow crops, and harvesting and packing them for marketing. . . .

The reclamation of swamp and overflowed lands in the Sacramento-San Joaquin Delta took place in two stages. During the first period, which lasted from the early 1850s to the early 1880s, the work was done mainly by "wheelbarrow brigades" of Chinese laborers employed by both individual landowners and land-reclamation corporations. During these early decades the mainland tracts around the periphery of the Delta—where the soil had a lower organic content, was less swampy, and therefore was easier to drain and dike—were reclaimed. In the later period, the clamshell dredge—a barge outfitted with a long boom at the end of which were two claw-like buckets—introduced in 1879, soon replaced the Chinese wheelbarrow gangs. The giant buckets scooped earth out of the riverbeds, and the boom then swung the buckets over to the levee to dump the fill. The introduction of heavy equipment made it possible to reclaim the swampier central peat islands. Almost all of this work was carried out by large land companies because reclaiming the central Delta was so costly that individuals could not afford to undertake the task.

Once it had been shown that Delta land could be reclaimed, people were eager to establish farms in the area. Swamp land proved to be so fertile that some farmers could grow four crops a year. Another advantage was that the Delta was one of the few areas in the state that did not suffer from summer drought. Furthermore, with its maze of waterways, crops could be transported in bulk on barges to outside markets fairly cheaply. Sacramento and Stockton each developed inland ports to handle the export of the area's products. From the late 1850s to the 1880s, Delta counties were major producers of wheat, which was loaded at the two ports and taken to San Francisco for . . . shipment overseas. Later, potatoes, onions, beans, various green vegetables, asparagus, and fruit were also exported by water and by rail. . . .

Today, most Delta farms specialize in one or two crops, but over the area as a whole there is great diversification. Chinese tenant farmers and farm laborers were intimately involved in every phase of the Delta's land reclamation and farm-making. In fact, it can be argued that without them the Delta would have taken decades longer to develop into one of the richest agricultural areas in the world.

From the Family Farm to Agribusiness

Donald Pisani

All of American history reflects conflict between rural and urban values, between an often idealized life close to the earth and the impersonal, complicated social and economic relationships intrinsic to a commercial and by the late nineteenth and twentieth centuries, industrial nation. Californians were both more and less nostalgic than other Americans. Since statehood, their lives had been dominated by "the city," San Francisco, so they never witnessed at close hand the waning influence of the country *vis-à-vis* the metropolis. Nevertheless, the absence of strong rural traditions made those traditions all the more attractive. From the 1870s to the 1920s, leaders of the irrigation movement in California agreed that the state's cities were overcrowded and . . . that the family farm was the foundation of a stable society and the wellspring of republican virtues. . . .

The year 1931 represents a convenient dividing line. By that time most Californians acknowledged, though not always directly, that the health of their economy and society did not depend on the existence, perpetuation, or proliferation of the family farm. They might exhibit a sentimental attachment to "small town America," if not farming itself, but they also recognized that the values represented by the small freehold—widespread property ownership, high wages, and economic independence, to name a few—were anachronistic, however attractive. Californians had begun to see their state as "the great exception," and even to revel in its exceptional qualities, rather than try to recapture the institutions and life-style that had prevailed in the Midwest or "back East." By the 1930s, and especially after World War II, irrigation entered a new phase. It was no longer an agent to transform society, but an ally of the agricultural establishment.

During World War I, the crusade for the family farm enjoyed a renais-

From Donald Pisani, *From the Family Farm to Agribusiness: The Irrigation Crusade in California and the West, 1850–1931* (Berkeley: University of California Press, 1984), pp. 440–43, 451–52. Reprinted by permission of the Regents of the University of California and the University of California Press.

sance in many parts of the United States, but nowhere more than in California. A special California commission on land colonization reported in 1916:

> Within the last five years questions of land tenure and land settlement have assumed a hitherto unthought of importance in the United States. The causes for this are the disappearance of free, fertile public land; the rising prices of privately-owned lands; the increase in tenant farming and a clearer recognition of its dangers; and the increasing attractions of city life which threaten the social impairment of rural communities by causing young people to leave the farms. . . . In [some] countries the state has taken an active part in subdividing large estates and in creating conditions which will enable farm laborers and farmers of small capital to own their homes. They have adopted this policy because experience has shown that nonresident ownership and tenant farming are politically dangerous and socially undesirable; that ignorant and nomadic farm labor is bad; and that the balance between the growth of city and country can be maintained only through creating rural conditions which will make the farm as attractive as the office or factory for men and women of character and intelligence.

The problem of land monopoly and nonresident ownership was particularly acute in California where 310 property owners held over 4,000,000 acres of prime farmland suited to intensive cultivation, land capable of providing 100,000 40-acre farms and sustaining 500,000 additional rural residents. The Southern Pacific Railroad owned over 500,000 acres; four Kern Country land companies owned over 1,000,000 acres, or more than half the county's privately owned land; and in Merced County, Miller and Lux owned 245,000 acres. Most midwestern states had relatively homogenous populations, few very large or very small farms, few nonresident owners, and few very rich or very poor farmers. However, in many parts of California, rural society was characterized by wealthy, nonresident land barons and migratory farm laborers or tenants who had no allegiance to place or sense of civic responsibility. The commission on land colonization argued that land monopoly undermined democratic values and political stability just as it retarded the state's agricultural development.

The interest in restoring the family farm also grew out of a deep fear of the "yellow peril." By 1920, the Japanese, who had begun to migrate to California in great numbers in the 1890s to replace the excluded Chinese as field hands, had acquired over 500,000 acres of land, most of it reclaimed swamps in the Central Valley. They dominated the production of rice and tomatoes, and their success raised the prospect that they would one day displace the white farmer. Given the prevailing belief in white supremacy, many Californians assumed that democracy was possible only in a homogeneous society. They favored measures to restrict Japanese immigration, segregate Japanese schoolchildren, and prohibit alien landownership. They also hoped to lure more white farmers onto the land.

Aside from the problem of land monopoly and the "yellow peril," Californians shared the assumption of many other Americans that a massive economic slump, if not depression, would follow hard upon the end of World War I. Helping returning soldiers acquire a farm could soften the economic impact of demobilization, reward faithful service to the nation, and reverse the migration from country to city. "Nothing short of ownership of the land one toils over," Elwood Mead remarked, "will suffice to overcome the lure of the city." Federal reclamation had tried, but failed, to turn the tide. The United States population increased from 76,000,000 to 106,000,000 from 1900 to 1920, but the Reclamation Bureau had managed to provide rural homes for only about 1 percent of these new people. To make matters worse, easier living conditions in the cities and high wages, especially during World War I, led to dramatic increases in tenant farming and the abandonment of many farms in the older agricultural regions. . . .

In California, one of the prime obstacles to taking up a farm was the high price of land and water. The average price of unimproved farmland increased from $27.63 an acre in 1900 to $116.84 in 1920. The value of the average farm tripled in the same period and the price of improved land increased even faster. By 1920, most irrigated land sold for $100 to $500 an acre. And the cost of setting up a new farm, including the price of livestock, machinery, barns, and fences, made farming even more expensive. . . .

The 1920s represented a turning point in the history of California agriculture. . . . The West had entered a new era. Little public land remained, successful farming required much more knowledge as well as equipment, and the city offered new opportunities as well as temptations. The economic hard times of the 1920s and 1930s increased the gulf between landownership and the act of tilling the soil—a gulf symbolized by the plight of migrant farm workers. Both the Central Valley Project and State Water Project contributed to the growth of agriculture as a business and the disappearance of farming as a way of life. Both the Central Valley Project and State Water Project represented a shift away from local control as water bureaucracies in Washington and Sacramento assumed vast new powers. . . . The rise of agribusiness coincided not just with the decline of the family farm ideal, but also with the virtual disappearance of the hope that California agriculture could be built on relatively autonomous, middle-class rural communities.

Of course, demographic shifts also contributed to the rise of agribusiness and the decline of "rural California." In 1900, 47.7 percent of the state's population resided in communities of 2,500 or fewer. This number decreased to 38.2 percent in 1910, 32.1 percent in 1920, and 26.7 percent in 1930. Though the state's rural population increased from 708,233 in 1900 to 1,099,902 in 1920, migration from out of state accounted for less than half the increase. Put in a broader perspective, California's urban population grew by 89 percent from 1900 to 1910 while the rural population increased by only 28.4 percent. In the next decade, the growth rate was 58.5 percent urban and 21 percent rural. During the 1920s, the rate was 78.8 percent to 37.9 percent. . . .

FIGURE 6. CENTRAL VALLEY IRRIGATION CANAL.
The federal Bureau of Reclamation's Delta-Mendota and Friant-Kern canals deliver water for irrigation to Central Valley farmers, allowing vast acres of land to be farmed year-round. Courtesy, The Bancroft Library, University of California,

California had contributed much to western economic development since statehood. . . . But urbanization, farm mechanization, the soaring price of land and water, and other trends could not be reversed. After 1930, irrigation became one of the foundation blocks of agribusiness. The dream of using it to reform California society was all but forgotten.

Further Readings

Chambers, Clarke. *California Farm Organizations: A Historical Study of the Grange, the Farm Bureau, and the Associated Farms, 1929–1941*. Berkeley: University of California Press, 1952.

Chan, Sucheng. *This Bittersweet Soil: The Chinese in California Agriculture, 1860–1910*. Berkeley: University of California Press, 1986.

Cockcroft, S. V., and O. B. Frank. *Outlaws in the Promised Land: Mexican Immigrant Workers and America's Future*. New York: Grove Press, 1986.

Daniel, Cletus. *Bitter Harvest: A History of California Farmworkers, 1870–1941*. Berkeley: University of California Press, 1981.

Daniels, R. *The Politics of Prejudice: The Anti-Japanese Movement in California and the Struggle for Japanese Exclusion*. Berkeley: University of California Press, 1977.

Dunne, John Gregory. *Delano, the Story of the California Grape Strike*. New York: Farrar, Straus & Giroux, 1967.

Galarza, Ernesto. *Farm Workers and Agribusiness in California, 1947–1960*. Notre Dame, IN: University of Notre Dame Press, 1977.

Gates, Paul W. *California Ranchos and Farms, 1846–1862*. Madison: State Historical Society of Wisconsin, 1967.

Gates, Paul W. *Land and Law in California: Essays on Land Policies*. Ames: Iowa State University Press, 1991.

Gonzales, Juan L. *Mexican and Mexican American Farm Workers: The California Agricultural Industry*. New York: Praeger, 1985.

Gregory, James N. *American Exodus: The Dust Bowl Migration and Okie Culture in California*. New York: Oxford University Press, 1989.

Hart, John. *Farming on the Edge: Saving Family Farms in Marin County, California*. Berkeley: University of California Press, 1991.

Jacobson, Yvonne Olson. *Passing Farms, Enduring Values: California's Santa Clara Valley*. Los Altos, CA: W. Kaufman, De Anza College, 1984.

Jelinek, Lawrence J. *Harvest Empire: A History of California Agriculture*. San Francisco: Boyd & Fraser, 1982.

Johnson, Stephen, Gerald Haslam, and Robert Dawson. *The Great Central Valley: California's Heartland*. Berkeley: University of California Press, 1993.

Kelley, Robert. *Battling the Inland Sea: American Political Culture, Public Policy, and the Sacramento Valley, 1850–1986*. Berkeley: University of California Press, 1989.

Liebman, Ellen. *California Farmland: A History of Large Agricultural Landholdings*. Totowa, NJ: Rowman and Allanheld, 1983.

Majka, Linda C., and Theo J. Majka. *Farm Workers, Agribusiness, and the State*. Philadelphia: Temple University Press, 1982.

McWilliams, Carey. *California: The Great Exception*. New York: Wyn, 1949.

McWilliams, Carey. *Factories in the Field: The Story of Migratory Farm Labor in California*. Santa Barbara: Peregrine Smith Books, 1935.

McWilliams, Carey. *Ill Fares the Land: Migrants and Migratory Labor in the United States*. Boston: Little, Brown, 1942.

Mitchell, Don. *The Lie of the Land: Migrant Workers and the California Landscape*. Minneapolis: University of Minnesota Press, 1996.

Pisani, Donald J. *From the Family Farm to Agribusiness: The Irrigation Crusade in California and the West, 1850–1931*. Berkeley: University of California Press, 1984.

Preston, William L. *Vanishing Landscapes: Land and Life in the Tulare Lake Basin*. Berkeley: University of California Press, 1981.

Reisler, M. *By the Sweat of Their Brow: Mexican Immigrant Labor in the United States, 1900*.

Chapter 9

PRESERVING PARKS

DOCUMENTS

Lafayette Bunnell Describes the "Discovery" of Yosemite, 1852

He came alone, and stood in dignified silence before one of the guard, until motioned to enter camp. He was immediately recognized by Pon-wat-chee as Ten-ie-ya, the old chief of the Yosemites, and was kindly cared for—being well supplied with food—after which, with the aid of the other Indians, the Major informed him of the wishes of the commissioners. The old sachem was very suspicious of [James] Savage, and feared he was taking this method of getting the Yosemites into his power for the purpose of revenging his personal wrongs. Savage told him that if he would go to the commissioners and make a treaty of peace with them, as the other Indians were going to do, there would be no more war. Ten-ie-ya cautiously inquired as to the object of taking all the Indians to the plains of the San Joaquin valley, and said: "My people do not want anything from the 'Great Father' you tell me about. The Great Spirit is our father, and he has always supplied us with all we need. We do not want anything from white men. Our women are able to do our work. Go, then;

From Lafayette Bunnell, "Discovery of the Yosemite," in Robert Leonard Reid, ed., *A Treasury of the Sierra Nevada* (Berkeley: Wilderness Press, 1983), pp. 36–40.

let us remain in the mountains where we were born; where the ashes of our fathers have been given to the winds. I have said enough!"

This was abruptly answered by Savage, in Indian dialect and gestures: "If you and your people have all you desire, why do you steal our horses and mules? Why do you rob the miners' camps? Why do you murder the white men, and plunder and burn their houses?"

Ten-ie-ya sat silent for some time; it was evident he understood what Savage had said, for he replied: "My young men have sometimes taken horses and mules from the whites. It was wrong for them to do so. It is not wrong to take the property of enemies, who have wronged my people. My young men believed the white gold-diggers were our enemies; we now know they are not, and we will be glad to live in peace with them. We will stay here and be friends. My people do not want to go to the plains. The tribes who go there are some of them very bad. They will make war on my people. We cannot live on the plains with them. Here we can defend ourselves against them."

In reply to this Savage very deliberately and firmly said: "Your people must go to the Commissioners and make terms with them. If they do not, your young men will again steal our horses, your people will again kill and plunder the whites. It was your people who robbed my stores, burned my houses, and murdered my men. If they do not make a treaty, your whole tribe will be destroyed, not one of them will be left alive." At this vigorous ending of the Major's speech, the old chief replied: "It is useless to talk to you about who destroyed your property and killed your people. If the Chow-chillas do not boast of it, they are cowards, for they led us on. I am old and you can kill me if you will, but what use to lie to you who know more than all the Indians, and can beat them in their big hunts of deer and bear.

Therefore I will not lie to you, but promise that if allowed to return to my people I will bring them in."

Ten-ie-ya was allowed to go. The next day he returned to the camp and convinced Savage that his people would soon follow. But when several days passed and none of Ten-ie-ya's followers showed up, Savage decided to go after them. The old chief agreed to accompany the battalion to the Indians' hiding place.

While ascending to the divide between the South Fork and the main Merced we found but little snow, but at the divide, and beyond, it was from three to five feet in depth, and in places much deeper. . . . To somewhat equalize the laborious duties of making a trail, each man was required to take his turn in front. The leader of the column was frequently changed; no horse or mule could long endure the fatigue without relief. To effect this, the tired leader dropped out of line, resigning his position to his followers, taking a place in the rear, on the beaten trail, exemplifying, that "the first shall be last, and the last shall be first." The snow packed readily, so that a very comfortable trail was left in the rear of our column.

Old Ten-ie-ya relaxed the rigidity of his bronze features, in admiration of our method of making a trail, and assured us, that, notwithstanding the depth of snow, we would soon reach his village. We had in our imaginations pictured it as in some deep rocky cañon in the mountains.

While in camp the frantic efforts of the old chief to describe the location to Major Savage, had resulted in the unanimous verdict among the "boys," who were observing him, that "it must be a devil of a place." Feeling encouraged by the hope that we should soon arrive at the residence of his Satanic majesty's subjects, we wallowed on, alternately becoming the object of a joke, as we in turn were extricated from the drifts. When we had traversed a little more than half the distance, as was afterwards proved, we met the Yosemites on their way to our rendezvous on the South Fork.

Seventy-two Indians gave up without a fight. Ten-ie-ya insisted that none of his braves were left in the valley, but Savage was unconvinced. Ten-ie-ya now joined his people on the march out from the mountains, while Savage and his men pressed on.

We found the traveling much less laborious than before, and it seemed but a short time after we left the Indians before we suddenly came in full view of the valley in which was the village, or rather the encampments of the Yosemites. The immensity of rock I had seen in my vision on the Old Bear Valley trail from Ridley's Ferry was here presented to my astonished gaze. The mystery of that scene was here disclosed. My awe was increased by this nearer view. The face of the immense cliff was shadowed by the declining sun; its outlines only had been seen at a distance. . . .

That stupendous cliff is now known as "El Capitan" (the Captain), and the plateau from which we had our first view of the valley, as Mount Beatitude.

It has been said that "it is not easy to describe in words the precise impressions which great objects make upon us." I cannot describe how completely I realized this truth. None but those who have visited this most wonderful valley, can even imagine the feelings with which I looked upon the view that was there presented. The grandeur of the scene was but softened by the haze that hung over the valley—light as gossamer—and by the clouds which partially dimmed the higher cliffs and mountains. This obscurity of vision but increased the awe with which I beheld it, and as I looked, a peculiar exalted sensation seemed to fill my whole being, and I found my eyes in tears with emotion. . . .

I hurriedly joined the Major on the descent, and as other views presented themselves, I said with some enthusiasm, "If my hair is now required, I can depart in peace, for I have here seen the power and glory of a Supreme being; the majesty of His handy-work is in that "Testimony of the Rocks." That mute appeal—pointing to El Capitan—illustrates it, with more convincing eloquence than can the most powerful arguments of surpliced priests." . . . When we overtook the others, we found blazing fires started, and preparations commenced to provide supper for the hungry command; while the light-hearted "boys" were indulging their tired horses with the abundant grass found on the meadow near by, which was but lightly covered with snow.

Congress Grants Yosemite to California, 1864

An Act authorizing a Grant to the State of California for the "Yo-Semite Valley," and of the Land embracing the "Mariposa Big Tree Grove."

Be it enacted by the Senate and House of Representatives of the United States of America in Congress assembled, That there shall be, and is hereby, granted to the State of California the "Cleft" or "Gorge" in the granite peak of the Sierra Nevada mountains, situated in the county of Mariposa, in the State aforesaid, and the headwaters of the Merced River, and known as the YoSemite valley, with its branches or spurs, in estimated length fifteen miles, and in average width one mile back from the main edge of the precipice, on each side of the valley, with the stipulation, nevertheless, that the said State shall accept this grant upon the express conditions that the premises shall be held for public use, resort, and recreation; shall be inalienable for all time; but leases not exceeding ten years may be granted for portions of said premises. All incomes derived from leases of privileges be expended in the preservation and improvement of the proper; or the roads leading thereto; the boundaries

United States Congress, "The Yosemite Grant," in Robert Leonard Reid, ed., *A Treasury of the Sierra Nevada* (Berkeley, CA: Wilderness Press, 1983), pp. 317–18.

to be established at the cost of said State by the United States surveyor-general of California, whose official plat, when affirmed by the commissioner of the general land office, shall constitute the evidence of the locus, extent, and limits of the said cleft or Gorge; the premises to be managed by the governor of the State with eight other commissioners, be appointed by the executive of California, and who shall have no compensation for their services.

SEC. 2. And be it further enacted, That there shall likewise be, and there is hereby, granted to the said State of California the tracts embracing what is known as the "Mariposa Big Tree Grove," not to exceed the area of four sections, and to be taken in legal subdivisions of one quarter section each, with the like stipulation as expressed in the first section of this act as to the State's acceptance, with like conditions as in the first section of this act as to inalienability, yet with same lease privilege; the income to be expended in preservation, improvement, and protection of the property; the premises to be managed by commissioners as stipulated in the first section of this act, and to be taken in legal sub-divisions as aforesaid; and the official plat of the United States surveyor-general, when affirmed by the commissioner of the general land-office, to be the evidence of the locus of the said Mariposa Big Tree Grove.

APPROVED, June 30, 1864.

The Yosemite Grant, from Yosemite Grant, U.S. Congress: Act of June 30, 1864 (13 STAT., 325). An Act Authorizing a grant to the State of California of the "Yo-Semite Valley," and of the land embracing the "Mariposa Big Tree Grove."

John Muir on Saving Hetch-Hetchy Valley, 1910

Most of our forests have already vanished in lumber and smoke, mostly smoke. Fortunately, the federal government is now faithfully protecting and developing nearly all that is left of our forest and stream resources: nor even in these money-mad commercial days have our beauty resources

From John Muir, "The Hetch-Hetchy Valley: A National Question," *American Forestry*, 16, no. 5 (1910): 263–69.

been altogether forgotten. Witness the magnificent wild parks of the west, set apart and guarded for the highest good of all, and the thousands of city parks made to satisfy the natural taste and hunger for landscape beauty that God in some measure has put into every human being. . . .

Nothing dollarable is safe, however guarded. Thus the Yosemite Park, the beauty, glory of California and the nation. Nature's own mountain wonderland has been attacked by spoilers ever since it was established, and this strife, I suppose, must go on a part of the eternal battle between right and wrong.

The Yosemite National Park is not only the greatest and most wonderful national playground in California, but in many of its features it is without rival in the whole world. It belongs to the American people and is among their most priceless possessions. In worldwide interest it ranks with the Yellowstone and the Grand Canyon of Colorado.

The Yosemite National Park was created in 1890 by Congress in order that this great natural wonderland should be preserved in pure wildness for all time for the benefit of the entire nation. The Yosemite Valley was already preserved in a state park, and the national park was created primarily to protect Hetch-Hetchy Valley and Tuolumne Meadows from invasion.

The Yosemite Park embraces the headwaters of two rivers—the Merced and the Tuolumne. The Yosemite Valley is in the Merced Basin and the Hetch-Hetchy Valley, the Grand Canyon of the Tuolumne, and the Tuolumne Meadows are in Tuolumne Basin. Excepting only the Yosemite Valley, the Tuolumne Basin is the finer and larger half of the park. Practically all of the Tuolumne Basin drains directly into Hetch-Hetchy Valley, which is a wonderfully exact counterpart of the great Yosemite, not only in its crystal river, sublime cliffs and waterfalls, but in the gardens, groves, meadows of its flowery parklike floor. . . .

The floor of the Hetch-Hetchy Valley is about three and one-half miles long and from one-fourth to one-half mile wide. The lower portion is mostly a level meadow about a mile long, with the trees restricted to the sides and partially separated from the upper forested portion by a low bar of glacier-polished granite, across which the river breaks in rapids.

Standing boldly out from the south wall is a strikingly picturesque rock called "Kolana" by the Indians, the outermost of a group 2,300 feet high corresponding with the Cathedral Rocks of Yosemite, both in relative position and form. On the opposite side of the valley facing Kolana there is a counterpart of the El Capitan of Yosemite rising sheer and plain to a height of 1,800 feet, and over its massive brow flows a stream which makes the most graceful fall I have ever seen. From the edge of the cliff

FIGURE 7. HETCH-HETCHY VALLEY BEFORE THE DAM.
"Hetch-Hetchy Valley," John Muir claimed, "is charmingly diversified with groves of the large and picturesque California live oak, and the noble yellow pine. . . . Beneath them spreads a sumptuous fern carpet." With the passage of the Raker Act in 1913, the valley was drowned. From John Muir, "The Hetch-Hetchy Valley: A National Question," American Forestry, *vol. XVI, no. 5 (May 1910), p. 262. Photo by J.N. LeConte.*

it is perfectly free in the air for a thousand feet, then breaks up into a ragged sheet of cascades among the boulders of an earthquake talus. It is in all its glory in June, when the snow is melting fast, but fades and vanishes toward the end of summer. . . .

So fine a fall might well seem sufficient to glorify any valley; but here, as in Yosemite, nature seems in no wise moderate, for a short distance to the eastward of Tueeulala booms and thunders the great Hetch-Hetchy fall, Wapama, so near that you have both of them in full view from the same standpoint. It is the counterpart of the Yosemite Fall, but has a much greater volume of water, is about 1,700 feet in height, and appears to be nearly vertical, though considerably inclined, and is dashed into huge outbounding bosses of foam on the projecting shelves and knobs of its fagged gorge. . . . Besides this glorious pair, there is a broad, massive fall on the main river a short distance above the head of the valley. There is also a chain of magnificent cascades at the head of the valley on a stream that comes in from the northeast, mostly silvery plumes, like the one between the Vernal and Nevada falls of Yosemite, half-sliding, half-leaping on bare glacier-polished granite, covered with crisp, clashing spray into which the sunbeams pour with glorious effect. . . .

The principal trees are the yellow and sugar pines, Sabine pine, incense cedar, Douglas spruce, silver fir, the California and goldcup oaks, Balm of Gilead poplar, Nuttall's flowering dogwood, alder, maple, laurel, tumion, etc. The most abundant and influential are the great yellow pines, the tallest over 200 feet in height, and the oaks with massive, rugged trunks four to six or seven feet in diameter, and broad, arching heads, assembled in magnificent groves. The shrubs forming conspicuous flowery clumps and tangles are manzanita, azalea, spirea, brier rose, ceanothus, calycanthus, philadelphus, wild cherry, etc.; with abundance of showy and fragrant herbaceous plants growing about them, or out in the open in beds by themselves—lilies, Mariposa tulips, brodiaeas, orchids—several species of each; iris, spraguea, draperia, collomia, collinsia, castilleia, nemophila, larkspur, columbine, goldenrods, sunflowers, and

FIGURE 8. HETCH-HETCHY VALLEY AFTER THE DAM.
O'Shaughnessy Dam and reservoir in Hetch-Hetchy Valley harnessed water and power for the City of San Francisco, delivering its first supplies in 1934 at an initial cost of $100 million. From Warren D. Hanson, San Francisco Water & Power: A History of the Municipal Water Department and Hetch Hetchy System. *San Francisco: Department of Water and Power, 1987, p. 34. Reprinted by permission.*

mints of many species, honeysuckle, etc., etc. Many fine ferns dwell here, also, especially the beautiful and interesting rock-ferns—pellaea, and cheilanthes of several species—fringing and rosetting dry rockpiles and ledges. . . .

In spite of the fact that this is a national property dedicated as a public park for all time in which every citizen of the Untied States has a direct interest, certain individuals in San Francisco conceived the idea that here would be an opportunity to acquire a water supply for the city at the expense of the nation.

But light has been brought to bear upon it, and everybody is beginning to see more and more clearly that the commercial invasion of the Yosemite Park means that sooner or later under various specious beguiling pleas, all the public parks and playgrounds throughout our country may be invaded and spoiled. The Hetch-Hetchy is a glaringly representative case, involving as it does the destruction of one of the grandest features of the Yosemite National Park, which if allowed, would create a most dangerous precedent.

Judging from the way that the country has been awakened to the importance of park preservation, it is incredible that the people will tolerate the destruction of any part of the great Yosemite Park, full of God's noblest handiwork, forever dedicated to beneficent public use.

Yosemite Indians Recount Their Origin Story, Recorded in 1927

The Coming of the Indians to the Great Valley

Untold moons ago, the Great Spirit, in kindly thought for his favourite people, gathered them together in bands and led them into the high mountains. Long and arduous were the trails, but finally they reached a high-set valley, grassy and all girt about with great cliffs. Streams of crystal were gleaming with fish, trees and bushes bent with the weight of acorns and berries and nuts, and deer were very plentiful. Here the Great

From Elinor Shane Smith, *Po-ho-no and the Legends of Yosemite* (Monterey, CA: Peninsula Printing Co., 1927), pp. 6–7.

Spirit bade them to stop and make themselves homes. "Ah-wah-nee," "Deep Grassy Valley," they called the new home and themselves "Ah-wa-nee-chees."

How the Name of Yosemite Came to Ah-wah-nee

Many, many seasons of flowers passed by, and the Ah-wah-nee-chees were now a strong and powerful tribe, dwelling in peace and in plenty. Perhaps they grew careless, forgot the Great Spirit and all his kindness and care for them, for he seemed to grow angry with them. He sent famine and earthquake, drought, and woe to the Valley. The rains ceased to fall, the acorn crop failed, and mighty tremblings hurled down huge rocks to the floor of the Valley, crushing the huddled and terrified people. The end of the world seemed at hand!

A pitiful remnant of the proud tribe fled across the Great Snowy Mountains to the land of the Monos, and dwelt with them there by their lake. One of the exiled young Ah-wah-nee-chee chieftains married a Mono maiden, and to them a sturdy wee son was born, named Te-nie-ja, who grew to be a valiant Chief. When his aged father had died, an ancient brave, one of his father's faithful friends, came to Te-nie-ja and besought him to gather the rest of the tribe and return to Ah-wah-nee, now rightly his own.

Te-nie-ja, longing to see the land of his father, having heard many tales of its fairness and plenty, assembled his people and such of the other adventurous folk as wished to go with them, and they fared forth together.

Over the lofty Snow Mountains they traveled old trails, half forgotten, grown over with bushes and flowers. At last, bearing their heavy burdens, weary but glad, they came to the Valley, and once more Ah-wah-nee-chee camp-fires flickered among the pines and cedars.

The returned wanderers and their children flourished in their ancestral home, but they were a small tribe at best, and not strong. Yet there were sturdy leaders, and one such gave the name of "Yosemite" to all the tribe.

One morning as he was on his way to spear fish in "Ah-wei-ja," "The Lake of the Mirror," he met a huge grizzly bear, who resented his coming and fiercely attacked him. Having no weapon, spear nor bow, and arrow tipped with flint, the brave youth seized the limb of a tree and waged a terrible battle with the huge beast. Long and fiercely they fought, but at last the young chief, sorely wounded, slew the great bear. Bleeding and exhausted, he crept back to camp, where the rest of the braves, greatly admiring this feat of courage and strength, gave him the name of the huge beast he had slain—"Yosemite"— the great Grizzly.

The proud title passed down to his children's sons, and finally became the tribal name of the erstwhile Ah-wah-nee-chees.

Huey Johnson on Long Live Mono Lake, 1983

Mono Lake is a national treasure whose continued survival is threatened by the diversion of four of its five input streams. The real tragedy is that Mono Lake does not need to die. A single decision, a single stroke of the pen by one political jurisdiction could reverse the Lake's fate. The Department of Water and Power of the City of Los Angeles has the opportunity—particularly in this extraordinarily wet year—to halt Mono Lake's slow death. Instead the City's response has been to redouble its lobbying against efforts to protect Mono Lake. . . .

Mono Lake should be considered an indicator of our interest in survival. The recent history of Mono Lake demonstrates how over-consumption and special interest manipulation can threaten the long-term survival of unique and productive resources that are important to a wide range of species—including our own. The arrogance implicit in the City's position is manifested by its recent assertion that the public trust doctrine—a protection as old as the Magna Carta—is "irrelevant" to the Mono Lake situation. If the alternatives were costly and harsh, the Mono Lake case study would be easier to understand. But the alternatives are reasonable. An extensive inter-agency study four years ago determines that water conservation, reclamation and alternative supplies could make up for a reduction in diversions from Mono Basin sufficient to save the Lake. Mono Lake is a victim of waste and mismanagement.

The Department of Water and Power is the captive of a cornucopian philosophy which assumes that there will always be more. Rather than face the realities of limited resources and turn to modern water conser-

From Huey Johnson, "Mono Lake: The Developing Tragedy," Testimony before Subcommittee on Public Lands and National Parks, U.S. House of Representatives, John Seiberling, Chairman, May 18, 1982, reprinted in Robert Leonard Reid, ed., *A Treasury of the Sierra Nevada* (Berkeley, CA: Wilderness Press, 1983), pp. 356–60.

vation practices which would provide reasonable and economic solutions to the City's problems, it steps up its public relations efforts. . . .

Because of the lack of cooperation on the part of the Department of Water and Power, things have reached a political stage. Some Los Angeles citizens have become convinced that institutional arrangements within the City of Los Angeles are themselves a major impediment to change. They are seriously considering a door-to-door petition drive that would overhaul the Department's approach to water and energy management.

As a scenic and recreational resource, Mono Lake has a national importance. If you have visited the Mono Basin, you will know what I am talking about; if you haven't, I would urge you to do so on your next trip to California. Mono Lake is a profound experience. Rimmed on the West by snow-covered granite peaks of Yosemite National Park, and on the east by the high desert and ranges of the Great Basin, Mono Lake offers a special sense of space and geologic transition to the thousands that visit each year. Recognizing its scenic value, John Muir and others sought to have Mono Lake included within Yosemite National Park.

Mono Lake is valuable as a seasonal and nesting habitat for over 70 species of migratory birds and waterfowl. Large proportions of the world's eared grebe and Wilson's phalarope populations stop at Mono Lake. Mono Lake supports the State's second largest inland population of snowy plover, an endangered shorebird. On many late Summer and Fall days, over a million individual birds of various species are present. They don't come for the scenery, but for the extremely productive brine shrimp and brine fly populations. They use Mono Lake as a rest and "refueling" stop en route to South America or other winter homes. For most of these birds, there is no alternative "refueling" area. The Wilson's phalarope, for example, flies directly from Mono Lake to South America—a 5,000 mile nonstop journey. The loss of the Mono Lake brine shrimp populations will mean the loss or drastic reduction in the number of these bird species. . . . Mono Lake is the summer habitat for 95 percent of the California Gulls that nest in California, or 20 to 25 percent of the world's population. . . .

Each of these values is threatened by the continued diversion of Mono Lake's tributary streams at current rates. Diversions have had three devastating impacts on Mono Lake.

First, diversions have reduced the Lake level by almost 50 feet since 1940. This lowering has exposed a broad band of former Lake bed to wind erosion. The fine alkali sediments which were formerly under the Lake are now lifted by the strong winds of the area, posing a health risk to the inhabitants of the area and to visitors.

Second, diversions have reduced nesting habitat by making peninsulas of what were once islands. Negit Island, the largest historical nesting area, supported 17,000 nests in 1978. After it became connected to the mainland in 1979, there has been no successful nesting on Negit. Although gulls have nested on smaller islands, there has been an overall decrease in nesting activity of 20 percent since 1978. . . .

Third, diversions are fundamentally altering the temperature characteristics of Mono Lake. Since 1940, the volume of the Lake has been cut in half. Because Mono Lake has no outlet, this has caused a dramatic increase in salinity and alkalinity. These changes stress both the bird species and the brine shrimp populations on which they depend. Last year, the early brine shrimp hatch, the first of two annual hatches at Mono Lake, was less than 10 percent of normal. This was a radical departure from past patterns. Although we may not know for sure until it's too late, the overwhelming likelihood is that salinity and temperature changes caused the poor brine shrimp hatch. . . .

These are warning signs that we should not ignore. Although scientists argue about what is happening at Mono Lake, the range of disagreement appears to be narrowing. Increasingly, observers are applying the phrase "ecological collapse!" to the situation at Mono Lake. If this is indeed what we are witnessing, the loss of Mono Lake's value as habitat and as a productive ecosystem will be sudden and it will be irreversible. . . .

The loss of Mono Lake—an incredibly rich and productive ecosystem—would be a tragedy of major dimension. It was John Muir, an early Mono Lake enthusiast, who pointed out that all things in nature are interconnected. We should not casually contemplate the destruction of this resource. We have a responsibility to save Mono Lake—a responsibility to ourselves and our children and grandchildren.

ESSAYS

Indian–White Conflicts over Yosemite

Rebecca Solnit

Lieutenant Moore's Second Infantry was better known as part of the Mariposa Battalion, after Mariposa County, where the group of about 200 men was organized. It was in pursuit of Indians that the Mariposa Battalion became the first party of whites to enter Yosemite Valley, on March 27, 1851. Most of what we know about the battalion's expedition comes from Lafayette Bunnell's *Discovery of the Yosemite and the Indian War of 1851 Which Led to That Event*, and it was Bunnell who gave most of the principal landmarks in the valley the names they still bear.

Bunnell's is a strange account, switching back and forth from the lushest romantic response to the land to cool, journalistic recounting of how the war was conducted. When I read the book, I was shocked to learn that Yosemite had been first explored in the course of a war, that a place always described in terms of its idyllic scenery could have such a brutal history, and shocked most of all by the way Bunnell could be lyrical and cold blooded at the same time. The views moved him to tears, he wrote, and the rocks reaffirmed his faith in the deity. For me the inadvertent climax of the book is a scene at Lake Tenaya, after the old chief and his people have been captured, just before they are marched to a reservation in the flatlands of the San Joaquin Valley.

"When Ten-ie-ya reached the summit, he left his people and approached where the captain and a few of us were halting," Bunnell recounts. "I called him up to us, and told him that we had given his name to the lake and river. At first he seemed unable to comprehend our purpose, and pointing to the group of glistening peaks, near the head of the

From Rebecca Solnit, "Up the River of Mercy," *Sierra* (November/December, 1992): pp. 52–57, 78. Reprinted by permission of the author.

lake, said 'It already has a name; we call it Py-we-ack.' Upon my telling him that we had named it Ten-ie-ya, because it was upon the shores of the lake that we had found his people, who would never return to it to live, his countenance fell and he at once left our group and joined his family circle. His countenance indicated that he thought the naming of the lake no equivalent for his loss of territory."

Annihilating a culture and romanticizing it are not usually done at the same time, but Bunnell neatly compresses two stages of historical change into one interaction. Bunnell is saying, in effect, that there is no room for these people in the present; instead, they will provide a decorative past in another culture's future. Pyweack means "shining rocks"; like most of Yosemite's original place names, it describes the landscape rather than memorializing a passing human figure. Tenaya is a name given from outside, a name that sheds light on neither the lake nor the man unless one knows its pathetic origin. . . .

Yosemite is a crucible of the American landscape, a catalyst for turning beliefs into tangible effects. It is, among other things, the subject of the first significant landscape photographs. The valley was the first piece of land recognized by the federal government as worthy of protection as a national park. It reigned supreme in John Muir's heart, and was central to the founding of the Sierra Club. It is the most famous park in the country, and the most photographed.

Those photographs of Yosemite portray, again and again, a sublimely empty wilderness; early authors compare the place to the Garden of Eden, emphasizing its tranquil purity. None of the material prepared me for the picture Bunnell presents, of an old man held captive by a rope around his waist being told by the U.S. Army that his culture was going to be obliterated. In art, Yosemite always looks like a virgin bride, not somebody else's mother.

If this history of Yosemite begins anywhere, it begins with a contrary young adventurer named James D. Savage, who was born in Illinois in 1823, on what was then the white frontier. . . . No solid account of him exists, only glimpses in dozens of memoirs of California at midcentury. Family legend has it that he was kidnapped by Indians as a child or ran away to join them as an adolescent; when he was 23 he joined a wagon train heading west—the wagon train that included the Donner Party, though he didn't join them on their unfortunate shortcut. He appears next in California, as a soldier under Fremont, helping to seize the territory from Mexico, and as a freelance looter raiding rancheros for the U.S. troops and for himself. He worked for John Sutter in the Sacramento Valley, rustled livestock, and was around when gold was discovered on the

American River in 1848. Savage was one of the early miners who explored further, finding gold on the Tuolumne, the river that flows out of Hetch-Hetchy Valley, and finding Indians to work his mines for him.

One pioneer ran into him at this time "under a brushwood tent . . . measuring and pouring gold dust into candle boxes by his side. Five hundred naked Indians . . . brought the dust to Savage, and in return for it received a bright piece of cloth or some beads." Another explorer remembered that "Jim Savage was the absolute and despotic ruler over thousands of Indians, extending all the way from the Cosumnes to the Tejon Pass, and was by them designated in their Spanish vernacular El Rey Guero—the blond king. He called himself the Tulare King."

Los Tulares was the Spanish name for what would later be called the San Joaquin Valley, and Savage achieved his sway over the tribes there through tricks—sleights of hand, electrical shocks—pretenses of supernatural power, and his gift for Indian languages: Contemporaries say he spoke several regional dialects, and he was often in demand as a translator. . . .

By some estimates Savage was extracting $10,000 to $20,000 a day during his brief reign in the Sierra foothills. He moved from the Tuolumne to the Merced watershed, about ten miles downstream from Yosemite Valley (whose existence was still unsuspected by the gold miners), and established two trading posts on the banks of the River of Mercy. You might call Savage the mountain man as diversified corporation.

The whites derisively referred to the California Indians as "Diggers," a name—still current in my elementary-school textbooks in the 1960s—supposedly descriptive of their root-gathering methods. Some of the Indians called the whites Gold Diggers, after their principal activity, as panning for gold quickly gave way to digging for it. In 1853 digging was abandoned for hydraulic mining. . . . Afterward came chemical refinement of the ore with mercury, then chlorine gas and finally cyanide (which remains the method of choice in open-pit mines all over the West). . . .

The landscape the Forty-Niners thus transformed was fully inhabited. . . . The miners' incursions displaced many foothill tribes, and the logging, sullying of streams, hunting, and grazing not only constituted the first wave of environmental damage, but devastated native food sources. During the Gold Rush, confrontations between Indians and whites were frequent and violent. Some whites believed extermination the only solution to the "Indian problem," and harbored few inhibitions about killing native people. (The indigenous population of California declined by two-thirds during the Gold Rush era.)

A few foothill tribes took to rustling livestock as a solution to their food problems, and attacked some of the settlements in their territory. . . . Savage reported that an all-out war against the whites in the region was brewing. His Indian sources told him that the attacks on his posts had been launched by a fierce people they called the Yosemites; the Mariposa Battalion was organized largely to exterminate or relocate them. It was Savage's war, and the troops elected him head of the expedition: Major Savage.

Tenaya came out to meet the battalion near what is now El Portal, at the head of Yosemite Valley. Savage told him that if he did not bring his people out and sign a treaty, they would be utterly destroyed. A snowstorm came, and perhaps because of it Tenaya's people did not, so the battalion set off for the Valley. Halfway between the main Merced River and its south fork they met Tenaya with 72 people—those, the chief said, who were willing to be relocated. Savage and part of the battalion pressed on into Yosemite Valley to get the rest. They got as far as El Capitan and, at dusk, made camp at the foot of Bridalveil Fall. . . .

Bridalveil Fall, under whose spray the Mariposa Battalion camped, had been called Pohono by the Miwok, meaning "a potent wind." At Bridalveil Fall that cold March night in 1851, the 50 or so men built a campfire and sat around it arguing about what to name the valley they had just marched into. Romantic and biblical names were brought forth; Paradise Valley was proposed by a man who cursed the Indians and their names. (The place was often compared to Paradise, and to Eden: A companion of the painter Albert Bierstadt wrote in 1863, "If report was true, we were going to the original site of the Garden of Eden." . . . Bunnell insisted that the valley have an Indian name, "that we give the valley the name of Yo-sem-ity, as it was suggestive, euphonious, and certainly American; that by so doing, the name of the tribe of Indians which we met leaving their homes in this valley, perhaps never to return, would be perpetuated."

"Yo-sem-i-ty" won by a voice vote; then Bunnell went to ask Savage what the word meant. Savage, who spoke many neighboring dialects, said it meant "grizzly bear" and that the name was given to Tenaya's band "because of their lawless and predatory character." Savage's translation is still almost universally accepted; it is the version on all the park signage. But Craig Bates, the Park Service's chief ethnologist in Yosemite, talked to the old Miwok speakers in the region and came to a very different conclusion. It is true that *uzumati* is the Miwok word for grizzly bear, but

Yosemite seems to be a version of another Miwok word, *yohemiteh*, which means "some among them are killers." Some-Among-Them-Are-Killers National Park, a place named by those who knew not what they did. . . .

Yosemite's untidy history continued. The Mariposa Battalion destroyed the acorn granaries and villages it found. Bunnell's division caught three of Tenaya's sons near a triple rock formation later named Three Brothers in honor of the coincidence, and then shot one of the brothers as he tried to escape. Tenaya and some of his people took refuge with the Mono Lake Paiutes for a while after they fled the uninhabitable reservation they had been sent to, and Tenaya is said to have been killed by some Paiutes in 1853 in a gambling dispute. The year before, Savage had been shot down by another white man in an argument over treatment of the Indians; Savage had taken the Indians' side.

In 1855 the first tourist party arrived in Yosemite. It had taken a little over three years for Yosemite to go from indigenous homeland to military zone to vacationland, where visitors rarely thought of Indians as they gazed up at granite walls and sighed over waterfalls. The valley the early tourists saw was dominated by broad meadows and spreading oak trees, a landscape beautiful in romantic terms and conducive to deer-hunting and acorn-gathering, a landscape that had been maintained by the torches of its former inhabitants. Later in the 19th century overgrazing damaged the meadows; suppression of fires allowed incense cedars to encroach on the grassland and unusual amounts of flammable undergrowth to build up. . . .

"Soon after its discovery in 1851, Yosemite attracted artists, writers, photographers, and lovers of natural landscape," reads the wall-text at the park's Visitor Center. "Their interpretations of such scenic splendor helped awaken the public to its natural heritage." . . . John Muir arrived in 1867 and ignored Yosemite's remaining Miwok villages; when the great landscape photographer Eadweard Muybridge came to the Valley in the 1870s, he took a lot of huge photographs of the land and a handful of stereoscope pictures of its people. In the 20th century, Ansel Adams carefully cropped out evidence of habitation to make his majestic, uninhabited landscape images, and argued that "people, buildings, and evidence of occupation and use will simply have to go out of Yosemite if it is to function as a great natural shrine."

Muir Woods and Hetch-Hetchy Valley

Roderick Nash

In March, 1868, a self-styled "poetico-trampo-geologist-bot. and ornith-natural, etc!-!-!-!" named John Muir arrived in San Francisco and, allegedly, immediately asked to be shown "any place that is wild." His search took him into the Sierra, and in the following decades Muir became the leading interpreter of these mountains as well as America's foremost publicizer of wilderness values in general. Three years after Muir came West, another young man, fleeing the great Chicago fire, arrived in California. William Kent also sought the wilderness—to such an extent that he could state without much exaggeration: "My life has been largely spent outdoors." Independently wealthy and of a reforming temperament, Kent pursued a political career in which conservation played a major part. Inevitably his trail crossed Muir's. Indeed in 1908 Kent insisted that the tract of virgin redwood forest he gave to the public be called the Muir Woods National Monument. But in the next five years San Francisco's attempt to secure the Hetch-Hetchy Valley in Yosemite National Park as a reservoir site created an issue that caused a major schism among conservationists. The friendship of John Muir and William Kent was one of the casualties of the Hetch-Hetchy fight. Faced with the need to choose between different definitions of "conservation," they ended in opposing camps. . . .

Rising north of the Golden Gate in Marin County, Mt. Tamalpais dominates San Francisco Bay, and its western slope affords a favorable environment for the towering coast redwood and a lush understory of alder, laurel, and fern. William Kent's home was near Tamalpais, and in 1903 it came to his attention that one of its last unlogged canyons was for sale. Kent knew the area and confessed that "the beauty of the place attracted me, and got on my mind, and I could not forget the situation." . . . The purchase of almost three hundred acres in Redwood Canyon followed. Kent had hopes of "a wilderness park for San Francisco and the Bay Cities." Commenting in September, 1903, on the plan, he explained

From Roderick Nash, "John Muir, William Kent, and the Conservation Schism," *Pacific Historical Review*, 36 (November 1967): 423–33. Reprinted by permission.

that "whatever occupation man may follow, there is planted within him a need of nature, calling . . . to him at times to come and . . . seek recuperation and strength." Crowded cities, he added, produced "physical, moral, and civic degradation" and, at the same time, the need to escape to wilder environments.

With this attitude John Muir was in full sympathy. "Civilized man chokes his soul," he noted in 1871. . . . He believed that centuries of primitive existence had implanted in human nature a yearning for adventure, freedom, and contact with nature which city life could not satisfy. Deny this urge, and the thwarted longings produced tension and despair; indulge it periodically in the wilderness, and there was both mental and physical reinvigoration. Steeped as he was in Transcendentalism, Muir never doubted that nature was a "window opening into heaven, a mirror reflecting the Creator." And wild nature, he believed, provided the best "conductor of divinity" because it was least associated with man's artificial constructs. Summing up his philosophy, Muir declared: "In God's wildness lies the hope of the world—the great fresh, unblighted, unredeemed wilderness."

Like Kent, Muir recognized the necessity of the formal preservation of wild country if future generations were to have any left. In 1890 he was a prime mover in the establishment of Yosemite National Park. Two years later he became president of the Sierra Club, an organization dedicated to wilderness enjoyment and preservation. After the turn of the century Muir emerged as a major figure in both the nature-writing genre and the conservation movement.

In 1907 William Kent returned to Marin County from a Hawaiian vacation to find that the North Coast Water Company was beginning condemnation proceedings against his land in Redwood Canyon for the purpose of creating a reservoir. Convinced that wilderness preservation took precedence over private development of water resources, Kent searched for a way of obtaining permanent protection for the area. When the Antiquities Act of 1906 came to his attention, he recognized its possibilities at once. The statute enabled the president to issue executive orders designating tracts of land in the public domain with exceptional historical or natural interest as national monuments. The federal government might also accept gifts of private land for this purpose. Taking the act at its word, in December, 1907, Kent informed Chief Forester Gifford Pinchot and Secretary of the Interior James R. Garfield that he wished to give Redwood Canyon to the government. . . .

On January 9, 1908, President Theodore Roosevelt proclaimed the land a national monument. . . . Although he did not know Muir personally, Kent had long admired him as an interpreter of wilderness values and determined to name the reserve in his honor. . . .

The wide publicity Kent received for his philanthropy pleased him on several counts. In the first place the land involved was still subject to condemnation; Kent wanted an aroused public opinion on his side. Second, his political ambitions, which carried him to the House of Representatives in 1910, were beginning to stir, and "conservation" was a potent, if vaguely defined, word in the Progressive vocabulary. The attention accorded Kent the donor of Muir Woods could not fail to help Kent the candidate. And praise poured in from all sides. Sunset, Collier's, and the Sierra Club Bulletin ran illustrated articles on the new national monument while newspapers throughout the country picked up the story. The Sierra Club made Kent an honorary member. . . .

Kent's gift and personal tribute deeply touched John Muir. On the day Roosevelt created the monument, Muir wrote that in view of the "multitude of dull money hunters" usually associated with undeveloped land, it was "refreshing" to find someone like Kent. Five days later he wrote Kent personally, calling Muir Woods "the finest forest and park thing done in California in many a day." "How it shines," Muir enthused, "amid the mean commercialism and apathy so destructively prevalent these days." . . .

The three years following the establishment of Muir Woods National Monument marked the zenith in the relationship of Kent and Muir. They even discussed the possibility of collaborating in "the general cause of nature preservation." But friction was already mounting within the conservation movement. Those who would preserve undeveloped land for its esthetic, spiritual, and recreational values as wilderness found themselves opposed to resource managers with plans for efficiently harvesting nature's bounties. In the fall of 1897 Muir abandoned his efforts to support professional forestry and, as a consequence, feuded with Gifford Pinchot, the leading exponent of the "wise use" school. Thereafter Muir poured all his energies into the cause of preservation, particularly the national park movement. Yet Pinchot, WJ McGee, Frederick H. Newell, Francis G. Newlands, and James R. Garfield among others were directing federal resource policy toward utilitarianism and even succeeded in appropriating the term "conservation" for their viewpoint. The Pinchot-dominated governors' conference on the conservation of natural resources held at the White House in 1908 revealed the depth of the schism. Spokesmen for the protection and preservation of nature, including John Muir, were kept off the guest lists in favor of practical men who interpreted conservation to mean the maintenance of an abundance of important raw materials. The frustrated advocates of wilderness preservation had no choice but to call Pinchot a "deconservationist."

Meanwhile William Kent was construing conservation in his own way. . . . If private interests took precedence over the people's voice in regard to natural resources, democracy was endangered as well as the land. . . . In this frame of mind Kent welcomed the idea of public control of natural resources as a panacea for land policy as well as for American government. And most importantly for his relationship to Muir, Kent's conception of conservation accorded greater value to democratic development of natural resources than to wilderness preservation.

The Hetch-Hetchy controversy . . . deepen[ed] the rift in conservation . . . bringing the friendship of John Muir and William Kent to an abrupt end. Situated on a dry, sandy peninsula, the city of San Francisco faced a chronic fresh-water shortage. In the Sierra, about one hundred and fifty miles distant, the erosive action of glaciers and the Tuolumne River had scooped the spectacular high-walled Hetch-Hetchy Valley. Engineers had long recognized its suitability as a reservoir and source of hydro-electric power, but in 1890 the act creating Yosemite National Park designated the valley and its environs a wilderness preserve. Undaunted, San Francisco applied for Hetch-Hetchy shortly after the turn of the century, and, riding a wave of public sympathy generated by the disastrous earthquake of 1906, obtained preliminary federal approval of its plans.

John Muir, however, determined to arouse a nation-wide protest over what he conceived to be a needless sacrifice of wilderness values and a betrayal of the whole idea of national parks. In the five years after 1908, while the Hetch-Hetchy question was before Congress, Muir labored to convince his countrymen that wild parks were essential, "for everybody needs beauty as well as bread, places to play in and pray in where Nature may heal and cheer and give strength to body and soul alike." As such a statement implied, nature, for Muir, was steeped in spiritual truth. Its desecration for material reasons was sacrilege. He had no doubt that he was doing the Lord's battle in resisting the reservoir. San Francisco became "the Prince of the powers of Darkness" and "Satan and Co." This conviction that he was engaged in a battle between right and wrong prompted Muir and his school of conservationists to issue vituperative outbursts against the opposition. In a popular book of 1912 Muir labeled them "temple destroyers" who scorned the "God of the Mountains" in pursuit of the "Almighty Dollar." Using such arguments, and playing upon the growing American enthusiasm for wildness in both man and nature, Muir succeeded in stimulating a remarkable amount of public concern for Hetch-Hetchy.

As a California congressman and well-known conservationist, William Kent could not ignore the Hetch-Hetchy question. On March 31, 1911, a few weeks after he arrived in Washington to begin his first term, Kent received a personal letter from John Muir. Assuming that Kent, the donor of Muir Woods, would champion the cause of wilderness preservation, Muir simply encouraged him to watch developments concerning Hetch-Hetchy closely and "do lots of good work." But for Kent the matter was not so simple. He realized Hetch-Hetchy was an extraordinary wilderness area and part of a national park. But he also knew that the powerful Pacific Gas and Electric Company wanted Hetch-Hetchy as a step toward tightening its hold on California hydro-electric resources. Municipal control of Hetch-Hetchy's water would block this plan and at the same time be a significant victory for the ideal of public ownership. The sacrifice of wilderness qualities, Kent concluded, was regrettable but in this case necessary for a greater good. Making this point in a letter to Muir's colleague, Robert Underwood Johnson, Kent stated his conviction that conservation could best be served by granting the valley to San Francisco.

In 1913 the Hetch-Hetchy struggle entered its climactic phase, and as a second-term congressman and key member of the house committee on the public lands, Kent was in a position to exert considerable influence. . . . When the Hetch-Hetchy bill was under consideration in the house, he rose to answer the preservationists' arguments: "I can lay claim to being a nature lover myself, I think that is a matter of record." He then proceeded to defend the reservoir plans as "the highest and best type of conservation." The same technique appeared in a letter to President Woodrow Wilson where Kent asserted that in the cause of protecting nature he had "spent more time and effort . . . than any of the men who are opposing this bill." And there was, in fact, much truth in this claim.

The final stages of the Hetch-Hetchy controversy revealed just how far apart Muir and Kent had drawn. "Dam Hetch-Hetchy!" cried Muir, "as well dam for water-tanks the people's cathedrals and churches, for no holier temple has ever been consecrated by the heart of man." Mustering the Sierra Club and wilderness advocates throughout the country, the elderly Californian threw his remaining energy into what he regarded as the most crucial conservation struggle of his lifetime. Kent's emphasis, on the other hand, was all on the beneficence of public ownership. Speaking in the house, he declared that "the ideal conservation is public social use of resources of our country without waste." Non-use, Kent explained, which the preservation of wilderness entailed, was waste. Searching for a dramatic illustration, he declared it his sentiment that if Niagara Falls could be totally used up in providing for humanity's need for water, he would be "glad to sacrifice that scenic wonder." . . . He had made up his mind that "real conservation meant proper use and not locking up of natural resources" and the furtherance of democracy through their public development.

It remained for Kent, as an acknowledged admirer of Muir to provide public explanation for their divergence over Hetch-Hetchy. He did so in the summer of 1913 in a series of letters to his congressional colleagues. To Representative Sydney Anderson of Minnesota he wrote: "I hope you will not take my friend, Muir, seriously, for he is a man entirely without social sense. With him, it is me and God and the rock where God put it, and that is the end of the story. I know him well and as far as this proposition is concerned, he is mistaken."

Similarly he wired Pinchot that the Hetch-Hetchy protest was the work of private power interests using "misinformed nature lovers" as their spokesmen. In October Kent told a public gathering in California that because Muir had spent so much time in the wilderness he had not acquired the social instincts of the average man.

The nearest Kent came to accounting directly to Muir was an undated memorandum to the Society for the Preservation of National Parks, of which Muir was a director. After commending the group for its statement on Hetch-Hetchy, Kent reiterated his conviction that the "highest form of conservation" called for a reservoir that would provide Californians with an abundant supply of cheap water. "I make these comments," Kent concluded, "with the utmost regard for your sincerity of purpose, and with a full understanding of your point of view."

Muir never responded directly to these remarks, but in the year of his life that remained after the reservoir plan received federal approval in December, 1913, he must have felt betrayed. The man who had done him his greatest honor in creating Muir Woods became an influential opponent in the Hetch-Hetchy fight. But it was not that Kent changed his mind about wilderness after 1908. At the very time he was helping draft the bill authorizing a reservoir in Hetch-Hetchy, he asked Gifford Pinchot for a statement in support of a state park on Mount Tamalpais. Specifically, Kent wanted Pinchot to show "the advantage of such a wilderness, particularly near San Francisco." And after Hetch-Hetchy, Kent went on to author the bill establishing the National Park Service (1916), participate in the founding of the Save-the-Redwoods League

(1918), and add more land to Muir Woods National Monument (1920). At his memorial service in 1928 one of the chief speakers was William E. Colby, president of the Sierra Club. Kent's problem was that the necessity to decide about Hetch-Hetchy left no room for an expression of his ambivalence. The valley could not simultaneously be a wilderness and a publicly owned, power-producing reservoir.

In spite of their common interest in wilderness, Kent and Muir ultimately gave it a different priority. The result was a bitter conflict. Yet both men were sincere and energetic proponents of conservation. Indeed, few Americans after their generation openly opposed it. But that hardly ended the controversy over the value and uses of nature in America. One man's conservation was frequently another's exploitation, which is another way of saying that conservationists neither were nor are a homogeneous interest group or political bloc. As the relationship of Kent and Muir revealed, the dynamics of the history of the American landscape in the twentieth century comes not so much from "conservationists" embattled against greedy, wasteful exploiters, but from the conflict of diverse interpretations of the meaning of conserving natural resources.

Sequoia and King's Canyon National Parks

Larry Dilsaver and Douglas Strong

In 1990, Sequoia National Park mark[ed] its centennial, [and] Kings Canyon National Park celebrate[d] its fiftieth birthday. These contiguous parks in the southern Sierra Nevada constitute one of the nation's finest wilderness regions. The history of their establishment represents a major success story in preservation efforts in the United States, and the account of their management adds a valuable chapter to the history of the National Park Service.

Such success did not come easily. The creation of Sequoia resulted primarily from the determined efforts of a few San Joaquin Valley residents,

From Larry Dilsaver and Douglas Strong, "Trees—Or Timber? The Story of Sequoia and Kings Canyon National Parks," *California History* (Summer 1990): 98–117. Reprinted by permission of the California Historical Society.

and the expansion of the park and establishment of Kings Canyon came only after extended battles and compromises. The history of their management reveals the transiency of policies that depend on changing public awareness, lobby groups, and leadership.

Between 1772, when Europeans first sighted the Sierra Nevada, and the discovery of gold at Sutter's mill in 1848, few white people set foot in the Sierra Nevada. Not until 1858 did cattleman Hale Tharp make the first known visit by a white person to the mountainous area east of the central San Joaquin Valley that became Sequoia National Park. Guided by local Indians, he traveled to Giant Forest, one of the finest concentrations of giant sequoias, where he later established a summer cattle camp. Although the Indians appealed to Tharp to protect their land, nothing could prevent the increasing influx of settlers into the Sierra foothills. Smallpox, scarlet fever, and measles devastated the Indians. The survivors retreated into the high mountains and crossed the Sierra to the east. The Indians of Kings Canyon met the same fate.

Soon sheepherders, prospectors, and lumbermen in pursuit of their trades entered the Kings-Kern-Kaweah watersheds in the Sierra east of Fresno and Visalia. Following the great California floods and drought of 1862–1864, sheepherders from the southern San Joaquin Valley drove their flocks north and east into the highest mountains in search of grazing land. Unfortunately, sheepherding practices at that time, combined with a complete lack of governmental control over the use of public land, resulted in widespread damage to the mountain watersheds. When sheep entered the mountains as the snow melted each spring, their sharp hoofs cut deeply into the moist soil, severely damaging the meadows. Sheepherders' fires, set in the fall to clear away brush and deadfall, ran unchecked over the mountain slopes.

Fires and overgrazing alarmed some explorers of the Sierra. In 1873, Clarence King noted that the Kern Plateau, which had numerous meadows and lush grass when he had visited earlier, now appeared as a "gray sea of rolling granite ridges." Two years later, John Muir vividly described the threat to Kings Canyon's fragile beauty. He urged that the forests be protected so the spring run-off would be sure to provide enough water for the San Joaquin Valley during the dry summer months. In Muir's opinion, "sheepmen's fires" did a great deal more damage than lumbermen's axes or mill fires.

Prospectors also participated in the early exploration and utilization of the Kings-Kern-Kaweah watersheds. Their extensive prospecting came to little except for one strike at Mineral King. The discovery of silver in

1873 touched off a rush to this high mountain valley. With the completion of a road into the isolated mining camp by the end of the decade, Mineral King reached its peak of development. The boom soon ended, however. With the failure of the mines, the toll road passed into the hands of the county and became a public highway. A few summer tourists, attracted by the cool mountain air, built cabins and continued to visit the valley each year.

In the meantime, discovery of the big trees elsewhere in the Sierra Nevada had attracted worldwide attention. . . . In the 1860s lumbermen entered the forests of the Kings and Kaweah watersheds. At first lumber mills served only local communities, but the completion of the Southern Pacific Railroad line through the San Joaquin Valley in the mid-1870s opened more distant markets. Although pine and fir trees provided most of the lumber, many giant sequoias were cut to provide shakes, fence posts, and grape stakes. The worst was yet to come. Log flumes, introduced in 1889, opened previously inaccessible timberlands to loggers. Perhaps the finest stand of giant sequoias, in the Converse Basin, fell quickly.

Danger to the General Grant Grove first stimulated interest in protection of the giant sequoias of the Kings-Kaweah watersheds. . . . Beginning in 1878, editorials in the Visalia Delta criticized the destruction of the forests, including the cutting of giant sequoias for exhibit.

In 1880 Theodore Wagner, the United States Surveyor General for California, wrote to the registrar of the United States Land Office in Visalia to request that four sections in the Grant Grove be suspended from entry, temporarily prohibiting anyone from claiming the land under existing land laws. He was responding to the concerns of Secretary of the Interior Carl Schurz, several scientists, and a growing number of local citizens who advocated protection of the big trees. In the following year, General John F. Miller of California introduced into Congress the first bill to establish a park. The measure died in committee, however, perhaps because the proposed park was so large and would be opposed by timber and grazing interests.

In 1885 fifty members of the Cooperative Land and Colonization Association, with the intention of founding a utopian community, filed ownership claims on extensive tracts of land in Giant Forest under the Timber and Stone Act. They next sought capital for a railroad to connect with a road they planned to build from the foothills to the forestlands. When the railroad plan failed, they formed a joint stock company, the Kaweah Cooperative Commonwealth Company, and constructed an 18-mile road through rugged country to the edge of Giant Forest. After setting up a portable sawmill, the colonists produced a small amount of lumber.

While a government land agent examined the Kaweah Company's land claims, local residents in Tulare County initiated a determined drive to protect the Sierra forests by having Congress permanently withdraw large tracts of land from the market. There were precedents for such action. Congress had granted Yosemite Valley to California in 1864 for "public use, resort and recreation." It was the first area in the country specifically set aside to be preserved for all future generations. As such, Yosemite marked the real beginning of the national park system, even though Yellowstone, created in 1872, was the first officially designated national park. Yosemite did not achieve this status until 1890, and the valley actually remained under state management until 1906.

Farmers in the San Joaquin Valley wanted to protect the watershed on which they depended for irrigation, and they also wished to preserve the groves of giant sequoia remaining in public ownership as scenic and recreational areas. With these goals in mind, George W. Stewart, editor and publisher of the *Visalia Delta*, spearheaded the movement to protect the southern Sierra Nevada. . . .

The *Delta*'s editorials soon attracted local attention, and members of the Tulare County Grange called a meeting for October 9, 1889, in Visalia. Prominent residents from Fresno, Kern, and Tulare counties who attended agreed unanimously to petition Congress to establish a national park. When Stewart and Tipton Lindsey, former receiver of the U.S. Land Office, drew a map of the proposed reservation, they expanded it to include the entire western slope of the Sierra from the present Yosemite National Park in the north to the southern end of the forest belt in Kern County. They wished to protect all major rivers flowing from the mountains into the San Joaquin and Tulare valleys.

A few months later, Stewart and his local supporters—including John Tuohy, Frank Walker, and Lindsey—became alarmed over rumors that the federal government was about to open the Garfield Grove—south of Giant Forest—to private land ownership under the Swamp and Overflow Act, Timber and Stone Act, and other statutes. Lumbermen coveted the timber, and sheepmen desired access to mountain meadows. Letters sent from the Delta office to interested groups and influential people from coast to coast warned of the danger to the world's largest trees, the giant sequoias. Lindsey notified Congressman William Vandever, initiating a full-scale campaign for a park.

By the end of July 1890, Vandever introduced a bill for a national park in the township that included the Garfield Grove. The Delta's campaign attracted the support of *Garden and Forest*, *Forest and Stream*, and other publications. The California Academy of Sciences, American Association for the Advancement of Sciences, and American Forestry Association also adopted resolutions favoring the park bill. Such support brought results; President Benjamin Harrison signed the bill on September 25, 1890. Sequoia National Park became the nation's second national park, created eighteen years after Yellowstone. The enabling act provided that two townships plus four sections be withdrawn from settlement, occupancy, or sale, and that they be set apart as a public park or pleasure ground for the enjoyment of the people.

Wanting to see all the forests of the Sierra preserved, Stewart declared in the Delta that the first important step in a great work had been taken. Believing that the Kings and Kern canyons and other desirable areas could be added later, park advocates had not clamored for a large park because of the imminent danger to the Garfield Grove. They felt that any effort to secure a larger park, which would have included much privately held land, would be sure to fail without an educational campaign that they had no time to conduct.

On October first, less than one week after creation of the park, Congress passed a second bill that established Yosemite National Park, tripled the size of Sequoia, and set aside the Grant Grove as a small separate national park. The measure came as a complete surprise to Stewart and others who had initiated the movement and worked so hard for the first bill. Daniel K. Zumwalt, a resident of Visalia and a land agent of the Southern Pacific Railroad who visited Vandever in Washington, has been credited with proposing both Grant National Park and the expansion of Sequoia, as well as for lobbying on behalf of establishment of Yosemite National Park. . . . More important, the Southern Pacific had long been concerned about protecting the water supply of the San Joaquin Valley, and it recognized that national parks would attract tourists and increase business. Support by the powerful Southern Pacific helped win the day for Sequoia, Yosemite, and General Grant national parks.

The Kaweah colonists, however, were shocked and dismayed by the news of the October first legislation. To make matters worse, a special land agent of the Department of the Interior reversed a previous report and ruled unfavorably on the colony's already-filed land claims in Giant Forest. Despite widespread support of the colonists by many residents of Tulare County, who respected the time and labor they had invested, Secretary of the Interior John Noble ruled against the colonists' land claims in April 1891. The government even denied compensation for the road they had built. . . .

[In 1940], the small General Grant National Park, created in 1890 and administered jointly with Sequoia National Park until 1933, was converted into a part of the new Kings Canyon National Park. Since 1943, the neighboring parks of Sequoia and Kings Canyon have been administered jointly.

Redwood National Park

Susan Schrepfer

During the battle to establish the Redwood National Park in the 1960s, the Sierra Club and Save-the-Redwoods League differed so adamantly over where and how large the proposed park should be that there were times when it was doubtful whether a national park could be established at all. Historical narration often neglects the fact that the politics of individuals and groups reflect their philosophic assumptions and that, when beliefs change, behavior changes. In the 1960s the League still based its political moderation and its interpretation of the function of a park on an earlier view of a progressive and benevolent world. The Sierra Club had once shared this assumption. By the time of the Redwood National Park battle, however, the Club was drawn to a modern premonition of environmental doom in a universe that lacked order and direction. As the organization adapted its politics accordingly, it found that its old ideological ally, the Save-the-Redwoods League, had become an antagonist.

By 1960 these two preservation groups had been headquartered across the street from each other in San Francisco for almost forty years. These had been decades of mutual support. The Club guarded the Sierra Nevada, and the League purchased redwoods for California state parks. This mutuality stemmed not only from similar goals but from a common philosophic heritage.

The Sierra Club had been formed in 1892, during a period when many

From Susan Schrepfer, "Conflict in Preservation: The Sierra Club, Save-the-Redwoods League, and Redwood National Park," *Journal of Forest History*, 24 (April 1980): 60–68, 75–76. Reprinted by permission.

sought to reconcile theism and Darwinian evolution. One Club founder Joseph LeConte, a University of California paleontologist, outlined for Americans a synthesis of direct agency and natural law. LeConte's world view was predicated upon the existence of a force external to the material world and upon man's ability to grasp the meaning of this force. He advanced a rational, scientific theology in which God has both a real, independent existence and a presence in nature. LeConte's idealism saw man as possessing two natures—the animal and the spiritual. With these dualities, LeConte perceived design in nature. Evolution was creation, "the conception of the one infinite, all-embracing design, stretching across infinite space." The very existence of man proved that evolution moves toward greater physical and spiritual development. From the late nineteenth century through the 1930s, such Sierra Club officers as naturalist John Muir, business leader Duncan McDuffie, professor of theology William Bade, and Stanford University President David Starr Jordan similarly influenced the organization with their conceptions of progressive designs in nature.

When the Save-the-Redwoods League was established in 1918, its leaders also held this vision of a moral universe, despite the ready evidence of global evil. John Campbell Merriam, paleontologist and League president, preached reconciliation of the religious impulse and evolution. To Merriam research was an act of reverence. Naturalists and League officers Henry Fairfield Osborn and Vernon Kellogg similarly wrote that science was verifying the existence of a central, moral wisdom in the universe. Scientists were proving not only that the world was progressive but that it was one in which man could control his destiny. If the human intellect is the highest product of directional evolution, our mastery of the world must be beneficial. . . .

These men, in the Club and in the League, projected their world view into preservation. Parks were temples within which to worship a universe of continuity and design. Parks would teach, as did the Museum of Natural History, that "Life developed continuously toward more complicated, more broadly comprehending, and more intelligent forms." Hence these men designed their parks to capture what they considered the highest examples of evolutionary momentum, such as the oldest of the coastal redwoods.

Not surprisingly, these individuals, who combined faith in human progress with faith in divine will, balanced their preservationist philosophy with respect for technology. Man had the right and ability to manipulate nature for his benefit. Parks were to be Laboratories that would aid this process. Most lands would be used. Only unique examples of the greatest scenic wonders would be saved. These tenets were reflected in the programs of both organizations.

For the Save-the-Redwoods League these attitudes dictated, and in turn were reinforced by, its financial base. Established in 1918 to purchase northern California coast redwoods, the League benefited from capital made available by industrialization and from the willingness of the newly wealthy to use philanthropy for reasons of conscience and taxes. The organization blossomed forth from the "stewardship of wealth," supplied by such figures as William Harkness and John D. Rockefeller, Jr. It is not surprising that its values were those of turn-of-the-century entrepreneurs. These included a genuine appreciation of economic and technological progress, as well as a desire to accommodate the lumber industry. . . .

Frederick Law Olmsted, Jr., acting as a consultant to the California State Park Commission, outlined in 1928 a series of state park projects, including several in the redwoods. Because the criteria for selection included ease of public access, the parks were located on the major north–south highway, U.S. 101, commonly known as the Redwood Highway. Limited funds as well as the legitimate needs of industry necessitated small reserves. Nonetheless, the state and the League together established Humboldt, Prairie Creek, Jedediah Smith, and Del Norte Coast redwoods state parks. With success in this venture, interest in a national park soon waned. . . .

Since the 1920s the Sierra Club had supported the League's campaign to establish the state park system and save the redwoods. The groups shared a preservationist perspective, and many Club leaders were officers in the League. In orientation and structure, however, the organizations differed. As a nonprofit, philanthropic organization, the Club depended upon tax-deductible dues and volunteer services. It offered members personal involvement in preservation campaigns and group hikes. Since most of the Sierra Nevada had remained in federal ownership, it was not necessary to raise funds. Instead, the group needed a large membership and coordination to influence government actions in the Sierra. Not surprisingly, in the 1930s and 1940s Club leaders generally had a more positive view of the federal government than did the officers of the League.

The Club leaders' philosophy did show, however, a teleological world view and strong faith in human progress. In their idealism the men of both groups believed parks could accommodate technological changes. The League cultivated park accessibility. Similarly, the Club's 1892 bylaws announced among its purposes: to "render accessible the mountain regions of the Pacific Coast." Members "proposed roads up Tenaya Canyon, into Little Yosemite, and across Kearsarge Pass and all the highest passes of the Sierra Nevada." . . . The establishment of parks was a final conquest of the West. Bordered, improved, and well-traveled, parks would complement a rapidly industrializing civilization in a universe that tended toward the good. Man was not the destroyer of worlds — rather, the enhancer of life. . . .

Between 1960 and 1963 proposals for a Redwood National Park came to the fore of national politics, and the Sierra Club was a catalyst. But the group had changed since World War II. In the 1950s and 1960s a philosophical naturalism had replaced the teleological view. . . .

The affinity of this world view and ecology was expressed by [Loren] Eiseley: "If I remember the sunflower forest, it is because from its hidden reaches man arose. The green world is his sacred center." But these words implied more than the reality that man's physical existence is impossible without the chain of life. The universe was not only man's physical prison but his spiritual prison. The teleological world view demanded movement beyond the physical order to grasp greater intelligence and purpose. The vision had been lost. . . .

To the Club, wilderness held answers to more questions than man yet knew or might ever know how to ask. The role of the scientist was to understand the world, not to seek to master it. Eiseley's random universe denied man a world priority. Man was a temporary adaptation to a specific environment and not innately superior to other species. To many Club leaders of the 1960s, nature had rights that should not be continually compromised for the benefit of man. . . .

The preservationist and developer were polarized by the former's loss of faith in progress. In the 1960s the Sierra Club rejected the leadership of those members active in industry, and the militant faction within the organization opposed cooperation with developers in the selection of alternate sites. Such cooperation connoted a sanction difficult for proponents of zero population growth. The flattery and pressure of communicating with business interests might seduce one from the truth. The organization publicly attacked both the actions and intentions of opponents. . . .

In 1960 David R. Brower, the group's executive director, publicized this apocalyptic vision that had begun to affect the Club. In an article titled "A New Decade and a Last Chance: How Bold Shall We Be?" he prophesied, "What we save in the next few years is all that will ever be saved." . . . He relished being called an extremist. Extremism meant distrust of authority, public attacks on misguided expertise, and avoidance of compromise. The Club stood ready to draw swords with whatever agents rationalized the compromise of wilderness. . . .

The Sierra Club launched a book to capture this desperate reality. In print by the winter of 1963, *The Last Redwoods* juxtaposed forest beauty and logging destruction. Critics countered there would always be redwoods in the state parks, but the Club retorted that these parks were too small and too close to the highways to tolerate heavy use or offer a wilderness experience. These were "see-through roadside strips" and "almost as tall as they were wide." . . .

In September 15, 1964, the National Park Service announced three plans for a Redwood Creek park, varying in size from 30,000 to 50,000 acres. . . .

The League gradually accepted the idea of a Redwood National Park. Its shift in thinking corresponded with the general postwar acceptance of big government by American liberals and conservatives alike. The accord was anticipated as early as 1934 when Newton Drury himself acknowledged the existence, "for good or for evil," of Washington's irresistible centripetal force. By 1960 the League advocated minimal, well-planned federal involvement that was respectful of industry and based upon the group's traditional program. . . .

[In 1968,] Congress passed a park of 58,000 acres, 27,500 acres of which represented the projected inclusion of three state parks. The bulk of the privately owned land acquired was on Redwood Creek, some 15,325 acres as opposed to only 4,600 acres on Mill Creek. (The remaining 4,100 acres comprised several coastal strips.) . . .

The national park battle was multifaceted, with this schism only one of the major skirmishes. The division significantly delayed resolution, however, and resulted in a park whose logic lay more in politics than ecology. The Redwood National Park's boundaries defied the integrity of both the Redwood Creek and the Mill Creek watersheds and would allow further erosion damage to the Tall Trees. The compromise would result as well in another ten-year battle over whether or not the park should be expanded. The division among preservationists in the 1960s was significant, therefore, because it helps to explain the fragmented nature of the park and illustrates the direction of organized environmental preservation in the twentieth century.

The Club's politics had mirrored a modern philosophic orientation. It had purposely made its issue one of the popular will versus the industry. Its confrontation politics precluded compromise with the timbermen. It demanded a large wilderness park, away from civilization and capturing the ongoing processes of varied ecosystems. Similarly, the League's ability to influence the executive, rather than Congress, its willingness to parley with the interests, and its vision of a park as a museum in which to study the culmination of progressive evolution all harkened back to the faith in human dominion and universal purpose that it had acquired in the 1910s and 1920s and which its closed internal governance had preserved.

At the turn of the century, many park preservationists had accepted a world view in which man's technological progress was compatible with the progressive, directional action of evolution. The League retained this older idealism, stoically adhering to its conservative politics and its philosopher, John Merriam. The Sierra Club abandoned the philosophy of men like Joseph LeConte for that of a Loren Eiseley, who found no will, no knowledge beyond the natural order. Evolution was random. A nondirectional universe complemented a technology that appeared as destructive as it was beneficent. Technology was not divinely programmed progress. Indeed, it may be no more than the random movement of the human parasite through an indifferent cosmos. By denouncing the League, the Club rejected an image that looked like an old photograph of itself.

Further Readings

Albright, Horace M., and Robert Cahn. *The Birth of the National Park Service*. Salt Lake City: Howe Bros., 1985.

Allin, Craig W. *The Politics of Wilderness Preservation*. Westport, CT: Greenwood, 1982.

Bunnell, Lafayette Houghton. *Discovery of the Yosemite, and the Indian War of 1851, Which Led to That Event*. Chicago: Fleming H. Revell, 1880.

Clary, Raymond H. *The Making of Golden Gate Park: The Early Years, 1865–1906*. San Francisco: California Living Books, 1980.

Cohen, Michael P. *The History of the Sierra Club, 1892–1970*. San Francisco: Sierra Club Books, 1988.

Cohen, Michael P. *The Pathless Way: John Muir and American Wilderness*. Madison: University of Wisconsin Press, 1984.

Dilsaver, Larry M., and William C. Tweed. *Challenge of the Big Trees: A Resource History of Sequoia and Kings Canyon National Parks*. Three Rivers, CA: Sequoia National History Association, 1990.

Engebeck, Joseph H. *State Parks of California from 1864 to the Present*. Portland: Charles H. Belding and Graphic Arts Center Publishing Co., 1980.

Farquhar, Francis P. *History of the Sierra Nevada*. Berkeley: University of California Press, 1966.

Fox, Stephen. *John Muir and His Legacy: The American Conservation Movement*. Boston: Little Brown, 1981.

Gilliam, Harold. *Between the Devil and the Deep Blue Bay: The Struggle to Save San Francisco Bay*. San Francisco: Chronicle Books, 1969.

Huth, Hans. *Nature and the American: Three Centuries of Changing Attitudes*. Berkeley: University of California Press, 1957.

Jones, Holway R. *John Muir and the Sierra Club: The Battle for Yosemite*. San Francisco: Sierra Club, 1965.

Muir, John. *Our National Parks*. Boston: Houghton Mifflin, 1901.

Muir, John. *The Yosemite*. New York: The Century Co., 1912.

Nash, Roderick. *Wilderness and the American Mind*. 3rd ed. New Haven, CT: Yale University Press, 1982.

Pomeroy, Earl. *In Search of the Golden West: The Tourist in Western America*. New York: Knopf, 1957.

Ranney, Victoria Post, ed. *The Papers of Frederick Law Olmsted. Vol. V: The California Frontier, 1863–1865*. Baltimore: The Johns Hopkins University Press, 1990.

Reid, Leonard, ed. *A Treasury of the Sierra Nevada*. Berkeley, CA: Wilderness Press, 1983.

Richardson, Elmo R. *Dams, Parks, and Politics: Resource Development and Preservation in the Truman-Eisenhower Era*. Lexington: University Press of Kentucky, 1973.

Runte, Alfred. *National Parks: The American Experience*. Lincoln: University of Nebraska Press, 1979.

Runte, Alfred. *Yosemite: The Embattled Wilderness*. Lincoln: University of Nebraska Press, 1990.

Runte, Alfred, and Richard Orsi, eds. *Yosemite and Sequoia: A Century of California National Parks*. Berkeley: University of California Press, 1993.

Russell, Carl P. *One Hundred Years in Yosemite*. Yosemite National Park: Yosemite Association, 1992 [1959].

Sax, Joseph. *Mountains without Handrails: Reflections on the National Parks*. Ann Arbor: University of Michigan Press, 1980.

Schrepfer, Susan R. *The Fight to Save the Redwoods: A History of Environmental Reform, 1917–1978*. Madison: University of Wisconsin Press, 1983.

Smith, Michael L. *Pacific Visions: California Scientists and the Environment, 1850–1915*. New Haven: Yale University Press, 1987.

Solnit, Rebecca. *Savage Dreams: A Journey into the Hidden Wars of the American West*. San Francisco: Sierra Club Books, 1994.

Strong, Douglas H. *Trees—Or Timber? The Story of Sequoia and Kings Canyon*

National Parks. Three Rivers, CA: Sequoia Natural History Association, 1968.

Worster, Donald. *Nature's Economy: A History of Ecological Ideas*. Cambridge: Cambridge University Press, 1977.

Chapter 10

Battles Over Energy

Documents

David Pesonen Describes a Visit to an Atomic Park, 1962

There are public atoms and there are private atoms. Public atoms are used widely in therapy, research and bombs. Until 1954 all atoms were public. In that year, however, at the urging of the Atomic Energy Commission, Congress passed the Atomic Energy Act. This was the official launching of the Government's sponsorship of peaceful uses of the atom. Its most pregnant provision was to stimulate development of nuclear-electric generation by the private utility industry.

The industry joined the AEC in lobbying for the bill and, anticipating its passage, enjoyed a running start the moment the bill became law. Among the front-runners was the Pacific Gas and Electric Company which, with four other utilities, began development work as early as 1951, under contract to the AEC.

PG&E is a multi-billion dollar corporation that in one way or another touches nearly every person in California. Like most utilities, it is a regulated monopoly. Its substations dot the landscape from Mount Shasta to the Mojave Desert. PG&E transmission lines are stitched the length

From David Pesonen, A *Visit to the Atomic Park* (Berkeley, CA: Author, 1962), pp. 1–2, 9, 11–12, 38. Reprinted by permission of the author.

and breadth of the state. To serve more than two million paying customers the company operates 76 generating plants, including hydroelectric dams on all of California's major river systems, geothermal plants in the state's principal geyser region, oil and gas fueled steam plants on the coast and in the delta, and increasingly, nuclear fueled generators—a small experimental one now operating at Vallecitos in the delta, and a larger one at Humboldt Bay near Eureka on the north coast, scheduled to start operation late in 1962. . . .

The company also enjoys the confidence of the Atomic Energy Commission which, despite company pronouncements to the contrary, has vigorously encouraged PG&E's expansion in the nuclear field. . . . California's third nuclear generator will be the largest in the world—a 325 thousand kilowatt giant reactor at PG&E's proposed "Bodega Bay Atomic Park." The utility turned its eye on the Bodega area soon after passage of the Atomic Energy Act. And the events surrounding its advance inward the Bodega headland give a fascinating glimpse into what the future of private nuclear power holds in store. They also paint a disturbing picture of corporate power at work when government puts the atom in its fist.

Bodega Head is a stubby peninsula, several miles long, hooking sharply into the Pacific Ocean on the coast of Sonoma County, about 50 miles north of San Francisco. It lies a long shout north of Tomales Point on the Point Reyes Peninsula—a brooding, fog-bound newcomer to the National Park System.

The Bodega peninsula resembles an arm raised in defense against the sea. It terminates in a blunt fist of granite called Bodega Head. Curled inside the bend of its elbow is Bodega Harbor, an anchorage for several hundred commercial and sport fishing boats. . . . Bodega Harbor is considered the safest among only five harbors of refuge along the 300 miles of forbidding coast between San Francisco and Coos Bay, Oregon. . . .

Looking down on the harbor is the town of Bodega Bay. With nets drying on the wharves and the salty odor of steaming crabs, fresh cod and salmon, the town breathes an atmosphere of old world fishing ports, like those strung along the Mediterranean from Spain to Yugoslavia. In fact, most of the fishermen who sail from Bodega Harbor, the fish processors, the shipwrights, and the mechanics are immigrants or their descendants from the Mediterranean region. . . .

Everywhere are the auspices of nature. Here State Route 1 alternately skirts a procession of esteros, drifts through fine dairyland bounded by dark groves of Eucalyptus then climbs to the lip of cliffs scanning the Pacific. It serves beauty more handily than commerce. In Spring it seems a thin gray wake through an ocean of wildflowers blue lupines, golden poppies, mustard and daisies, splashing yellow over the hills rolling to the sea.

Because north–south traffic runs mainly on US 101, thirty miles inland from Bodega Bay, the town, the harbor and their beautiful peninsula are little known to most Californians. The headland is more subtle than Yosemite, the town more remote than Carmel. And the headland is little used except by grazing cattle and wandering hikers.

But in 1957 a whiff of rumor suggested that the world's largest gas and electric company had found an industrial use for Bodega Head—involving a use for the private atom. . . . [T]he park agency abruptly withdrew its interest in Bodega Head and the University followed in the Fall. At this time, no public evidence of the company's plans existed; but tenacious rumors persisted in Sonoma County. And in the Spring of 1958 the company confirmed them. . . .

The best data available suggest that the western edge of the San Andreas Fault Zone is almost exactly one-quarter mile from the centerline of the proposed reactor at Bodega Bay. It is difficult to avoid the inference that PG&E's investment in Bodega Head, rather than public safety or a desire to eliminate ambiguities, may have played the largest part in the AEC's newest earthquake criterion.

Some geophysicists have postulated that the San Andreas is part of a much longer rift, circling the globe, mostly beneath the ocean. In the Bodega-Point Reyes region it climbs out onto land and shakes itself periodically like a creature of the deep.

Whether this theory is true or not, there is no doubt that the fault is a very real and ornery feature of the earth's crust. It is the same great rift that heaved in San Francisco's 1906 catastrophe. Near Point Reyes it threw a locomotive and several cars off the track and onto their sides. A roadway nearby was displaced 20 feet. The granite of both Point Reyes and Bodega Head is unduplicated anywhere to landward. It appears to have migrated to its present location through slippage along the fault by fits and starts over eons of geologic time. Reliable computations show that the Point Reyes Peninsula may have moved 67 feet northward since Sir Francis Drake careened his Golden Hind on its beaches in 1579.

The fault is as much a part of people's thinking in this region as is Vesuvius to those who live in its shadow. Despite, or perhaps because of solemn assurances from the company that the AEC will review the reactor's design, some people feel that putting a giant nuclear reactor on the shifty headlands of Bodega would be like pouring radium down the throat of Krakatoa. . . .

The town of Bodega Bay will be two miles from the reactor in a generally 90 degree direction from the prevailing wind. The city of San Francisco will be about 50 miles in a generally downwind direction and the trailer camps on Doran Park will be 1600 feet directly downwind. The centerline of the channel entrance to Bodega Harbor—through which the fishing boats of the local citizenry, the yachts from San Francisco, and other vessels seeking a harbor of refuge must pass—will be exactly 800 feet straight downwind from the centerline of the reactor core. Directly downwind from a three-hundred-foot stack, emitting such radioactive wastes as the thyroid-loving Iodine-131, will be the dairy ranches and poultry farms on which Sonoma County's economy depends. . . .

All the risks which have been swept under the rug . . . should be fully aired before the people who must, in the last analysis, run [the reactor]. Although the fact has been somewhat obscured in Sonoma County, local government is the backbone of Democracy.

William Bronson on Electrical Transmission-Line Blight, 1968

There is nothing wrong with wires and power poles as such. . . . But the land is draped and defiled by hundreds of thousands of miles of power lines; our city streets are spoiled by the same, ever-increasing blight. One wire, or even one great transmission system, is not necessarily bad in itself. However, when the lines grow so thick and when the view is obstructed and the face of the otherwise untouched earth is obscured to the point we have reached in California then something must be said in protest.

The means-to-ends-at-the-lowest-cost philosophy, which guides so much of our technological and commercial life, is the basic culprit here again. That, and plain numbers. The wire mess is growing at an even faster rate than the population. As the economy grows, the per capita consumption of electricity—in homes and industry—grows with it. And if that factor were not in itself enough cause for alarm, we will at some point begin to consume electricity in an ever increasing proportion to the other source of heat and power, for no other reason than that the smog problem will force greater curbs on uncontrolled and partially controlled burning of fossil fuels.

There are several remedies that we could employ if and when we demand it, but it should be clearly understood that the public and not the utility companies will bring about the change. The cost of undergrounding and rerouting is going to be passed on the user if it is to be done at all, and it is folly to think otherwise. The problem, of course, is to find a way of amortizing the cost over a period long enough to spread the burden. . . .

Ultimately, I would like to think, we plan our ground transportation system not only for private and high-speed public conveyances, but for electrical power transmission, and perhaps even our communications and bulk fuel transportation systems in great inter-city corridors. But that is a long way off, and in the meantime, we have our work cut out for us.

What are the alternatives to the mess we suffer with today?

In the case of the high-tension transmission lines that link our power sources to the distribution centers, there are two obvious alternatives. One is to underground where no other alternative will work, and this method should be used universally in urban regions as it now is in "downtown" sections.

Undergrounding is probably too expensive a method for the cross-country transmission lines, but the utilities could route their wires in such a way as to create a minimum of visual pollution for the greatest number. This, of course, implies higher construction cost, perhaps higher maintenance costs, greater power loss, and hence a slightly higher power unit cost, but in my eyes, it would be money well spent.

In our cities, we can underground not only high-tension lines, but all distribution lines. There is evidence that public opinion will eventually force all of our overhead wires underground. In a Gallup Poll in 1965, it was shown that seventy percent of the American adult population favored undergrounding in urban areas, and more persons than not were willing to pay extra for it. . . .

We long ago gave away our rivers, which belong to all of us, to the great power companies, and we have given away too much of our skyline in the process. Carey McWilliams wrote in 1949 in *California, The Great Exception*, that he believed that the state would have to take over the PG&E. I would think there is less likelihood of this today than there was then, although communities, such as Redding, Palo Alto, and Sacramento, have had the good sense to set up publicly owned electric systems. The threat of converting to public ownership is perhaps the most powerful weapon in the anti-blight arsenal, but it isn't being exploited and we can't count on it to achieve the goal we seek.

Clearly, the great power companies and the publicly owned systems must be made to respond to the growing need for relief from visual pollution. And to achieve this, public clamor must grow until the question becomes a statewide political issue. We have quite a way to go yet.

A Writer on the 1969 Santa Barbara Oil Spill, 1972

[Robert Easton's *Black Tide*] is the first full account of an ecological crime—a crime without criminals but with many victims—and a community's response to it. The eruption of the oil well on Union Platform A off Santa Barbara on January 28, 1969, has had profound effects, and could be described as the blowout heard around the world.

The blowout shook the industry and the federal bureaucracy, whose rules and safeguards had failed to prevent it, and is gradually forcing the reform of those rules and safeguards. It triggered a social movement and helped to create a new politics, the politics of ecology, which is likely to exert a decisive influence on future elections and on our lives. It brought to a head our moral and economic doubts about the American uses of energy and raised the question of whether we really have to go on polluting the sea and land and air in order to support our freeway philosophy of one man, one car.

The oil operations on the Rincon were not pretty, in spite of the palm trees that were brought in to decorate the artificial island. Still, few Santa Barbarans realized that such operations might some day constitute a threat to the city itself. There had been an oil strike just east of Santa

From Ross Macdonald, "Introduction," in Robert Easton, *Black Tide: The Santa Barbara Oil Spill and Its Consequences* (New York: Delacorte, 1972), ix–xvi. Reprinted by permission of Harold Ober Associates Incorporated.

Barbara some fifty years before, and the Summerland shore was still a wasteland being cleaned up at state expense. But such enormities seemed a thing of the past." . . .

It was an ecologically aware citizen force that held the city together during the oil crisis. Though the city is quite conservative politically, and backed Nixon and Reagan in recent elections, it is far from devoid of intellectual activism. This activism can be reminiscent of a New England town meeting, where citizens are a functioning and vocal part of government. . . .

The oil crisis crept up on us in near silence. Concentrating on protecting the city and the surrounding lands, we didn't fully realize that the sea could be in danger. In my early years in Santa Barbara, I had made the narrator of a novel say: "I turned on my back and floated looking up at the sky, nothing around me but cool clear Pacific, nothing in my eyes but long blue space. It was as close as I ever got to cleanliness and freedom, as far as I ever got from all the people. They had jerrybuilt the beaches from San Diego to the Golden Gate, bulldozed super-highways through the mountains, cut down a thousand years of redwood growth, and built an urban wilderness in the desert. They couldn't touch the ocean. . . ."

A series of oil drilling platforms had been erected in the state tidelands southeast of Santa Barbara. Now there was pressure for further platforms in the deeper and more treacherous federal waters outside the three-mile limit. The pressure was intensified by the federal government's need for money to finance the war in Indochina, and local government was unable to withstand it. A number of citizens, including Fred Eissler and Bob Easton tried to open public discussion on the matter, but the decision had already been made in Washington, based on Interior Department findings which were rather loosely related to local reality. Without a public hearing, or any serious examination of the dangers of deep-water drilling in the earthquake-prone channel, oil rights in large sections of the channel were auctioned off to the oil companies. After the damage was done, we learned why there had been no public hearings: a permanent official in the Interior Department had vetoed it on the grounds that it might "stir up the natives." . . .

The great oil spill that began on January 28, 1969, failed to interrupt our walks, but it displaced other topics of conversation. I remember a day, about two weeks after the eruption, when we stopped on Mountain Drive and looked out over the contaminated sea. The flowing oil had been partly choked off, but it was still leaking up through the ruptured sea floor. Thousands of diving birds had died, and the quality of human life in the area was being threatened. The beaches were black for forty miles along the coast, and reblackened every day as the tides came in. The odor of crude oil reached us like the whiff of a decaying future. . . .

The people of Santa Barbara had hoped that an enlightened response to the city's near tragedy would be made by other agencies of the federal government. But the Department of the Interior, staffed by many of the same permanent officials whose recommendations and decisions precipitated the disaster persist[s] in its determination to convert the Santa Barbara Channel into an oil field.

The issue, as it appears to local residents, is whether their environment can be bought and sold over their objections, or whether there is an inalienable right to the use and enjoyment of air and water. It is a test case, between a pervasive new form of tyranny and an ancient freedom, which will help to determine the future conditions of life throughout the United States.

Martin Goldsmith on Siting Nuclear Power Plants, 1973

[Nuclear] power plant sites [can] be grouped into four classes. The first is coastal, where the plant is sufficiently close to the ocean that its cooling requirements are met by the once-through flow of ocean water. The second class includes plants built in proximity to lakes or rivers whose waters may be used for once-through cooling. The third class is a variant, where the body of cooling water is a small lake or pond whose prime purpose is plant cooling. The fourth class employs cooling towers, which need not be in proximity to any natural body of water. . . .

In California, coastal and inland siting has been employed. The nuclear plants at Humboldt, San Onofre, and Diablo Canyon are immediately adjacent to the ocean, and use its waters for cooling. On the other hand, the Rancho Seco plant, nearing completion near Sacramento, will

From Martin Goldsmith, *Siting Nuclear Power Plants in California: The Near Term Alternatives* (Pasadena, CA: Environmental Quality Laboratory, 1973), prologue, 2–5, 7, 10, 41. Reprinted by permission of the California Institute of Technology, Environmental Quality Laboratory.

use canal-fed cooling towers, and numerous fossil-fuel plants now in operation employ cooling towers at inland sites.

There are few large rivers or lakes in California, and their use, or misuse, is already a subject of concern. The regulations of the State Water Resources Control Board effectively preclude once-through cooling with natural inland waters. Even in other states, whose natural bodies of fresh water greatly exceed California's in number and size, power plant cooling is under sharp scrutiny, and restrictions on the use of once-through cooling are increasing. Thus, this category of site is not a realistic possibility in California except, perhaps, at a very few locations. . . .

The [Coastline] Controversy

The controversy in this instance can be fairly simply drawn. Coastal siting is resisted because nuclear plants must be situated on remote portions of the coastline, thus altering the natural state there. The concern of California's citizens for their coastline was reflected in the passage of the Coastline Initiative in 1972. Moreover, some suspect that the use of sea water as a coolant will have a detrimental effect on the ocean. On the other hand, coastal siting is preferred by many because remote sites can be found, and the cooling requirements are expected to be met inexpensively, as compared to other alternatives.

The other side of the coin is inland siting using towers. Some feel that sufficient areas of lesser import, environmentally, than coastal sites can be found and should be used. However, water for cooling towers is required, and a substantial body of opinion holds that this is an inefficient and improper use of a precious resource.

Thus, the question finally comes down to the relative worth and proper utilization of two finite and precious and overworked resources—California water and California coastline. . . .

Visual

It is, of course, difficult to evaluate the visual impact of any construction because of the subjective nature of the aesthetic judgments required. In a setting of great natural scenic value, any man-made structure is an intrusion—usually an unwelcome one. On this basis then, a power plant will in some cases be of obvious impact. In some areas, however, man-made structures already exist, or the presence of such structures may be acceptable. However, not all structures are equally acceptable. In some cases, one might even argue that the grandeur of the natural setting may be enhanced by a man-made structure. . . . Nuclear power plants . . . are not usually considered to be of inspired design. . . . On the other hand, they generally represent good examples of industrial architecture, and tend to be simple of form, of regular geometry, and relatively free of functional or decorative excrescence. In point of fact, the largest visual impact is often due to the switchyard structures which are almost invariably constructed adjacent to the nuclear power plant itself. . . .

Recreational

The power plant need not physically interfere with use of the beach itself. Indeed, this is the case at several Southern California Edison (SCE) plants where the actual construction is set back from the beach, with water intakes and outlets passing under the beach and into the ocean. Thus, there is no effect on beach recreation. . . . The presence of power plant structures need not restrict camping, walking, birdwatching, fishing, and so forth. There is no doubt, however, that the presence of a power plant definitely interferes with contemplation of unspoiled nature for some distance away from the power plant. The nuclear power plants produce no loud noises, smells, or other interferences with noncontemplative recreational pursuits. . . .

Terrestrial Wildlife

Once constructed, a nuclear power plant has relatively little effect on neighboring wildlife. Plant operations are nearly silent, and relatively few personnel are required to conduct operations. Permanent effects are caused, of course, by the fact that habitat has been removed or disturbed over the actual area of the plant. The more significant impact may occur during construction of the plant, where noise, dust, transport of goods, and personnel may drive away most wildlife, which generally avoid human company. The importance of these effects can be gauged only on a case-by-case basis. If the plant is otherwise acceptable, this effect would prove to have major impact only if rare or endangered species occupied the general area of the plant construction zone. . . .

Oceanic Effects

The discharge of waste heat from the power plant represents the largest effect on the ocean, and is qualitatively the same for both fossil-fuel and nuclear power plants. However, nuclear power plants (per unit capacity) discharge up to 50 percent more heat to the ocean than a fossil-fuel plant. . . . To put the problem in perspective, it requires over one million acre-feet per year (af/yr) of cooling water (once-through) for a nuclear power plant of 1000-megawatt capacity, if the temperature rise is restricted to 20°F as the water passes through the condenser tubes. Such a flow is equivalent to that of a very substantial river, and could become very serious in a stream even the size of the Sacramento River, the largest in the state of California, or in estuaries, such as some parts of San Francisco Bay, San Diego Bay, or other enclosed areas. . . .

Greater concern has been expressed as to the entrainment of larger organisms, fish larvae, and even small fish which might be swept into the passages. It is possible to prevent the passage of large fish by the inclusion of racks, traveling screens, etc. This serves to protect not only the fish, but also the power plant. In some instances, the installations protect only the power plant; hence, the fish can be sucked against the racks and destroyed. What is being done on the West Coast to protect against this problem is to maintain the intake velocity of the water at very low levels and in the horizontal direction; thus, fish are capable of swimming against the intake suction and can keep themselves from being trapped in the collection mechanism. . . .

Seismicity

Another serious physical constraint, especially in California, is the seismic risk at a site. The AEC has set no absolute exclusion criteria based on seismic or geologic factors, but the siting guidelines in some ways effectively make some areas far less likely for siting. For example, the draft "Seismic and Geologic Siting Criteria" do not preclude the siting of plants within zones which may be subject to surface faulting. The criteria simply set down a methodology for determining the geographic extent of the zone that might be subject to faulting, and call for a body of investigation to accompany applications for siting in such zones. The criteria also state that plant designs in such zones must be able to accommodate the faulting, with safety.

Although the language of the criteria does not preclude siting in highly seismic regions, even in the zone of faulting, in practical fact there are no accepted design techniques for plants which might be subject to the differential displacements of surface faulting. Several such sites have been considered in California. After lengthy investigations, and often acrimonious hearings and proceedings, the sites have been dropped from consideration. Thus, although no specific approved criteria can be cited which preclude plant sites in highly seismic areas, it is our judgment that such areas realistically should be excluded from consideration, for the near-term at least. At the very best, lengthy delays in licensing will be encountered in many zones in California, and recent history would indicate that sites in zones subject to faulting will be rejected.

Moreover, the seismic situation does not seem to be tending toward less restrictive regulation. On the contrary, some predictive theories for magnitude of ground shaking formerly accepted are now being reconsidered. Much of the data from the San Fernando earthquake of February 1971 have tended to confound simplistic relationships, and in the face of uncertainty, the normal and proper attitude of a regulatory body is to require greater margins of safety. . . .

Summary and Conclusions

Present electrical demand growth in California and the various problems associated with alternate means of generation suggest that nuclear power plants will continue to be favorably considered by the electric utilities. Increased concern for plant safety and for environmental values is reflected in ever more stringent siting requirements by the various licensing bodies. Although coastline siting has been emphasized by utility planners, inland siting has been chosen in selected cases. Availability of cooling water has been a prime constraint for inland siting, while environmental and seismic constraints have strongly influenced coastal choices.

This study concludes that the unavoidable environmental impact of coastal siting is mostly associated with the visual presence of the plant. The effects of the discharged cooling water can be minimized by straightforward (but expensive) engineering methods. The environmental impact of inland plants is largely associated with the prominent cooling towers and their discharges.

ESSAYS

Energy and the Making of California

James Williams

The give-and-take between technology and the environment mingles over time, and in various ways, with the distinctively regional factors in California's energy experience. Together they encourage or retard exploitation of energy resources. This combination of factors, in turn, coalesces with the influence of energy myths, the tension between authoritarian and democratic technology, the value of time, and concerns for efficiency, reliability, and societal structure. This final mixture comprises the crude calculus of ever-changing advantages by which Californians make their energy-related choices. It provides, at any given time, the *Weltanschauung* fundamental to understanding the historical evolution of energy technologies, systems, and use patterns.

Energy Models: Problems and Solutions

California's . . . energy landscape, human exploitation and use of energy resources, and the interplay between technology and the environment all [evolved] in the context of a steadily growing population. In the nineteenth century . . . people use[d] muscle power, wood and coal, hydraulic and wind power, and other natural energy resources. After 1850, coal became the primary fuel in America, outpacing wood in industrial and household use and undermining waterpowered manufacturing. But Californians discovered they had few coal reserves. Although many of them dreamt of replicating eastern manufacturing development on the Pacific

From James Williams, *Energy and the Making of Modern California* (Akron, OH: University of Akron Press, 1997), pp. 9–11, 350–51, 353–54, 356–57. Reprinted by permission.

Coast, inconvenient water power sites and expensive steam fuels stifled such efforts, and most people remained dependent on wood for fuel. As the national energy pattern shifted, only in agriculture, which relied on human and animal muscle power, did Californians adhere to the same energy model as the rest of the nation.

During the 1890s, entrepreneurs earnestly began developing California's abundant petroleum deposits, and they pioneered hydroelectric power development, drawing on the rich knowledge of hydraulic engineering which they had gleaned from four decades of gold mining. Petroleum, hydroelectricity, and, after 1910, natural gas formed an energy triad that virtually eliminated all other resources from the state's energy budget. Moreover, oil, gas, and hydroelectricity freed Californians from dependence on imported energy resources and opened enormous opportunities for economic and industrial development. Oil replaced coal and wood as steam fuel in industry and transportation, and a hydro-based electric power network delivered energy throughout the state for domestic, agricultural, and manufacturing use. Motor vehicles and electricity transformed both landscapes and lifestyles, and, as energy companies came to be among the largest and most influential in the emerging urban/industrial world, issues of power and control over energy resources surfaced in the region.

World War II wrought an economic transformation in California that played itself out during the cold war era. The war stimulated enormous military and industrial investment in the state and attracted tens of thousands of new immigrants. The explosive population and economic growth, which it spawned, continued unabated into the 1980s. Southern California's decentralized urban environment, in part attributable to the petroleum industry, spread across the southland and to other parts of California. An electronics industry, partly grown from regional electric power research and development, provided a foundation on which Californians built an industrial society quite unlike the eastern model of which so many of them long had dreamed. . . .

Yet the state's enormous growth brought equally immense energy problems. Oil and gas imports exceeded domestic production, and hydroelectricity became secondary to giant steam-turbine generating plants. Metropolis-dominated and energy-hungry in the second half of the twentieth century, California joined the nation in facing an uncertain energy future. Air pollution, offshore oil drilling, nuclear power, and other energy-related issues increasingly fell afoul of the rising environmental movement that was itself prompted by the good life engendered in part by abundant, cheap energy. . . .

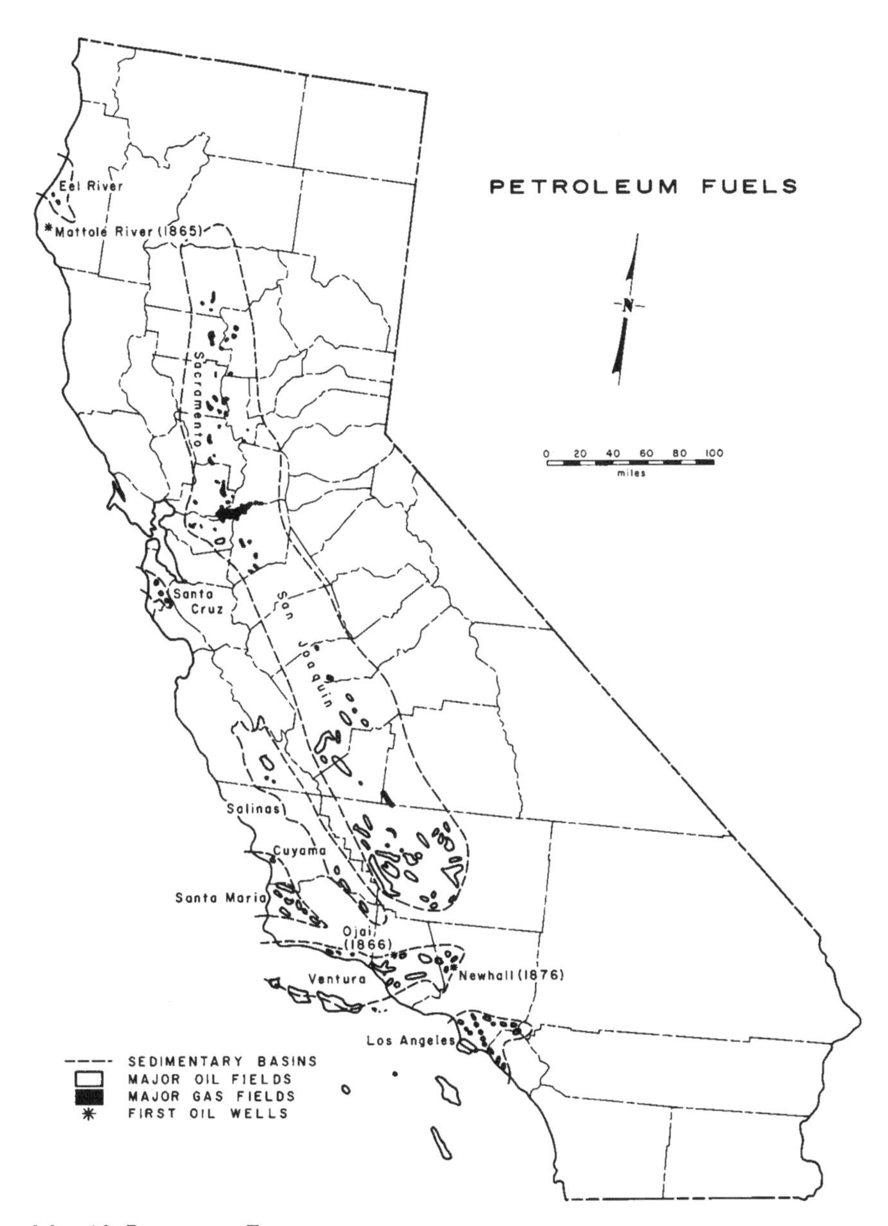

MAP 10. PETROLEUM FUELS.
Petroleum, or black gold, began to replace metallic gold as a major source of California wealth in the 1890s, supplying energy for industrial development and fuel for transportation. From Warren A. Beck and Ynez D. Haase, Historical Atlas of California. *Norman: University of Oklahoma Press, 1974, map 89. Reprinted by permission.*

By 1970, California's energy condition seemed to be in line with that of the rest of the nation. Its prosperity appeared inextricably linked to energy development. Its energy regime followed the same grow-and-build strategy found elsewhere in the country. Its energy resource budget, save having no coal, followed the national pattern. But its citizens' concern about air pollution, offshore oil drilling, and nuclear power exposed an undercurrent of environmentalism that came into direct conflict with the energy regime. As energy shortfalls struck the nation, the state once again diverged from the national course. Californians and their policy makers balked at national energy plans. Acting on a profound concern for the deteriorating state of their natural environment, they took an independent course. Grassroots activists and legislators halted installation of nuclear power plants, urged development of renewable energy resources, and took up energy conservation. Resources used and then abandoned in past years were rediscovered, and new technologies harnessed them in different ways. Federal and state legislation helped to stimulate investment in conservation and development of these and other untested energy resources and technologies.

In taking up appropriate technology and the soft energy path, California forged a new energy paradigm. Hard energy industries and the oil, gas, and waterpower triad were not replaced, but the introduction of alternative energy resources, new production methods, and conservation strategies substantially refashioned them. California became a laboratory for environmentally astute energy strategies and technologies. There soon trickled from it to the world a stream of new energy ideas and techniques, and the California Energy Commission undertook an export technology program, in 1986, to support the new renewable energy industries. In effect, the CEC came to act as an international economic development agent for renewable energy companies, among other things, granting funds to assist them in competing against German, Japanese, and other foreign firms. The resulting good health of California's new energy industries elevated the state as a leader in the battle against the worsening greenhouse effect of the earth's atmosphere caused, in large part, by carbon dioxide gas emissions. Because of its extensive research into solar panels, fuel cells, and advanced gas turbines for clean, efficient, and cost-effective electric power generation, the state made possible viable alternatives to the exclusive use of fossil fuels. . . .

Some lessons from energy history are plain. Fossil fuels are exhaustible. Just as California's coal reserves were spent in some fifty years and its oil and gas reserves depleted substantially in another fifty years, fossil fuels everywhere on earth constantly are being used up. We inevitably will reach the time when worldwide reserves will near exhaustion. . . . Reasonable people probably would be safest in assuming it will be sooner rather than later, if for no other reason than to be prepared for the inevitable. Such a position is especially important for societies that persist in opting for economic growth as the way to the good life and as the solution to social ills.

Most insights from history, however, are not as straightforward. For example, general economic betterment over time correlates, to some extent, with increased energy consumption. Per capita energy consumption in California followed an upward trend for 150 years, moving from between 150 million and 200 million BTUs in 1850–1860 to some 280 million BTUs in 1970–1980. Significant dips in per capita consumption appear to coincide with periods of economic depression (the 1870s, 1890s, and 1930s). Other declines not explained entirely by economic slowdowns can be attributed to improved efficiency in production and energy use. A variety of technological improvements between 1870 and 1900 contributed to more efficient energy conversion in steam boilers, water wheels, kerosene lamps, and agricultural field equipment. Similarly, since 1900, steady technological improvements to steam turbines, electric power generating and transmission systems, electric lights and appliances, and internal combustion engines have slowed dramatically the growth of per capita consumption despite an enormous and almost unbroken climb in overall energy use. In all this, California's experience is not unlike that of the United States as a whole.

This apparent linkage between economic growth and energy use still governs some energy planning, but, as California's experience during the 1970s and 1980s reveals, economic growth and energy consumption are not linked in some immutable fashion. Therefore, if our ideological position remains opposed to a relatively static economic state, one which might force a redistribution of wealth among our citizens, this does not mean we need to continue using more and more energy in order to achieve the economic growth and opportunity needed to stave off redistribution. . . . Indeed, California provides a model of energy conservation and diverse resource development. But, it remains very difficult for us not to place too much emphasis and hope on some single known resource, such as solar energy or nuclear fusion, or dream unrealistically of some magic, unknown resource just waiting to be discovered, that will let us get on with lives as usual. . . .

Concern for the natural environment, the source of much regulatory activity in recent years . . . will be of continuing importance for policy makers, particularly if we fully meld environmental planning and policymaking with that of energy. Would the environmental regulations and labor laws in place today financially or otherwise have prevented the construction of many, if not most, of the pre–World War II hydroelectric plants on which Californians greatly depend today? Likewise, what might have been the impact on earlier developments in California's oil industry under current regulatory restrictions? Policy makers might find it prudent to consider that environmental concerns have the potential to prevent tapping sorely needed resources and to slow other societal developments which might depend on the experiences gained in resource development. Conversely, it might be wise to remember that enacting regulations against [problems of] flaring natural gas connected to oil production helped to harness its beneficial use, and entrepreneurs do adjust to changes over time, as the builders of minihydro projects during the 1980s revealed in their successful dealings with the modern bureaucracy.

Finally, in a world shrinking in terms of time and distance because of advances in communication and transportation technologies, California's regional differences with other parts of the world now may appear less important as determinants of its energy future than in the past. Yet the state's present leadership in solar, wind power, and geothermal energy production seems inextricably linked to regional geographic characteristics, much as its hydroelectric system developed as it did because of the location of rivers and streams. Therefore, policy makers also must consider that regional peculiarities will continue to play a role in California's quest to achieve a mixed energy resource base for the twenty-first century. What the energy past plainly reveals is that the constant interplay between technology and the environment, along with a host of other factors, has always made up a crude calculus of ever-changing advantages by which we make energy-related choices. California's recent energy experience leaves little reason to believe this has changed and every reason to challenge our creative thinking for the future.

Oil and the Environment

Nancy Quam-Wickam

"A paradise of sunshine, fruit, and flowers," the true-life embodiment of "Oz, the Garden of the West," a land where "smokeless and cloudless skies have let the sunshine stream into human hearts"—such were typical characterizations of the Los Angeles basin in the early twentieth century. But this Eden-like landscape was also to become, in the decade of the 1920s, the site of what was then the most intensive oil field development in history, producing "the greatest outpouring of mineral wealth the world has ever known." Southern California's tremendous petroleum industry accounted for over 20% of the world's output of crude oil during part of the 1920s. Petroleum exploitation, in fact, not only provided a basis for southern California's economic growth during that decade, but also, in the words of contemporary economist John Ise, "so dominated the oil industry" that the history of southern California's oil development is "almost the history of the oil industry of the entire country."

The oil industry changed southern California's landscape forever. Located within the fastest growing metropolitan area of the country, oil development radically changed existing land-use patterns, encouraged industrialization and suburbanization, and contributed to real estate speculation in the region. But more than this, the particular nature of oil exploitation in the Los Angeles basin—the preponderance of small landholdings ("town-lots"), its intense and unplanned development, and the unusual geological formations encountered in drilling—contributed to production practices which devastated the physical environment and severely strained the economic vitality of the industry. So serious was oil pollution that one observer found the region to be a "stygian landscape" in 1924.

The shift of oil production from the San Joaquin Valley to the Los Angeles basin during the 1920s also profoundly transformed the industry. This new context of oil development in an urban area presented indus-

From Nancy Quam-Wickam, "Cities Sacrificed on the Altar of Oil": Popular Opposition to Oil Development in 1920s Los Angeles," *Environmental History*, 3, no. 2 (April 1998). Reprinted by permission.

FIGURE 9. SOUTHERN CALIFORNIA OIL FIELDS.
Oil derricks, erected by Standard Oil of California, dotted Huntington Beach along the state's southern coast in this 1939 photograph. Courtesy, The Bancroft Library, University of California, Berkeley.

try management with new problems: public concerns over unprecedented levels of oil pollution, a changed political setting offering greater regulatory opportunities to municipal and state governments, and the prospect of a revived union movement that explicitly based its rebirth on the social control of industry.

Pollution, overproduction, and profligate waste were the consequences of unchecked oil development in the Los Angeles basin, consequences that reached crisis proportions in the 1920s. The few previous studies of the politics of oil pollution control have generally ignored the local political context of oil's operations, focusing instead on the few relatively weak federal regulatory achievements of the 1920s. In California, state and local oil politics diverged significantly from federal policies. At all levels of politics in the state—from local referendums in working-class communities to the chambers of the legislature—the crisis of California's oil industry animated the efforts of policymakers to control the industry. The 1920s was a decade in which members of the Assembly and State Senate entertained more than fifty separate bills proposing to regulate virtually all oil industry practices: from bills establishing maximum hours of labor in the oil fields to measures calling for outright public ownership of important branches of the industry. Although many of these legislative measures ended in failure after hard-fought battles, local initiatives were more successful. In particular, the actions of residents of the mostly blue-collar communities of the southland to regulate and restrict oil development, and consequently lessen the devastating effects of oil pollution on their immediate environment, convinced oil management that political control of industry was indeed imminent.

The crisis of oil was rooted in the political economy of its production; opponents to oil development advanced political, not technological or economic, alternatives most frequently to resolve that crisis. There were good reasons for this emphasis on political solutions: economic control of the industry seemed impossible, as the example of the oil workers' strike of 1921 had just demonstrated. Technological control of the industry was likewise removed from the realm of public endeavor: in the late 1910s, engineers from the State Division of Oil and Gas had assumed responsibility for mandating technological improvements in oil development work. . . .

Oil was not new to Californians in the 1920s. California had been an important oil-producing state from the 1890s on, but before the 1920s, oil production was concentrated in the San Joaquin Valley, with some additional production along the coastal sections of Orange, Los Angeles, Ventura, Santa Barbara, and San Luis Obispo counties, as well as in some inland areas of Monterey and San Benito counties. Los Angeles had first experienced an oil boom in the early 1890s, with the development of the Los Angeles City Field, in what is the present-day Westlake District. The City Field developed at a frenzied pace, creating conditions that would be greatly magnified some thirty years later: rapid development of the field, overproduction without an accompanying increase in demand, a drastic drop in the price per barrel of crude oil, and waste. Yet production in the City Field declined rapidly, and although several new oil pools were located in Los Angeles and Orange counties in subsequent years, by 1913, the Los Angeles basin accounted for only 14% of the state's oil production. . . .

These rich new oil fields of the 1920s challenged the technological capabilities of the oil industry. Geological conditions in the Los Angeles basin posed enormous problems to those engaged in oil extraction; to varying degrees, each oil pool was blanketed with a reservoir of pressurized natural gas. These gas pressures were so great that special drilling technologies had to be employed, or the drilling crew would risk "blowing out" the well. . . .

Oil development was unplanned; the rush to get down the first well produced an enormous waste of oil and gas. During the southern California oil booms of the 1920s, dozens of gushers rained oil over the landscape, ruining orchards, vegetable fields, and grazing lands. Oil well fires—numbering well into the hundreds by the end of the decade—blackened the sky, shooting "livid towers of flames" a hundred or more feet high. Well fires, in fact, often remain the most vivid and enduring images of people who spent their childhoods near the oil fields. Oil wells released so much pressurized gas that the land sometimes buckled, caved in, and sank. Some contemporary observers warned frantically that the oil fields and surrounding communities might be blown apart in one gigantic firestorm if the gasses underground ignited. Oil exploitation in the Los Angeles basin was environmental destruction on a grand scale. . . .

The region's landholding patterns contributed directly to oil pollution. In the Huntington Beach and Long Beach fields, where small leases were prevalent, there was often not enough space on the property for the drilling rig, its machinery and the sumps necessary to separate the emulsified oil (pumped from the well) from its water and to collect waste oil. On the many leases lacking these sumps, a method of separating the oil from water and sediment was improvised: a length of casing, twelve to eighteen inches in diameter and 50 to 100 feet long, was elevated at one end, with the oily mixtures being fed into the pipe at mid-length. Ideally, the lighter clarified oil would then migrate to the elevated end of the casing, where it was removed, which left the heavier water, tar-based sludge, and muddy sediments in the lower end of the pipe. These oily mixtures from the bottom end of the casing were then allowed either to flow freely over the ground or to collect in small sumps . . . [where it] remains until the [sump] walls break down or it is washed out by the rains and allowed to flow over the surrounding territory, eventually finding its way to drainage systems and thence to the ocean. It was estimated that in the southern Los Angeles basin area, "several hundred barrels" of oil flowed into the Pacific ocean every day from this oil-and-water separating practice alone. Indeed, these everyday oil discharges, from a variety of transportation, refinery, and field practices, were likely responsible for the majority of the region's oil pollution.

Contemporary government officials recognized the pollution problem. Earle Downing, California's official in charge of the Department of Commercial Fisheries of the state Fish and Game Commission, blamed land sources—refineries, gas plants, oil fields—for the state's "worst pollution." By 1923, numerous commercial fisheries in the state had already been damaged by oil pollution. Testifying before one Senate Committee, fisheries biologist (and former president of Stanford University) David Starr Jordan noted that the "worst and vilest destruction of fishes, clams, and sea birds occurs when crude oil is poured into the sea . . . along the California coast." Another authority stated that California's commercial fish catch had declined by 37,500 tons since extensive oil development had begun a decade earlier; much of the loss, he alleged, was directly attributed to oil pollution. At mid-decade, one newspaper reported that market prices for fish had increased dramatically during the early 1920s, as oil pollution contributed to declining stocks. Bird life was endangered. Fish were reported to be unable to survive in any of the "natural waterways" emptying into the Los Angeles/Long Beach harbor. In the mid-1920s, much of the energy of the Bureau of Pollution (later renamed the Bureau of Hydraulics) of the state Department of Fish and Game was expended in investigating and prosecuting oil polluters in the Los Angeles area. The secretary of Santa Ana's Chamber of Commerce charged that the "entire Southern coast of this state is now suffering from . . . this nuisance." By 1927, Long Beach's citizens were said to have "long complained" of the city's polluted conditions. . . .

The role of the oil industry in precipitating local political crises in suburban Los Angeles during the 1920s suggests one aspect about the development of the region that has eluded the attention of most historians. Despite its image as a region of unparalleled boosterism, residents of suburban Los Angeles—especially those areas in and near oil districts—did indeed have deep-seated anxieties about unchecked development; people transformed their concerns into concrete proposals for controlling, regulating, and restricting oil operations. The issue of environmental protection thus became a struggle between contending forces which each sought to determine the political direction of oil production. At the local level, hard-won working-class initiatives proved successful in restraining oil's operations in many locations throughout southern California. Although local limitations on oil development plagued oil operators throughout the 1920s, they rarely resulted in any long-lasting governmental reforms on more comprehensive levels. But these proposals did provide much of the controversy which informed the development of an intensely local political culture throughout suburban Los Angeles that had as its core debates over the pace and character of industrial development in the region. In the years following the oil booms of the 1920s, other matters—the situating of a factory or the location of a rock quarry—would displace oil as the most contentious issues in local political struggles. Further, the local politics of oil demonstrates that conservationists, middle-class reformers, and oil producers themselves held no monopoly on the political process and regulatory mechanisms of government. Working-class residents clearly expressed their concerns about environmental destruction, creatively using the political process to achieve environmental justice and determine for themselves the future shape of their communities.

The Battle for Bodega Bay

Thomas Wellock

In the spring of 1958, Dr. Edgar Wayburn, chairman of the Sierra Club's conservation committee, received a confidential letter. The sender wished to remain anonymous because he did not "know how far the long arm of PG&E [Pacific Gas and Electric Co.] reaches in this matter." The nation's largest utility had approached Rose Gaffney, the owner of a large tract of land on California's Bodega Head peninsula, to inquire into the purchase of her property for an electric generating facility. This news disturbed Wayburn.

Fifty miles north of San Francisco, Bodega is an isolated area of scenic charm that the State Division of Beaches and Parks had planned to acquire. The PG&E representatives told Gaffney that the parks division had withdrawn its interest. Although Gaffney responded that she had no desire to sell to PG&E, the aging woman actually had little choice. States had long ago delegated their power of eminent domain to electric utilities, since power plants were considered to be in the public interest. On May 23, N. R. Sutherland, the utility's president, announced that land acquisition was underway.

PG&E was about to embark on the construction of the nation's first commercially-viable nuclear power plant. Their efforts, however, became mired in concerns for the environment, questions of safety, and accusations of a conspiracy to subvert democratic institutions. PG&E emerged from the first significant citizens' battle over nuclear power with its reputation scarred and a new respect for the power of the conservation

From Thomas Wellock, "The Battle for Bodega Bay: The Sierra Club and Nuclear Power, 1958–1964," *California History* 71 (Summer 1992): 192–211, 289–91. Reprinted by permission of the California Historical Society.

movement that was coming of age politically in the 1960s. The controversy illuminates the growing dissatisfaction with the power relationships inherent in the existing political order and the identity crisis the environmental movement faced when confronted with social and internal expansion.

The issue emerged at a pivotal time in the Sierra Club's history. The protests in the early 1960s against the Bodega Head plant led directly to the notorious Diablo Canyon negotiations with PG&E that split the club and forced Executive Director David Brower's resignation in the late 1960s. Even before the Diablo Canyon controversy, Bodega clarified the issues of nuclear power and wiped away any ambivalence some club members felt about opposing nuclear energy. At the same time, other club members persisted in the post–World War II faith in nuclear energy as a relatively benign alternative to using fossil fuels and building more hydroelectric dams in wilderness areas. The earlier controversy thus established the terms of debate that, in the subsequent project at Diablo, proved impossible for the club to solve. Bodega marked the beginning of a transition in the traditional conservationist agenda of preserving a remnant of scenic lands for public use favored by older Sierra Club members, to a new generation's concern for the influence of economic growth and technology on the environment generally. The old guard believed their position best preserved the club's nonpartisan, tax-exempt status and cooperative relations with government and industry. In abandoning this position, the Bodega Head insurgents created an ad hoc organization that was committed to political activism. . . .

After being alerted to PG&E's plans for Bodega Bay, Edgar Wayburn spent the Labor Day weekend in 1958 camping near the head. He met with Gaffney and viewed the Indian artifacts she had collected from the area. He strolled among the ecological communities interspersed among its high granite cliffs and sand dunes. . . . Convinced that Bodega Head's splendor needed to be saved, Wayburn attended the first meeting of the new Redwood Chapter of the Sierra Club and encouraged them to spearhead the opposition to construction of any power plant. The club passed a resolution at the summer 1958 board of directors meeting in Yosemite National Park directing that "action be taken for [Bodega's] immediate acquisition" by the state.

The club's quick moves gave the appearance that it would fight for Bodega as it had on issues of aesthetics and wilderness preservation in the past decades. . . .

Yet it was PG&E who found allies in local business groups and politicians. Hostile to environmental concerns, the Sonoma County Board of

Supervisors held to the traditional conviction that scenic areas looked best when developed. Supervisor E.J. "Nin" Guidotti remarked that "it just didn't make sense . . . to have 'a beautiful area' [like Bodega] and just leave it undeveloped forever." With a power plant Bodega would "be developed a lot faster than if the PG&E plant were not located there." County Administrator Neal Smith thought the plant's power lines through Doran Park would look "artistic." He derided opposition to the plan as "just a lack of understanding." To cultivate their support, PG&E treated the supervisors to junkets at plants near Morro Bay and Hunter's Point. Officials declared that "a public hearing was not . . . necessary." In November 1958 the board ignored 1,300 petitions from angry residents and approved a use permit for the Bodega plant by a vote of four to one.

PG&E refused to disclose whether the plant would be coal or nuclear, but it was clear that the utility favored nuclear. . . . Further indication of their plans to go nuclear came when the utility moved the site of the plant on the head a sufficient distance away from the San Andreas fault to comply with Atomic Energy commission regulations. . . . The AEC [seemed] more interested in accelerating atomic construction than investigating earthquake hazards. . . .

At the time, even to friends of wilderness preservation, nuclear power seemed to be an answer, not a problem. The hazards of nuclear power were little understood by the public, and it was considered a "clean" alternative to fossil fuels. Environmentalists particularly embraced the new technology as a way of saving the wilderness from large dam projects. In the club's successful fight to save the Dinosaur National Monument from a dam at Echo Park, David Brower, the club's executive director, claimed that wilderness-preserving alternatives such as nuclear power could serve as a substitute. . . .

In the summer of 1961, PG&E removed any remaining doubt regarding what type of plant Bodega would be. In simultaneous announcements the Atomic Energy Commission declared that nuclear fuel costs would be cut 34 percent, and PG&E revealed that Bodega would be a 325 mega-watt Boiling Water Reactor. Bodega was to be a ground-breaking facility. It would be the first economically-competitive commercial reactor in the industry. The peaceful atom had reached maturity. . . .

For the moment, the way seemed clear for approval of the Bodega plant. The Public Utilities Commission's hearings held in San Francisco in early 1962 attracted little interest and were sparsely attended. Even the Sierra Club was absent. Joel Hedgpeth, who had placed his faith in the club, denounced its failure to act. In a letter to the PUC he charged that the club had "betrayed the memory of its patron saint, John Muir, who fought Hetch-Hetchy on his deathbed." Alfred Hitchcock's filming of "The Birds" in Bodega Bay was as much of interest in local circles as was the nuclear power issue. That both events illustrated the theme of man's relationship to nature seems to have gone unnoticed.

Help came from other quarters, however. Karl Kortum, director of the San Francisco Maritime Museum, used his friendship with the *Chronicle*'s editor, Scott Newhall, to have a letter on the Bodega plant prominently displayed on the editorial page. Kortum accused PG&E of subverting the democratic process and of being callous "demigods." He appealed for concerned citizens to "take five minutes to write a letter" to the PUC. The public response was so overwhelming that the PUC rescheduled additional hearings for May 1962. Rose Gaffney, whose property had been seized by PG&E, enlivened the new proceedings with a colorful slide show of Bodega Head, and frequent outbursts from her seat in the audience aimed at public officials. . . .

Young David Pesonen typified the new activists in the 1960s. Articulate, motivated, idealistic, and committed to environmentalism, he earned a degree in biological sciences from Berkeley in 1960. . . . A self-described "non-violent anarchist," he held to a philosophy that reflected the New Left ideology of decentralization, participatory democracy, and deep distrust of elites controlling technology. . . .

In conservation matters, Pesonen feared the speculative suburban development that threatened the proposed Point Reyes National Park and Bodega Head. He scoffed at government agencies that contended "it is not growth itself that is the problem, but the pattern of growth." . . . In his battle against the nuclear plant, Pesonen became the executive secretary of the foundering Northern California Association to Preserve Bodega Head and Harbor. . . .

But the fight did seem lost. . . . The state PUC gave its approval to PG&E subject to thermal pollution and radiation studies. Pesonen's association responded to the crisis in late 1962 by devising a strategy that followed three paths—citizen protest, legal intervention, and impugning the integrity of public officials. Pesonen and other members filed four appeals to the PUC requesting a new hearing; all were denied. . . .

By early 1963, this message of public activism and decentralized decision making espoused by the Association to Preserve Bodega Head drew in a broad array of activists. The movement's organizers ranged from far-right libertarians to former members of the IWW. Through leafleting and door-to-door visits, association activists were able to tap into a general discontent with a local political system that was so secretive that even boards of education held closed meetings. Sonoma State College students held a sympathy march in May to protest the "travesty" perpetrated by the supervisors. They called for the "spontaneous thinking and the liberal and individualistic freedom of all mankind." The students wanted each citizen to "have the privilege and also the responsibility to voice his own reasoned opinion in any public matter."

Political considerations alone could not have drawn a majority of the residents to the cause. Many of the local ranchers and farmers were not interested in politics, but were concerned with their health and incomes. . . . Fear of radioactive contamination served the association's purposes perfectly. In January 1963 they disseminated an analysis of the growing dangers of technology. Using the "food-chain argument" popularized by Rachel Carson's recent publication of *Silent Spring*, they warned that cows nearby Bodega could ingest radioactive contamination from grass. . . .

The Bodega Head plant received its death blow in August 1963, at the hands of Pesonen's association. Pesonen hired geologist Pierre Saint-Amand to consult the organization on the suitability of the excavated site at Bodega, which lay perilously close to the San Andreas Fault system. As Doris Sloan recalls "it was one of the high points of my life" when she escorted the geologist to Bodega Head on a cold rainy day. "I couldn't believe my eyes . . . we came around the corner and the gate was open and there was no PG&E person in the little gatehouse to keep you from walking in." As the two slogged through the mud toward the empty site, Saint-Amand exclaimed "'ooh' and . . . 'ah,'" as his eyes came to rest on a "spectacular" earthquake fault slicing through the excavation. At a press conference, Amand described the fault and the heavily fractured nature of the site, which, in his opinion, was sufficient cause to abandon the project. "A worse foundation condition," Amand concluded, "would be difficult to envision." . . .

The end was fast approaching. PG&E suspended construction in October 1963, when the fault controversy erupted. As Pesonen's association celebrated in October 1964 its first annual "Empty Hole in the Head Day," the AEC released four separate reports from various government bureaus, three of which found the site unsuitable. Even the AEC staff recommended against construction. The staff concluded that PG&E's new design was untried and could not provide reasonable assurance against earthquake hazards. The utility took Pesonen's advice to "bow out gracefully" and cancelled construction on October 30, 1964. The Sierra Club's president, Will Siri, praised the utility's "public spirited" decision and called for the establishment of a state park at the site. . . .

The changes wrought in the 1960s were also structural, and here Bodega foreshadowed the decentralization of club authority. While some dismiss the New Left of the 1960s and its accomplishments, single-issue activism has come to dominate the political landscape. Historians consider the environmental movement to be palpable evidence of a resurgent desire for participatory democracy. Bodega was a precursor of this trend.

Further Readings

Bartley, Ernest R. *The Tidelands Oil Controversy: A Legal and Historical Analysis*. Austin: University of Texas Press, 1953.

Bottles, Scott L. *Los Angeles and the Automobile: The Making of the Modern City*. Berkeley: University of California Press, 1987.

Brigham, Jay. *Politics and Power: The Fight for Electricity in the West, 1902–1932*. Lawrence: University Press of Kansas, 1997.

Coleman, Charles M. *PG&E of California: The Centennial Story of Pacific Gas and Electric Company, 1852–1952*. New York: McGraw-Hill, 1952.

Easton, Robert. *Black Tide: The Santa Barbara Oil Spill and Its Consequences*. New York: Delacorte Press, 1972.

Engelbert, E. A., and A. F. Scheuring, eds. *Competition for California Water*. Berkeley: University of California Press, 1982.

Ford, Daniel. *Meltdown: The Secret Papers of the Atomic Energy Commission*. New York: Simon & Schuster, 1982.

Fowler, Frederick H. *Hydroelectric Power Systems of California and Their Extensions into Oregon and Nevada*. Washington, DC: United States Geological Survey, 1923.

Fradkin, Philip. *A River No More: The Colorado River and the West*. Berkeley: University of California Press, 1996.

Freudenburg, William R., and Robert Gramling. *Oil in Troubled Waters: Perception, Politics, and the Battle over Offshore Drilling*. Albany: State University of New York, 1994.

Hay, Duncan. *Hydroelectric Development in the United States, 1880–1940*. Washington, DC: Task Force on Cultural Resource Management, 1991.

Holdren, John, and Philip Herrera. *Energy: A Crisis in Power*. San Francisco: Sierra Club Books, 1971.

Hutchinson, W. H. *Oil, Land and Politics: The California Career of Thomas R. Bard*. Norman: University of Oklahoma Press, 1965.

Jackson, Donald C. *Building the Ultimate Dam: John S. Eastwood and the Control of Water in the West*. Lawrence: University Press of Kansas, 1995.

Lillard, Richard G. *Eden in Jeopardy: Man's Prodigal Meddling with His Environment: The Southern California Experience*. New York: Alfred Knopf, 1966.

Lotchin, Roger. *Fortress California, 1910–1961: From Warfare to Welfare*. New York: Oxford University Press, 1992.

Mazuzan, George T., and J. Samuel Walker. *Controlling the Atom: The Beginnings of Nuclear Regulation, 1946–1962*. Berkeley: University of California Press, 1984.

Misrach, Richard. *Bravo 20: The Bombing of the American West*. Baltimore: Johns Hopkins Press, 1990.

Mosier, Dan L. *Harrisville and the Livermore Coal Mines*. San Leandro, CA: Mine Road Books, 1978.

Mumford, Lewis. *Technics and Civilization*. New York: Harcourt, Brace, 1934.

Novick, Sheldon. *The Electric War, The Fight over Nuclear Power*. San Francisco: Sierra Club Books, 1976.

Rubin, David M., and David P. Sachs. *Mass Media and the Environment: Water Resources, Land Use, and Atomic Energy in California*. New York: Praeger, 1973.

Salzman, Ed, ed. *California Environment and Energy: Text and Readings on Contemporary Issues*. Sacramento: California Journal Press, 1980.

Scott, Allen. *Technopolis: High Technology Industry and Regional Development in Southern California*. Los Angeles: University of California Press, 1993.

Sinclair, Upton. *Oil!* Berkeley: University of California Press, 1997.

Smeloff, Ed, and Peter Asmus. *Reinventing Electric Utilities: Competition, Citizen Action, and Clean Power*. Washington, DC: Island Press, 1997.

Solnit, Rebecca. *Savage Dreams: A Journey into the Hidden Wars of the American West*. San Francisco: Sierra Club Books, 1994.

Teller, Edward. *Energy from Heaven and Earth*. San Francisco: Freeman and Co., 1979.

Tygiel, Jules. *The Great Los Angeles Swindle: Oil, Stocks, and Scandal During the Roaring Twenties*. Berkeley: University of California Press, 1996.

White, Gerald T. *Formative Years in the Far West: A History of the Standard Oil Company of California and Predecessors through 1919*. New York: Appleton-Century-Crofts, 1962.

White, Gerald T. *Scientists in Conflict: The Beginnings of the Oil Industry in California*. San Marino, CA: The Huntington Library, 1968.

Williams, James C. *Energy and the Making of Modern California*. Akron, OH: The University of Akron Press, 1997.

Chapter 11

SECOND NATURE: CALIFORNIA CITIES

DOCUMENTS

Samuel Wood and Alfred E. Heller on California Going, Going, . . . , 1962

Over the years the tunes change, the rhythms vary, but the message remains: at the western shore of the American continent there lies a temperate land of unlimited beauty and unlimited bounty, which may be shared by all who choose to follow the sunset.

The message remains today; we still sing in praise of the golden state, notwithstanding the smog, the water pollution, the crowded roads, the dirty blighted cities, the disappearing open space.

Perhaps we sing out of nostalgia—for the old uncrowded ways of life. Most likely, and hopefully, we sing out of belief—California is a unique bright land—and somehow or other we must keep her so.

There is, however, a limit to our credulity: how polluted can a bright land become, and still be bright? The answer to that question is being written right now, across the surface of this chaotically growing state.

From Samuel Wood and Alfred E. Heller, *California Going, Going, . . . , : Our State's Struggle to Remain Beautiful and Productive* (Sacramento: California Tomorrow, 1962), pp. 5–6.

The evidence is mounting that Californians are beginning to recognize that the great asset of their state, the very goose that has laid and will lay the golden eggs of their pleasures and profits, is their golden land. This land, our bright land—the charm of its open spaces, the vitality of its soils—is the true economic base of our state, its attraction as a place to live.

Within the past decade, at all levels of government, there have been marked efforts to control the development and the uses of California land, in order to conserve it and protect it from unnecessary encroachments of new towns, new people, new roads, new sewage—to protect it, that is, from us. In other words, some plans have been laid, some laws and policies made.

In certain ways, California is America's most progressive state. Each county in California is now required under law to have a planning commission and to develop for itself a "master" plan for the use of its lands. Each city planning commission must do the same. Dozens of California cities and counties are carrying on active planning programs and adopting policies which say: this area is especially suitable for homes and gardens, this for industry, this for park land, this for schools, this for agriculture, and so on.

The various departments of state government are hard at work planning for land uses. The Division of Highways, for example, in fighting out where freeways must go, is trying to plan to meet the highway needs of millions of people decades from now. On its part, the federal government is giving generous financial aid to California communities willing to plan for the future of their lands. . . .

In spite of all efforts to the contrary, California's unique bright land is increasingly defiled by badly located freeways and housing subdivisions and industries which needlessly destroy beautiful scenery and entomb agricultural land; by reservoirs and water courses which unwittingly encourage the growth of mislocated communities; by waste products; by cars and jeeps and cycles which pre-empt our very living and breathing space. Already, the state's nose is bloody. How long before its whole magnificent body is beaten to deformity? How long before the bright lands are dead lands?. . . .

The story we tell is not new. But this is a great and progressive state, and we think its citizens are ready to face the compelling facts.

Harold Gilliam on the Devil and the Deep Blue Bay, 1969

Tossing the contents of the slop pail out the living room window is a practice not generally looked upon with favor these days, but that is almost exactly what local communities are doing when they heave their accumulated refuse into San Francisco Bay.

As a result, many parts of the Bay have been converted into a series of monumental garbage dumps. The most familiar, and obnoxious, of these dumps is the infamous "Candlestink Cove" area along the principal highway approach to San Francisco. Here, some years ago, near the city of Brisbane the freeway was extended across the mouth of the cove for two miles. The several hundred acres of water inshore, which might have become a prime water-recreation center, were preempted by the refuse interests and turned into a fuming sump alongside the highway.

By 1965 the filled-in cove was mercifully screened by vegetation; the aromas of refuse assaulted the nostrils of motorists less frequently than before. . . . In that same year, however, the dump was being extended across the road. The main highway entrance to San Francisco was about to be flanked by stinking garbage dumps on both sides when residents of Brisbane, resentful that their offshore waters had become San Francisco's refuse dump, revolted and voted to cancel contracts their own city officials had signed with the garbage firms. Eventually a compromise was reached, permitting completion of the filling of certain diked areas on the Bay side of the freeway but leaving most of the shoreline in open water.

There are 31 other refuse disposal sites around the shores of the Bay, "sanitary land fills" where dirt is spread over the refuse to cover it up—theoretically. The dirt has to be hauled from hills by trucks roaring through business and residential areas, creating traffic hazards, dust, and general uproar—not to mention the scarring of the hills themselves. . . .

Admittedly the problem is staggering. At the rate of four-and-a-half pounds per person per day, the city of San Francisco produces nearly

From Harold Gilliam, *Between the Devil and the Deep Blue Bay: The Struggle to Save San Francisco Bay* (San Francisco: Chronicle Books, 1969), pp. 59–67. Reprinted by permission.

1800 tons of garbage every day of the year. The entire Bay Area, with a population of some four million, yields up about 10,000 tons each day. With expected population growth, the continued accumulation would be enough, within the lifetimes of most people now living, to bury the city of San Francisco six feet deep in garbage.

It is quite possible to handle garbage in civilized ways. Incineration, the method in use in many cities, has been eliminated in the Bay Area because of the smog problem. New techniques may minimize air pollution, but incineration still leaves residues and unburnables amounting to as much as half of the original volume. It also wastes valuable organic materials. The same may be said for most schemes to dump garbage into the ocean, even if there were a fool-proof method of ensuring that the garbage would not return to litter the shores. Dumping it in remote areas is equally wasteful; the notion that unprocessed urban refuse can "fertilize" the desert is nonsense. . . . Refuse should be regarded as a resource to be conserved for use. Organic wastes came from the earth and should be returned to the earth to enrich the soil.

Richard A. Peters, of the California Department of Public Health, points out that there are at least two methods of reclaiming garbage. Both require separation of organic from inorganic refuse, which can be done either by the householder or by the processor. Metals and other inorganic material can be reclaimed for use in manufacturing.

"One method of handling organic garbage is destructive distillation," Peters reports. "It involves converting the refuse to carbon products, such as charcoal briquettes, with such side products as chemicals, coal tar, therapeutic drugs, and methane gas, which can be used to dilute natural gas.

"Another method is to convert the organic refuse into aerobic compost, which is good humus-like material and can be combined with other elements to make a complete soil additive. The product is a clean, odorless substance that does not attract insects or rodents. It can be baled and occupies one-fourth to one-third the space required for unprocessed refuse." . . .

The biggest obstacle to this method is a temporary one—the development of markets for the compost as a soil additive. Even if it were not initially possible for a city or region to sell its compost material on a commercial basis, it would be socially advantageous to distribute it below cost in order to preserve for higher uses the areas where the garbage might be dumped—whether the Bay or the canyons and ravines used by some communities for disposal.

The technicians have done their job in developing disposal processes.

It is now up to the economists and public officials to do theirs. It is time our garbage disposal methods were brought into the 20th century. . . .

Immediately before and during World War II, the volume of sewage going into the Bay had reached such proportions that fish were killed and the stench of the pollution spread for many miles beyond the shores. University of California students during those years, whiffing the aromas drifting across the campus, five miles from the Bay, dubbed the fragrance the "East Bay Stink."

The situation became such a menace to health that the State Department of Public Health and the San Francisco Bay Regional Water Quality Control Board, created in 1949, insisted that local governments build sewage treatment plants. As a result the water quality has improved; fish are more abundant; and the "East Bay Stink" has almost entirely disappeared. Yet the sewage problem is still serious.

Increases in population, industry, and home garbage-disposal units have resulted in large volumes of sewage, requiring continual expansion of facilities. Local officials are reluctant to spend money for this purpose and have to be prodded continually by the Water Quality Control Boards to do so. The consequent lag keeps the Bay dirty.

A further source of trouble is the Sacramento River, and other tributary waters; the board has no control over pollution from upstream. And raw sewage is still dumped freely into the Bay from ships and small boats. The city of San Francisco is possibly the worst offender. Its sewers are combined with storm drains. During heavy rains, the runoff overloads the treatment plants, and raw sewage of every description litters the beaches of the Golden Gate.

Another pollution problem is that filling of the shallow areas further reduces the Bay's capacity to handle the sewage that is dumped into it, processed or otherwise. Because of stronger sunlight, a large expanse of shallow water oxidizes sewage much more quickly than does the same volume of deep water. Developers often claim that even if all the shoreline waters were filled there would still be plenty of Bay left. The argument is spurious. These shoals are essential to the Bay's chain of life, and their necessity for health and sanitation constitutes an additional reason for their preservation. Eliminate the shallows and the Bay would become a stinking cesspool.

Still another problem is the plan to dump waste agricultural waters from the San Joaquin Valley into the Bay through the San Luis Drain. These runoff waters, high in salts, fertilizers, and the residue of pesticides sprayed on crops, would be transported down the west side of the valley and poured into the Bay at Antioch.

The Federal Water Pollution Control Administration has concluded that this proposed drain would be harmful to the Bay and has recommended that no discharge from the drain be permitted until 1972. Meantime sewage treatment plants could be constructed and tested. . . .

In the long run, in this region of water shortages, it would seem that the only prudent way to dispose of sewage would be to reclaim it for use. It has long been evident that sewage can be "purified" into fresh drinking water and the residues used for soil enrichment. Like solid waste, liquid sewage is a potential resource to be recycled back into the economy. With the growing threat that the human population will outrun available resources, it is no longer feasible to waste any element that may be vital to human survival.

In any case it is possible to build sewage facilities that would make all parts of the Bay safe for swimming and other water contact sports. The cost would doubtless be high. But the issue is whether this Bay is to continue to be used as a convenient sewer or whether it has higher uses for which residents should be willing to pay a price.

The California Environmental Quality Act, 1973

Introduction: Overview of the California Environmental Review and Permit Approval Process

The California Environmental Quality Act (CEQA) was enacted in 1973 as a system of checks and balances for land-use development and management decisions in California.

Environmental review is characterized by an Environmental Impact Report (EIR). Without an EIR, CEQA would be ineffectual. The EIR records the scope of the applicant's proposal and analyzes all its known environmental effects. Project information is used by state and local permitting agencies in their evaluation of the proposed project.

In 1977, the California Legislature passed the California Permit Streamlining Act (CPSA) and established the Governor's Office of Per-

From CEQA: *California Environmental Quality Act: Statutes and Guidelines* (Sacramento: Governor's Office of Planning and Research, 1992), pp.1–3. Reprinted by permission.

mit Assistance (OPA). The creation of both OPA and CPSA sought to remedy a complicated and often unresponsive permitting processes. The Permit Streamlining Act addressed some of CEQA's shortcomings: namely, that it lacked a calendar by which applicants and the public could expect the prompt review of a given project. The CPSA added time-lines and deadlines to expedite government review of proposals. While this did not guarantee the approval of projects or their favorable review, it did give applicants and the public an orderly, standardized process for filing reports and actions.

California's environmental review is rigorous by anyone's standards. In most cases it extends beyond federal statutes established under the National Environmental Policy Act (NEPA).

Cities and Counties Regulate Land Use by Way of Planning, Zoning, and Subdivision Controls. There are currently 58 counties and 468 incorporated cities in California, EACH with the same authority for land use regulation. Local government authority is granted by State law. Cities and counties have legislative power to adopt local ordinances and rules consistent with state law.

State Agencies Regulate the Private Use of State Land, Resources and Certain Activities of Statewide Significance. There are at least 21 state agencies which are or may be directly involved in the approval of development projects. The permitting authority of each state agency is established by statute, usually with additional administrative rules promulgated by the agency.

Federal Agencies Have Permit Authority over Activities on Federal Lands and over Certain Resources which have been the subject of congressional legislation: i.e., air and water quality, wildlife, and navigable waters. The U.S. Environmental Protection Agency generally oversees the federal agencies. In addition, the EPA regulates activities such as the disposal of toxic wastes and the use of pesticides. The responsibility for implementing some federal regulatory programs, such as those for air and water quality and toxics management, has been delegated to specific state agencies.

The Development Permit Process

In California, the development permit process is coordinated with the environmental review process under CEQA. Every development project, which is not exempt from CEQA, must be analyzed by the lead agency to determine the potential environmental effects of the project. This analysis is required by state law. . . .

Once the lead agency is identified, all other involved agencies, whether state or local, become responsible or trustee agencies. Responsible and trustee agencies must consider the environmental document prepared by the lead agency and do not, except in rare instances, prepare their own environmental documents. The procedure for issuing each particular development permit is governed by the particular law which establishes the permit authority and by the California Permit Streamlining Act.

Summary of the CEQA and Permit Application Process

There are three major phases in the development process as provided by CEQA: the Pre-Application Phase, the Application Phase, and the Review Phase.

I. Pre-Application Phase

The Pre-Application Phase begins when the developer-applicant has completed the conceptual and preliminary design work for a project and is ready to prepare a project proposal. At this point, enough information should be available to describe project activities and to identify the project's proposed location. The primary objective of this phase is to identify the appropriate permitting agencies and to collect as much relevant background information possible. . . .

By the end of the pre-application phase, the developer-applicant should have a good understanding of the detailed project information required, a list of probable permitting agencies, and an indication of the degree of environmental analysis required by the agencies.

At this point, the applicant will learn which agency (if there will be more than one permitting agency) will be the "lead agency." The lead agency is the single agency responsible for determining the type of environmental analysis CEQA requires. In addition, the lead agency must prepare the environmental review document it calls for.

The agency with the greatest authority over the project will usually assume the lead agency role. Criteria for determining the lead agency are provided in the CEQA Guidelines at Section 15051. In the event of a dispute over the lead agency status between or among agencies, the Office of Planning and Research may designate the lead. However, once the lead agency is identified, all other involved agencies, whether state or local, become responsible or trustee agencies. Responsible and trustee agencies must consider the environmental document prepared by the lead agency and do not, except in rare instances, prepare their own environmental documents.

II. The Application Phase

The Application Phase begins with the filing of the necessary permit application forms and/or a request for a land use decision (if required) along with a detailed project description. Supporting documents must also be filed, where CEQA requires, with the respective agencies. . . .

Once an application is accepted as complete, the lead agency has one year to approve or disapprove a project for which an Environmental Impact Report (EIR) will be prepared. The time limit in all other cases, including all actions by responsible Agencies, is six months.

III. Review Phase

The Review Process begins immediately with the completion of the specific application. In recognition of §65941 of Chapter 4.5 of the Permit Streamlining Act, the lead agency will simultaneously review the project under the applicable permit rules and conduct the necessary environmental analysis. Permit rules vary depending on the particular permit authority in question, but the process generally involves comparing the proposed project with existing statutes. The procedure usually results in a public hearing followed by a written decision by the agency or its designated officer. Typically, a project may be approved, denied, or approved subject to specified conditions.

Esther Gulick, Catherine Kerr, and Sylvia McLaughlin on Saving San Francisco Bay, 1988

Kay Kerr, Sylvia McLaughlin and I [Ester Gulick] are very grateful for the kind words expressed in [the] introduction [to the University of Califor-

From Esther Gulick, Catherine Kerr, and Sylvia McLaughlin, "Saving San Francisco Bay: Past, Present, and Future," in *The Horace M. Albright Lectureship in Conservation* (Berkeley, CA: College of Natural Resources), no. 28, April 14, 1988, pp. 1–10. Reprinted by permission.

nia Albright Lecture]. We have had other names bestowed upon us, such as enemies of progress, impractical idealists, do-gooders, posy pickers, eco-freaks, enviromaniacs, little old ladies in tennis shoes and almond cookie revolutionaries.

We were very much concerned about what was happening to our Bay, but it was a happenstance that we came together. Kay and Sylvia were talking at a tea and they discussed the Army Corps of Engineers' map that had been printed in the *Oakland Tribune* in December, 1959. It showed that the Bay could end up being nothing but a deep water ship channel by the year 2020 because of the enormous amount of fill being planned. The Corps had made a study for the United States Department of Commerce and it had just been released. Shortly before Christmas in 1960 I took some almond cookies over to Kay's house. . . . Of course, we talked about the Corps' map and Kay asked me if I thought I would have time to help do something about it. Little did I know what that was going to mean!

The three of us discussed how we could start. Berkeley's City Council and the business and industrial sector, supported by the *Berkeley Gazette*, were anxious to put over 2,000 acres of fill in the Bay so that Berkeley could almost double its size. There were grandiose plans of having an airport, industrial and commercial buildings, houses, apartments, a hotel or two, a parking lot and so on. Other cities had similar plans. We decided we would present the problems to some of the leaders of the larger conservation organizations. We asked a group of them to meet at my home in January 1961. They were very interested and thought it was an excellent idea but, unfortunately, they all said they were much too busy to take on anything more. They thought it was essential to form another group. . . . It was clear that if anything was going to be done we were "it," so we went to work, green as grass as we were.

Acquiring members was probably the most important first step so we composed a letter and sent it to people we knew, as well as to the Berkeley names on the lists we had received. We had a wonderful response—more than 90 percent. People could not believe that the public did not own the Bay. Most citizens did not know, just as we did not until a short time before, that approximately 70 percent of the Bay was less than 20 feet deep and very susceptible to being filled. As Mel Scott, who was with the University's Institute of Governmental Studies, wrote: "To attorneys, developers, title insurance companies, manufacturers of salt and cement, innumerable government officials, members of the state legislature and many others, it is some of the most valuable real estate in California."

The State much earlier had sold a great deal of the Bay for no more than a dollar an acre. . . .

In 1850, when California was admitted to the Union, the Bay was approximately 680 square miles. A little more than 100 years later this had been reduced by dikes and fills to about 430 square miles. There were at least 40 garbage dumps on the Bay shoreline. In Berkeley we have seen discarded furniture, old automobile tires and batteries, old mattresses and many other things, in addition to garden clippings and kitchen garbage. It was a common sight to see these dumps burning and smoking. Much of the refuse also spilled over into the water. . . .

In 1962 Berkeley secured an amendment to its tideland grant. This allowed fills in the Bay to be used "for all commercial and industrial uses and purposes, and the construction, reconstruction, repair and maintenance of commercial and industrial buildings, plants and facilities, as may be specified by the City Council after public hearing." Save San Francisco Bay Association tried to prevent the passage of the amendment, but unfortunately failed. However, we got much favorable publicity and more members. At the end of 1962 our total membership was 2,500.

It became very clear that numerous cities and counties around the Bay had plans in various stages of development to fill many more square miles. In 1963 and again in 1964 Assemblyman Nicholas Petris introduced bills in the state legislature to protect the Bay with a moratorium on fill while the problem was studied. Although these bills lost, they alerted the legislature to the destructive threats to the Bay. . . .

We talked with many professional people about the proposed Berkeley fill. We spoke with engineers, economists, city planners, architects and others. Throughout the years, many of the University's faculty had donated their time and expertise to the Association. Mel Scott's book *The Future of San Francisco Bay*, written for the Institute of Governmental Studies, was an invaluable resource and the definitive book about San Francisco Bay. . . . We learned a great deal and took some of this knowledge to the City Council. They began to listen to us. Also, they were impressed at the great concern so many citizens had over the concept of substantial fill and so-called balanced development.

Experts speaking for Save-the-Bay were hard to refute. The master plan became a political issue and more and more people spoke against it. The City Council made a complete change in policy at the end of 1963. A waterfront advisory committee was set up and existing plans were replaced by an interim waterfront plan which would greatly limit fill and development. It also defined the lines of how far out the fill would extend. The composition of the City Council had changed after an election and the new members were much more conscious of what many of their constituents wanted with regard to the preservation of the Bay.

Early in 1964 Kay, who knew Senator Eugene McAteer, went to see him in San Francisco and talked about the need to protect the Bay. . . . McAteer thought that the legislature should study the problem before any action was taken. In 1964 a bill was passed to create the San Francisco Bay Conservation Study Commission. The new commission had four months and a budget of $75,000 to study the Bay fill. . . . The commission had nine members and McAteer was chairman. Twelve public hearings were held and the commission heard the views of many diverse people. There were those who had plans for expansion by dredging and filling as well as those who worried about the Bay and the lack of access to its water.

The press and the public began to pay attention. Don Sherwood, the Bay Area's most popular disc jockey . . . was highly successful in getting people to write, as were the conservation groups, and many cartons of mail were delivered daily to the legislators.

The findings and recommendations were completed on schedule and sent to the legislature in January, 1965. These recommendations and the resulting bill that was sponsored by McAteer in the Senate and Petris in the Assembly led to a major conservation battle. The study commission recommended and the bill proposed that a San Francisco Bay Conservation and Development Commission be created. The new Bay Commission would have two major responsibilities: it would have four years to prepare a plan for the Bay and its shoreline and would have the authority to grant or deny permits for all Bay filling. With the astute leadership of Senator McAteer and Assemblyman Petris, the bill passed in June of 1965 and became the McAteer-Petris Act.

That same year, in Alameda, a serious battle was lost. Utah Construction and Mining Company planned a massive development on Bay Farm Island. . . . The state . . . granted the permit and the Corps therefore immediately did the same. The Corps also approved two permits for large fills for the Port of Oakland. This took place immediately prior to the effective date of the moratorium required in the McAteer-Petris Act.

A big success story for the Bay was the Westbay lawsuit. Westbay Community Associates was a group made up of David Rockefeller, Crocker Land Co., Ideal Cement Co., and the investment banking firm Lazard Freres & Co. In 1968 Westbay intended to fill and develop an area of the South Bay as large as Manhattan Island. This was to be done along 27 miles of the San Mateo County shoreline. There was to be a port, restaurants, convention, commercial and education centers, apartment buildings, hotels and light industry. Also, there were to be parks and recreation areas to be paid for by the public. . . .

Crocker Land Company owned San Bruno Mountain and the intent was to remove a large part of it for fill. The remainder of the mountain would then be level and would be used for real estate purposes. Fortunately, the State Lands Commission saw Westbay Community Associates as a direct threat to the tidelands. . . .

In April of 1968, the State Lands Commission authorized the filing of the lawsuit. . . . One lady wrote a letter to Senator Randy Collier, who was holding the bill hostage in his committee in the Senate. She wrote, "Senator Collier, if you don't release that bill, I promise you that every bus, filled with conservationists, that has been going to Sacramento this year is going to go to your district and campaign for your defeat." Greg says apparently Senator Collier believed the lady because the next time the committee met he made himself a co-author of the bill. He pushed the bill through the Senate so fast you would not believe it.

The lawsuit went on for over nine years. What was ultimately worked out was a tripartite negotiation among the developers, the State Lands Commission, and the environmentalists. This negotiation was over the redrawing of the property lines in the Bay within San Mateo County. The result is that there has been virtually no further encroachment on these tidelands. That was a very major victory. . . .

In January, 1969, the Bay Conservation and Development Commission sent its plan to the Governor and the state legislature. A second major political battle was started. Many of the same people were involved who had fought for BCDC earlier. There was the same worry as in 1965. If the BCDC bill was not accepted within ninety days after the legislative session ended, it would automatically die. Again the letters and the telegrams poured into the legislature. Telephones were constantly busy. There were stickers on car bumpers, buttons on lapels, and Don Sherwood again went into action. Concerned citizens and their friends drove to Sacramento. Many chartered buses were used.

We were worried about Governor Reagan. Most of his support, financial and otherwise, came from Southern California, and he was not known for his interest in the environment. However, he was running for reelection the next year. Some of his people were in favor of BCDC, but others were against it. This time the people who wanted to preserve the Bay were fortunate. . . . After Reagan supported the Bay, the issue in the legislature became bipartisan, and there were Democrats and Republicans on both sides. Probably the most important sponsors were Senator Petris and Assemblyman Knox.

We helped organize a temporary coalition, the Citizens Alliance to Save San Francisco Bay, which had its office in downtown San Francisco. One of the main reasons was to have as much unanimity as possible among the environmentalists. Some were adamant that we must not compromise on anything. Others feared we would lose BCDC altogether or else get a weakened BCDC. However, the Alliance worked well toward our common goal and was successful in giving information, coordinating activities, responding to undesirable amendments, and so forth. . . . Not everything that the conservationists wanted was achieved, but it was far beyond what it could have been. . . .

One committee meeting that will never be forgotten was the hearing on John Knox's Bill #AB 2057. It was the same as the McAteer-Petris Bill. KQED telecast this hearing to the Bay Area. The meeting room was packed and the large room next to it where one could hear, but not see what was going on, was also filled. People stood out in the hall. The lawyer for Westbay spoke passionately against the bill. Finally, John Knox asked him if he had read it. He said no.

The final vote was the climax of years of hard work. The developers' highly-paid lobbyists had also been very active. During the last session some legislators were still speaking against the bill. Sylvia and I were watching this dramatic event. Under Howard Way's very astute leadership the bill passed by one vote and the Bay Conservation and Development Commission became permanent. For us, and for our thousands of supporters, this was the ultimate victory.

John McPhee Depicts a Debris Slide, 1989

In *Los Angeles versus the San Gabriel Mountains*, it is not always clear which side is losing. For example, the Genofiles, Bob and Jackie, can claim to have lost and won. They live on an acre of ground so high that they look across their pool and past the trunks of big pines at an aerial view over Glendale and across Los Angeles to the Pacific bays. The setting, in cool dry air, is serene and Mediterranean. It has not been everlastingly serene.

On a February night some years ago, the Genofiles were awakened by a crash of thunder-lightning striking the mountain front. Ordinarily, in their quiet neighborhood, only the creek beside them was likely to make much sound, dropping steeply out of Shields Canyon on its way to the Los Angeles River. The creek, like every component of all the river systems across the city from mountains to ocean, had not been left to nature. Its banks were concrete. Its bed was concrete. When boulders were running there, they sounded like a rolling Freight. On a night like this, the boulders should have been running. The creek should have been a torrent. Its unnatural sound was unnaturally absent. There was, and had been, a lot of rain.

The Genofiles had two teen-age children, whose rooms were on the uphill side of the one-story house. The window in Scott's room looked straight up Pine Cone Road, a cul-de-sac, which, with hundreds like it, defined the northern limit of the city, the confrontation of the urban and the wild. Los Angeles is overmatched on one side by the Pacific Ocean and on the other by very high mountains. With respect to these principal boundaries, Los Angeles is done sprawling. The San Gabriels, in their state of tectonic youth, are rising as rapidly as any range on earth. Their loose inimical slopes flout the tolerance of the angle of repose. Rising straight up out of the megalopolis, they stand ten thousand feet above the nearby sea, and they are not kidding with this city. Shedding, spelling, self-destructing, they are disintegrating at a rate that is also among the fastest in the world. The phalanxed communities of Los Angeles have pushed themselves hard against these mountains, an aggression that requires a deep defense budget to contend with the results. Kimberlee Genofile called to her mother, who joined her in Scott's room as they looked up the street. From its high turnaround, Pine Cone Road plunges downhill like a ski run, bending left and then right and then left and then right in steep christiania turns for half a mile above a three-hundred-foot straightaway that aims directly at the Genofiles' house. Not far below the turnaround, Shields Creek passes under the street, and there a kink in its concrete profile had been plugged by a six-foot boulder. Hence the silence of the creek. The water was now spreading over the street. It descended in heavy sheets. As the young Genofiles and their mother glimpsed it in the all but total darkness, the scene was suddenly illuminated by a blue electrical flash. In the blue light they saw a massive blackness, moving. It was not a landslide, not a mudslide, not a rock avalanche; nor by any means was it the front of a conventional flood. In Jackie's words, "It was just one big black thing coming at us, rolling, rolling with a lot of water in front of it, pushing the water, this big black thing. It was just one big black hill coming toward us."

In geology, it would be known as a debris flow. Debris flows amass in stream valleys and more or less resemble fresh concrete. They consist of water mixed with a good deal of solid material, most of which is above sand size. Some of it is Chevrolet size. Boulders bigger than cars ride long distances in debris flows. Boulders grouped like fish eggs pour downhill in debris flows. The dark material coming toward the Genofiles was not only full of boulders; it was so full of automobiles it was like bread dough mixed with raisins. On its way down Pine Cone Road, it plucked up cars from driveways and the street. When it crashed into the Genofiles' house, the shattering of safety glass made terrific explosive sounds. A door burst open. Mud and boulders poured into the hall. We're going to go, Jackie thought. Oh, my God, what a hell of a way for the four of us to die together.

The parents' bedroom was on the far side of the house. Bob Genofile was in there kicking through white satin draperies at the panelled glass, smashing it to provide an outlet for water, when the three others ran in to join him. The walls of the house neither moved nor shook. As a general contractor, Bob had built dams, department stores, hospitals, six schools, seven churches, and this house. It was made of concrete block with steel reinforcement, sixteen inches on center. His wife had said it was stronger than any dam in California. His crew had called it "the fort." In those days, twenty years before, the Genofiles' acre was close by the edge of the mountain brush, but a developer had come along since then and knocked down thousands of trees and put Pine Cone Road up the slope. Now Bob Genofile was thinking, I hope the roof holds. I hope the roof is strong enough to hold. Debris was flowing over it. He told Scott to shut the bedroom door. No sooner was the door closed than it was battered down and fell into the room. Mud, rock, water poured in. It pushed everybody against the far wall. "Jump on the bed," Bob said. The bed began to rise. Kneeling on it—on a gold velvet spread—they could soon press their palms against the ceiling. The bed also moved toward the glass wall. The two teen-agers got off, to try to control the motion, and were pinned between the bed's brass railing and the wall. Boulders went up against the railing, pressed it into their legs, and held them fast. Bob dived into the muck to try to move the boulders, but he failed. The debris flow, entering through windows as well as doors, continued to rise.

Escape was still possible for the parents but not for the children. The parents looked at each other and did not stir. Each reached for and held one of the children. Their mother felt suddenly resigned, sure that her son and daughter would die and she and her husband would quickly follow. The house became buried to the eaves. Boulders sat on the roof. Thirteen automobiles were packed around the building, including five in the pool. A din of rocks kept banging against them. The stuck horn of a buried car was blaring. The family in the darkness in their fixed tableau watched one another by the light of a directional signal, endlessly blinking. The house had filled up in six minutes, and the mud stopped rising near the children's chins.

A metropolis that exists in a semidesert, imports water three hundred miles, has inveterate flash floods, is at the grinding edges of two tectonic plates, and has a microclimate tenacious of noxious oxides will have its priorities among the aspects of its environment that it attempts to control. For example, Los Angeles makes money catching water. In a few days in 1983, it caught twenty-eight million dollars' worth of water. In one period of twenty-four hours, however, the ocean hit the city with twenty-foot waves, a tornado made its own freeway, debris flows poured from the San Gabriel front, and an earthquake shook the region. Nature's invoice was forty million dollars. Later, twenty million more was spent dealing with the mountain debris.

There were those who would be quick—and correct—in saying that were it not for the alert unflinching manner and imaginative strategies by which Los Angeles outwits the mountains, nature's invoices at such times would run into the billions. The rear-guard defenses are spread throughout the city and include more than two thousand miles of underground conduits and concrete-lined open stream channels—a web of engineering that does not so much reinforce as replace the natural river systems. The front line of battle is where the people meet the mountains—up the steep slopes where the subdivisions stop and the brush begins.

Strung out along the San Gabriel front are at least a hundred and twenty bowl-shaped excavations that resemble football stadiums and are often as large. Years ago, when a big storm left back yards and boulevards five feet deep in scree, one neighborhood came through amazingly unscathed, because it happened to surround a gravel pit that had filled up instead. A tungsten filament went on somewhere above Los Angeles. The county began digging pits to catch debris. They were quarries, in a sense, but exceedingly bizarre quarries, in that the rock was meant to come to them. They are known as debris basins. Blocked at their downstream ends with earthfill or concrete constructions, they are also known as debris dams. With clean spillways and empty reservoirs, they stand ready to capture rivers of boulders—these deep dry craters, lying close above the properties they protect. In the overflowing abundance of urban nomenclature, the individual names of such basins are obscure, until a day when they appear in a headline in the *Los Angeles Times*: Harrow, Englewild, Zachau, Dunsmuir, Shields, Big Dalton, Hog, Hook East, Hook West, Limekiln, Starfall, Sawpit, Santa Anita. For fifty miles, they mark the wild boundary like bulbs beside a mirror. Behind chain links, their idle ovate forms more than suggest defense.

They are separated, on the average, by seven hundred yards. In aggregate, they are worth hundreds of millions of dollars. All this to keep the mountains from falling on Johnny Carson.

The principal agency that developed the debris basins was the hopefully named Los Angeles County Flood Control District, known familiarly through the region as Flood Control, and even more intimately as Flood. ("When I was at Flood, one of our dams filled with debris overnight," a former employee remarked to me. "If any more rain came, we were going to have to evacuate the whole of Pasadena.") There has been a semantic readjustment, obviously intended to acknowledge that when a flood pours out of the mountains it might be half rock. The debris basins are now in the charge of the newly titled Sedimentation Section of the Hydraulic Division of the Los Angeles County Department of Public Works. People still call it Flood. By whatever name the agency is called, its essential tactic remains unaltered. This was summarized for me in a few words by an engineer named Donald Nichols, who pointed out that eight million people live below the mountains on the urban coastal plain, within an area large enough to accommodate Philadelphia, Detroit, Chicago, St. Louis, Boston, and New York. He said, "To make the area inhabitable, you had to put in lined channels on the plain and halt the debris at the front. If you don't take it out at the front, it will come out in the plain, filling up channels. A filled channel won't carry diddly-boo."

To stabilize mountain streambeds and stop descending rocks even before they reach the debris basins, numerous crib structures (barriers made of concrete slats) have been emplaced in high canyons—the idea being to convert plunging streams into boulder staircases, and hypothetically cause erosion to work against itself. Farther into the mountains, a dozen dams of some magnitude were built in the nineteen-twenties and thirties to control floods and conserve water. Because they are in the San Gabriels, they inadvertently trap large volumes of debris. One of them—the San Gabriel Dam, in the San Gabriel River—was actually built as a debris-control structure. Its reservoir, which is regularly cleaned out, contained, just then, twenty million tons of mountain.

The San Gabriel River, the Los Angeles River, and the Big Tujunga (Bigta Hung-ga) are the principal streams that enter the urban plain, where a channel that filled with rock wouldn't carry diddly-boo. Three colossal debris basins—as different in style as in magnitude from those on the mountain front—have been constructed on the plain to greet these rivers. Where the San Gabriel goes past Azusa on its way to Alamitos Bay, the Army Corps of Engineers completed in the late nineteen-forties a dam ninety-two feet high and twenty-four thousand feet wide—this to stop a river that is often dry, and trickles most of the year. Santa Fe Dam, as it is called, gives up at a glance its own story, for it is made of boulders that are shaped like potatoes and are generally the size of watermelons. They imply a large volume of water flowing with high energy. They are stream-propelled, stream-rounded boulders, and the San Gabriel is the stream. In Santa Fe Basin, behind the dam, the dry bed of the San Gabriel is half a mile wide. The boulder-strewn basin in its entirety is four times as wide as that. It occupies eighteen hundred acres in all, nearly three square miles, of what would be prime real estate were it not for the recurrent arrival of rocks. The scene could have been radioed home from Mars, whose cobbly face is in part the result of debris flows dating to a time when Mars had surface water.

The equally vast Sepulveda Basin is where Los Angeles receives and restrains the Los Angeles River. In Sepulveda Basin are three golf courses, which lend ample support to the widespread notion that everything in Los Angeles is disposable. Advancing this national prejudice even further, debris flows, mudslides, and related phenomena have "provided literary minds with a ready-made metaphor of the alleged moral decay of Los Angeles." The words belong to Reyner Banham, late professor of the history of architecture at University College, London, whose passionate love of Los Angeles left him without visible peers. The decay was only "alleged," he said. Of such nonsense he was having none. With his "Los Angeles: The Architecture of Four Ecologies," Banham had become to this deprecated, defamed, traduced, and disparaged metropolis what Pericles was to Athens. Banham knew why the basins were there and what the people were defending. While all those Neurasthenic literary minds are cowering somewhere in ethical crawl space, the quality of Los Angeles life rises up the mountain front. There is air there. Cool is the evening under the crumbling peaks. Cool descending air. Clean air. Air with a view. "The financial and topographical contours correspond almost exactly," Banham said. Among those "narrow, tortuous residential roads serving precipitous house-plots that often back up directly on unimproved wilderness" is "the fat life of the delectable mountains."

People of Gardena, Inglewood, and Watts no less shall have Azusa and Altadena pay for the defense of the mountain front, the rationale being that debris trapped near its source will not move down and choke the channels of the inner city, causing urban floods. The political City of Los Angeles—in its vague and tentacular configuration—actually abuts the San Gabriels for twenty miles or so, in much the way that it extends to touch the ocean in widely separated places like Venice, San Pedro, and Pacific Palisades. Los Angeles County reaches across the mountains and far into the Mojave Desert. The words "Los Angeles" as generally used here refer neither to the political city nor to the county but to the multinamed urban integrity that has a street in it seventy miles long (Sepulveda Boulevard) and, from the Pacific Ocean at least to Pomona, moves north against the mountains as a comprehensive town.

The debris basins vary greatly in size—not, of course, in relation to the populations they defend but in relation to the watersheds and washes above them in the mountains. For the most part, they are associated with small catchments, and the excavated basins are commensurately modest, with capacities under a hundred thousand cubic yards. In a typical empty reservoir—whatever its over-all dimensions may be—stands a columnar tower that resembles a campanile. Full of holes, it is known as a perforated riser. As the basin fills with a thick-flowing slurry of water, mud, and rock, the water goes into the tower and is drawn off below. The county calls this water harvesting.

Like the freeways, the debris-control system ordinarily functions but occasionally jams. When the Genofiles' swimming pool filled with cars, debris flows descended into other neighborhoods along that part of the front. One hit a culvert, plugged the culvert, crossed a road in a bouldery wave, flattened fences, filled a debris basin, went over the spillway, and spread among houses lying below, shoving them off their foundations. The debris basins have caught as much as six hundred thousand cubic yards in one storm. Over time, they have trapped some twenty million tons of mud and rock. Inevitably, sometimes something gets away.

At Devils Gate—just above the Rose Bowl, in Pasadena—a dam was built in 1920 with control of water its only objective. Yet its reservoir, with a surface of more than a hundred acres, has filled to the brim with four million tons of rock, gravel, and sand. A private operator has set up a sand-and-gravel quarry in the reservoir. Almost exactly, he takes out what the mountains put in. As one engineer has described it, "he pays Flood, and Flood makes out like a champ."

ESSAYS

Los Angeles Air

Barry Commoner

For teaching us a good deal of what we now know about modern air pollution, the world owes a great debt to the city of Los Angeles, California. There are few cities in the world with climates so richly endowed by nature and now so disastrously polluted by man. . . . The first air pollutant to cause trouble in Los Angeles was an ancient one—dust. Between 1940 and 1946 dustfall in the city increased from about 100 tons to nearly 400 tons per day. The sources were easily located: industrial smokestacks and incinerators. Beginning in 1947 control measures—the installation of dust-precipitators and prohibition of open burning—were instituted. Within two years dust-fall was down to 200 tons per day and has since declined to its prewar level.

However, in 1943 the residents of Los Angeles noticed something new in the air: a whitish haze, sometimes tinged with yellow-brown that brought with it eye-smarting and tears. This condition rapidly worsened and spread throughout the mountain-ringed basin that comprises Los Angeles County.

The Angelenos came to call the new pollutant "smog," adopting a term invented in England to describe the thick clouds which in five days in 1952 killed 4,000 Londoners—the world's worst air pollution disaster. The dangerous component in London smog was sulfur dioxide, a gas that attacks the cells lining the lungs' air passages, reducing their natural self-protective action against other air pollutants such as dust and bringing on serious respiratory distress. . . . Sulfur dioxide emissions had in fact increased sharply in the area as a result of wartime industrialization (sulfur dioxide is produced by the burning of sulfur-containing coal and fuel oil). . . . [B]eginning in 1947 sulfur dioxide emissions were gradually reduced, reaching the prewar levels by 1960. . . .

Later research revealed the whole story. It begins with nitrogen oxides, which are produced whenever air becomes hot enough (as it does in high-temperature power plants and high-powered gasoline engines) to cause its natural nitrogen and oxygen to interact. Activated by sunlight, nitrogen oxides combine with organic compounds such as waste gasoline, producing the material which is the chief visible—and noxious—final product of photochemical smog, peroxyacetylnitrate (PAN). Thus Los Angeles smog is different from London smog (named from smoke plus fog), and a strict etymology would forbid its use to describe Los Angeles air. It is some kind of tribute to the growing importance of the Los Angeles type of smog that London's word has now been taken over by this newcomer among air pollutants. As now used, at least in the United States, the word "smog" (more precisely, "photochemical smog") refers to the Los Angeles variety.

With this chemical information in hand, Los Angeles authorities were quick to seek smog-control methods. An obvious approach was to reduce emissions of hydrocarbons into the air. A ready target were the numerous oil fields, refineries, and other activities of the large petroleum industry that had sprung up in the area. Rigid controls were imposed on open venting of oil wells and refineries. Hydrocarbon emissions from the petroleum industry were sharply reduced, from about 2,100 tons per day in 1940 to about 250 tons per day in 1957.

Nevertheless, Los Angeles smog conditions continued to deteriorate. Year by year high smog levels become more frequent. In 1959 eye irritation was reported in Los Angeles County on 187 days of the year; in 1960 there were 198 such days; in 1961, 186 days; in 1962, 212 days. Despite a vigorous and successful effort to control industrial emissions of hydrocarbons, Los Angeles was still in the grip of smog.

In 1953 new evidence made the situation—if not the air—much clearer. A survey showed that while the petroleum industry was emitting about 500 tons of hydrocarbons per day, about 1,300 tons per day were being emitted by automobiles, trucks, and buses. By 1957 motor vehicles were responsible for about 80 per cent of a total emission of about 2,500 tons of hydrocarbons per day. The real culprit had been found: the automobile industry.

This was an adversary worthy of even as extravagant a city as Los Angeles. The United States automobile industry sells about $15 billion of its products each year. It is closely tied to the petroleum industry, which produces about $10 billion worth of goods annually. The automobile industry is a major segment of the nation's military and financial complex. Industry executives have served in high national office. The industry's advertising budget amounts to hundreds of millions of dollars annually, a major part of the support of news media. The industry's technical facilities are enormous; it employs thousands of highly trained engineers and scientists.

Pitted against this formidable power were a few Los Angeles officials. Probably the most single-minded among them in the battle over automotive smog has been Los Angeles County Supervisor, Kenneth Hahn. The record of his long encounter with the giants of Detroit is illuminating.

On February 19, 1953, Mr. Hahn wrote to the president of the Ford Motor Company to ask whether the company "has conducted, or is conducting, research or experimentation designed to eliminate or substantially reduce exhaust vapors." The reply, from a member of the company's News Department, asserted that "the Ford engineering staff, although mindful that automobile engines produce exhaust gases, feels these vapors are dissipated in the atmosphere quickly and do not represent an air pollution problem." . . .

By the end of 1953 the automobile manufacturers acknowledged that an industrywide study of the problem was under way. Nineteen months later Mr. Hahn wrote again, asking if an exhaust control had been developed. The reply from one of the Detroit engineering giants: "We will soon be in a position to make recommendations which should point the way to reduction of hydrocarbons in automotive exhaust gases." . . .

The correspondence continues, until on October 18, 1960, Mr. Hahn is informed by the president of the General Motors Corporation: "I am gratified to be able to report that positive crankcase ventilation is available on all 1961 General Motors passenger cars being delivered to California. We believe that this relatively uncomplicated, inexpensive device will perform a major job of reducing air pollution." Since such a crankcase device eliminates only 25 per cent of the emitted hydrocarbons (most coming from the engine exhaust), Mr. Hahn replied to express his disappointment in the action. By 1965 no action to control exhaust emissions had been taken and Mr. Hahn's correspondence concludes with an appeal to the president of the United States for congressional action. In 1966 exhaust-control devices appeared on new cars in California and the emissions of hydrocarbons from that source began a downward trend in Los Angeles.

Between 1965 and 1968 emission of waste hydrocarbons from motor vehicles was reduced from 1,950 to 1,720 tons per day (without controls, emissions would have risen to 2,400 tons per day by 1968). Eye irritation was also reduced. At the same time, the levels of another important pollutant emitted by vehicles, carbon monoxide, were also reduced, for the new devices acted on this pollutant as well as on hydrocarbons. . . .

It might appear, then, that by 1968 Los Angeles would be ready to celebrate the end of a long and frustrating search for a solution to the smog problem. But now the situation took a new and ominous turn: the improvements in exhaust emissions had themselves brought on a new problem. For, coincident with the 12 per cent drop in hydrocarbon emissions from 1965 to 1968, the burden of nitrogen oxides in Los Angeles air increased by 28 per cent. . . .

This was cause for serious concern. While nitrogen oxide is relatively innocuous (except as an ingredient of the smog-forming reaction), nitrogen dioxide is highly poisonous, with a long history as the cause of serious industrial hazards. The gas destroys the cells of the lung, tends to enlarge lung blood vessels, and at a sufficiently high concentration, causes accumulation of fluid in the lungs, which may lead to death. Such serious effects are observed only at nitrogen dioxide concentrations that are well above the levels encountered in urban air. . . . The acceptable level for brief (one hour) exposure to nitrogen dioxide in urban air has been set at 2 parts per million (ppm). This value is uncomfortably close to the peak values recorded in Los Angeles—1.3 ppm.

Nitrogen dioxide is a colored gas which tinges the air a kind of whisky brown. As the concentration has increased in Los Angeles air, serious visibility problems have arisen—in the air lanes and along the high-speed freeways. In addition, nitrogen dioxide is toxic to plants; at levels less than 1 ppm the growth of tomato plants is reduced about 30 per cent.

There are two reasons why nitrogen dioxide levels would be expected to increase as a result of efforts to control other exhaust pollutants. One is the simple ecological rule that "everything must go somewhere." The nitrogen oxide–nitrogen dioxide mixture, formed in the air by the action of sunlight, combines with waste hydrocarbon to produce PAN. The latter eventually condenses and comes down to earth as a gummy precipitate, which is easily detected on a car windshield after a short time on a Los Angeles freeway. Hence, if hydrocarbon emissions are reduced so that the rate of smog formation declines, nitrogen oxides necessarily accumulate at higher levels.

Another reason for the unexpected rise in nitrogen oxides is that in devising the present pollution controls the automobile manufacturers considered only the demand for reduced emission of hydrocarbons and carbon monoxide. This goal led them to engine modifications designed to increase the extent of fuel combustion in the cylinders by increasing the engine's air intake. This also enhances the combustion of the major constituent of the air, nitrogen, to nitrogen oxides. Thus the engine modification introduced to reduce hydrocarbon emission increases nitrogen oxide emission. In enforcing the new automobile engine modifications, Los Angeles had simply traded one pollution problem for another. . . .

The paradigm of the enormous benefits of modern technology to the common man—as distinct from astronauts and generals—is the automobile. The dividing line between its success and failure is the factory door. So long as the automobile is being constructed, technology is admirably successful. However, once the automobile is allowed out of the factory and into the environment, it reveals itself as an agent which has rendered urban air pathogenic, burdened human bodies with nearly toxic levels of carbon monoxide and lead, embedded carcinogenic particles of asbestos in human lungs, and killed and maimed many thousands annually. The human value of the automobile is created by technology and diminished by its environmental failure.

Air pollution is not merely a nuisance and a threat to health. It is a reminder that our most celebrated technological achievements—the automobile, the jet plane, the power plant, industry in general, and indeed the modern city itself—are, in the environment, failures.

Earthquakes and Freeways

William P. McGowan

For Bill Harp, the traffic that Tuesday evening in October was light. Driving on the Bay Bridge, the San Francisco investment banker remembered that the first game of the 1989 World Series was scheduled to begin at 5:15 that night; most of his colleagues had left work early to watch the game at home or at their favorite bar. For Harp, who never liked baseball,

From William P. McGowan "Fault-lines: Seismic Safety and the Changing Political Economy of California's Transit System," *California History* (Summer 1993): 170–93. Reprinted by permission of the California Historical Society.

the game provided an escape from the usual bumper-to-bumper traffic he confronted every afternoon on the way to his East Bay home. Driving south on Interstate 880, Harp ascended to the top deck of the "Cypress Structure," a double-decked section of freeway that ran through downtown Oakland. At 5:08 P.M., Bill Harp was halfway down the Cypress when the Loma Prieta earthquake struck.

As the ground shook, the Cypress began to collapse. In his rearview mirror, Harp could see freeway sections behind him falling, and hoped that he could speed his way to safety. "The freeway was literally falling apart. . . . I would dip down, go into a gully and be airborne a couple of seconds. . . . I could see that the section ahead might remain intact, and I was praying: 'Dear God, just please get me there.'" When he reached that last section, the roadway lurched upward as the supports behind his car collapsed, transforming the freeway into a jump-like ramp. Moving at approximately ninety miles an hour, Harp's car shot off the freeway like a rocket and flew through the air, crash landing on one of the last sections of the Cypress that remained standing. The impact broke Bill Harp's arm, several of his ribs, and fractured his shoulder. But, he was alive. . . .

The collapse of the Cypress Structure forced many Californians to question the safety of their freeway system. In spite of assurances by the Department of Transportation (CalTrans) that the Cypress was "earthquake safe," it took less than twenty seconds of shaking to reduce the two-level, elevated freeway into a pile of twisted steel and broken concrete. . . .

The Cypress Structure was just one of thousands of state-owned bridges considered "seismically safe" by CalTrans engineers. . . . After the earthquake, Californians wanted to know why a structure deemed "earthquake safe" crumbled so quickly.

The answer came in June 1990. In a report to Governor George Deukmejian, a board of engineering experts found that the reasons behind the Cypress Structure's collapse were largely political. Exonerating CalTrans from blame, the board found that the transportation policies of three recent state governors, including Deukmejian, had resulted in a seismic retrofit program that was constantly neglected and underfunded. . . .

Insufficient funding critically affected seismic research, because no one knew if the seismic restraining cables actually worked. The board concluded that had more money been spent on research as originally planned, CalTrans could have predicted the collapse of the Cypress Structure. . . .

How could California, a state with both a history of earthquakes and the largest transportation budget in the country, have difficulty finding enough money to ensure that its freeways were seismically safe? An answer to this question lies in California's recent political and economic history. Between 1972 and the collapse of the Cypress Structure in 1989, the "political economy" (or the inter-relationship between economic and political processes) of California's transportation system changed with the election of every new governor. . . .

The era from 1947 until 1972 has been called the "Golden Era" of freeways in California. During that quarter of a century, political and economic conditions in the state proved conducive to an expansive transportation program. The state had an abundance of transportation dollars and plenty of politicians eager to spend them. A former secretary of Business and Transportation said that during the Golden Era, a legislator "could [always] bring back a piece of concrete—whether it [was] a highway or an off-ramp," to prove his political efficacy to the folks back home. These conditions first changed in 1972, just as the seismic retrofit program was beginning.

Beginning with Ronald Reagan, the political economy of the transportation system underwent a series of radical changes. By creating the Department of Transportation, Reagan concentrated transportation policy-making power in the governor's office. Over the next three administrations, CalTrans would be used by each governor in a different way. Under Reagan, CalTrans continued an aggressive highway building campaign. But because of declining transportation revenues, the governor focused on projects that would yield the greatest political advantage. During Jerry Brown's term, the agency almost stopped building freeways altogether, and concentrated instead on maintenance and the development of mass transit systems. Finally, George Deukmejian tried to begin a new capital spending program. However, hampered by the Gann Initiative spending limit and his own conservative, small-government philosophy, Deukmejian would not endorse new taxes to pay for it.

Throughout the administrations of these three governors, engineers at CalTrans tried to complete a seismic retrofit program designed to make the freeways "earthquake safe." The seismic retrofit program (hereinafter referred to as "retrofit") was designed to prevent total structural failure in bridges and viaducts built before 1971. In that year, during the San Fernando earthquake, CalTrans engineers discovered a crucial weakness in a commonly used freeway design that made structures particularly vulnerable to earthquakes. Prior to 1971, California's freeways had been built with no connection between a bridge's road-surface and its supports;

engineers had believed that gravity alone was sufficient to keep the two attached.

The San Fernando earthquake exposed this design flaw, as several overpasses literally shook themselves apart. While no one died when the freeways collapsed, the implications were ominous: California had well over 15,000 bridges and overpasses with this design. Because it was prohibitively expensive to rebuild every bridge or overpass that posed a seismic risk, the retrofit program was developed. The program was never intended as a permanent solution to the freeway earthquake safety problem, but only as a "quick fix" that would work until the state could replace its older, seismically vulnerable structures with new ones. The engineers who developed the program based their designs on the assumption that over the next twenty years the state would replace all freeways, ramps, and bridges built with the older, gravity bonding design. . . . Why [was] the retrofit program stalled over the course of . . . three gubernatorial administrations?

The Legacy of Ronald Reagan

Ronald Reagan changed the political economy of the state's transportation system twice: once as governor, and once as president. During his second term as governor, from 1970 to 1974, Reagan created a new transportation agency that gave the governor more direct control over the policies and direction of the state's transportation system. As president, however, in order to make the federal deficit look smaller, Reagan refused to spend billions of dollars in transportation funds. . . .

Before Reagan's second term as governor, the political economy of California's transportation system was oriented toward long-term projects. From the passage of the Collier-Burns freeway bill in 1947 until the creation of CalTrans in 1972, transportation projects remained unaffected by the vicissitudes of gubernatorial leadership. Freeways remained popular with California voters; population and the use of automobiles increased rapidly; and the revenues generated by a fuel tax provided the state with a relatively painless means of funding an ever-expanding freeway network. . . . What was most interesting about this Golden Era of California freeways was that the legislature and the governor had very little direct control over the state's freeways. With the exception of a legislative review every four years, the transportation system was controlled by the California Highway Commission, a board appointed by the governor.

Created in 1947 under the Collier-Burns bill the non-partisan California Highway Commission formulated transportation policies. . . . During its first year, many observers hoped the CalTrans would bring reform to the state's transportation policies. . . . While Reagan seemed to be continuing freeway construction with an intensity equal to or greater than that formerly displayed by the Highway Commission, the revenue base supporting his transportation policies was shrinking. To accommodate this decline, Reagan's priorities throughout his second term focused on projects that would be completed before he left office. The political economy of the transportation system was changing, moving away from long-term goals and focusing on those projects that promised rapid political payback. Many long-term projects faced extremely limited funding, as money was channeled into projects that could reap the most political advantage. The structural retrofit program met Reagan's political agenda perfectly. . . .

The Ice-Plant Lady Cometh: The Impact of Jerry Brown

While Ronald Reagan changed the political economy of California's transportation system for political reasons, Jerry Brown sought change for ideological ones. The OPEC fuel crises convinced Brown that mass transit, and not freeways, was the answer to the state's transportation problems. . . . After being drained of funds by the Reagan administration, CalTrans could not have continued an aggressive freeway construction program regardless of Brown's wishes. The fiscal imbalance left by Reagan allowed Brown to accelerate the shift away from construction and toward a greater emphasis on maintenance and mass transit. Unfortunately, the budget could not afford both. . . .

The person Jerry Brown chose to implement his transportation policies was the embodiment of his anti-freeway sentiment. Adriana Gianturco—dubbed "the Ice-Plant Lady" by her political enemies because of her affinity for planting the succulent ground cover along freeways—came to CalTrans as a complete outsider, a contradiction of the transportation official stereotype. First, she was a woman running an agency staffed predominantly by middle-aged white men. Second, she was the first director of any state's freeway system who was not a civil engineer. . . . Finally, Gianturco had little tolerance for political bargaining. . . .

In spite of her abrasiveness, Gianturco seemed the director most likely to support maintenance programs like retrofit. In her first budget, she stated her priorities clearly: "This budget year [we] shift away from construction of new freeways and [turn] towards more use of existing capacity. . . . Construction will be limited to projects vital to the operational integrity of the existing system." . . .

The most logical place to find the missing transportation funds was the federal government's matching funds program. Since the 1950s, the federal government had played a large role in financing California's freeway system: every dollar that the state spent building an interstate freeway was "matched" by the federal government on a nine-to-one basis. . . . The matching funds programs fit Gianturco's agenda perfectly. It allowed her to maintain freeways, while at the same time providing liberal reimbursement for creating mass transit systems.

Although Gianturco's emphasis on matching funds projects allowed her to stretch transportation dollars, federal programs limited her options. Matching funds programs were heavily weighted toward new construction, and maintenance projects that qualified for matching funds emphasized interstate freeways. These restrictions meant that retrofit did not qualify for federal funds, since it was a relatively inexpensive process that delayed new construction, and it was aimed at both inter- and intrastate freeways. . . . Instead of proceeding with such programs, Gianturco took the money and "banked" it, holding it in reserve for future projects that might qualify for federal aid. . . .

The budget crisis at CalTrans had an especially debilitating impact on the retrofit program. Just as the crisis developed, retrofit was supposed to be completed. Retrofit was supposed to take seven years to finish, yet in its seventh year, 1979, retrofit was less than half done. The program crept along undersupervised and underfunded. . . . At the same time, Gianturco completely eliminated all CalTrans research on partially completed programs, in order to save money. With retrofit cables and bolts in well over five thousand structures, no one knew how well they would work, or if they would work at all. Seemingly, the only reason for these delays was that neither Gianturco nor the governor was willing to raise the fuel tax. They had good political reasons. Encouraged by the passage of Proposition 13 in 1978, the California public was in the throes of tax revolt. . . .

In 1981, Jerry Brown broke the budgetary impasse by accepting a fuel-tax-increase proposal from the legislature. To compensate for inflation, the legislature proposed a six-cent increase; Brown accepted only two. Recognizing the ever-growing backlog of maintenance programs, Gianturco went along with the increase, but only on the condition that maintenance projects would have first claim to the new revenues. A year later, Gianturco regretted her concession. "The legislature raised the fuel tax and claimed that the first priority should be maintenance—yet when the funds were allotted . . . over forty-nine percent went to new construction." The battle over the fuel tax would be Gianturco's last fight in Sacramento. As the 1982 elections loomed, even Brown conceded that his director of transportation had lost more battles than she won, and even he openly questioned some of her tactics. . . .

From the early 1970s, it was clear that highways alone could not solve California's transportation needs. But, with a limited budget, Brown's attempts to refocus the state's transportation goals took a heavy toll on California's existing freeway system. One of the many casualties of the Brown administration was the seismic retrofit program. Since it did not qualify for federal matching funds, retrofit was constantly stalled, while its research budget was eliminated altogether. . . . Just as Ronald Reagan had done eight years earlier, Jerry Brown left the retrofit program for the next governor.

CalTrans, Retrofit, and the "Iron Duke"

Like Jerry Brown, George Deukmejian came into the governor's office in January 1983 with a transportation agenda radically different from that of his predecessor. Once again, the political economy of the state's transportation system changed, moving away from Brown's anti-freeway approach and toward a freeway-oriented policy. Deukmejian promised California voters that he would "steer the State away from exotic alternate transportation schemes" and refocus CalTrans on its traditional tasks: building and maintenance of roads. But, once in office, the new governor was handicapped by his own "no new taxes" campaign promise, by changing federal matching funds policies, and by the legacy of Jerry Brown. . . .

Unlike governors before him, George Deukmejian could not rebuild CalTrans with huge infusions of new money because of a new amendment to the state's constitution. In 1979, California voters overwhelmingly supported Proposition 4, an initiative known as the "Gann appropriations limit," to prevent runaway spending by state government. It was the Gann limit that proved to be the most damaging limitation that Governor Deukmejian inherited. The Gann limit was based on a simple principle: the state should not be able to increase appropriations at a rate greater than population growth or the national Consumer Price Index (C.P.I.), whichever was greater. . . . For Deukmejian, who was committed to an expansive transportation program, the Gann limit was especially restrictive. . . .

Deukmejian spent his first two years in office focusing transportation money on maintenance projects that were either under way or had been stalled by the Brown administration. It was during this period that the retrofit program began its second life. . . . In 1984, $191,000 was given to engineers at UCLA in order to study how "the [retrofit] cable restrainers behaved." This was the first full-scale research project in the program's twelve-year history. To the alarm of researchers, two out of three restrainers failed. Retrofit cable design was immediately altered for future projects. But since money was scarce, CalTrans never considered replacing the numerous retrofit cables of the older, inadequate design. All of the cables on the Cypress Structure, which served the fifth busiest intersection in the United States, were of this older design. . . .

By 1987, Governor Deukmejian was forced to make some tough decisions. Handicapped by the Gann limit, denied access to federal matching funds, and refusing to raise the fuel tax, Deukmejian had to decide which programs to fund, and which programs to kill. Maintenance programs faced the most serious liability, for as Adriana Gianturco had put it, "no politician was ever praised for fixing a 'pothole.'" The retrofit program was especially vulnerable, since it lacked the high visibility that could quickly be translated into votes for its supporters. Congressman Dominic Cortese knew this from experience. In 1987, after another earthquake in Sylmar confirmed suspicions raised about retrofit's inadequacy, the San Jose Democrat pushed for special funding to complete the fifteen-year-old retrofit program. Deukmejian vetoed the request in the 1988 budget, claiming that the state lacked the necessary funding. . . .

When the dust finally settled after the Cypress Structure's collapse, there was much finger pointing in Sacramento. Initially, CalTrans bore the majority of the blame for the Cypress collapse, but as information about the retrofit program began to surface, attention became focused on the governor. Deukmejian's chief of staff claimed that "had the Governor known how dangerous the Cypress structure was, he would have closed the road." But industry officials took another view of the situation. Said one, "[Deukmejian] didn't have to look any further than his own bathroom mirror" to find the one responsible for the structure's collapse.

Actually, if the governor ever did look into a magic mirror to see who was responsible for the Cypress collapse, he would have seen three faces: his own, Jerry Brown's, and Ronald Reagan's. Reagan found a way to concentrate transportation policy-making power in the hands of the governor, and thus opened a Pandora's box of temptation to the governors who would succeed him. Jerry Brown was the first governor to make full use of CalTrans power in his attempt to re-orient the direction of the state's transportation system. While Brown's objectives of mass-transportation and fewer freeways were admirable, the actual results were less commendable. . . .

With the creation of CalTrans in 1972, California's transportation policy became increasingly influenced by gubernatorial politics. The three governors who served California between 1967 and 1991 set transportation policy based on politics, often ignoring real long-term transportation problems. For Ronald Reagan and Jerry Brown, delay on the seismic retrofit project cost little politically. Even George Deukmejian escaped the collapse of the Cypress Structure largely unscathed. . . .

Freeway-bridge retrofit is only one aspect of the state's transportation system that has been woefully neglected for well over two decades. After the collapse of the Cypress Structure, a great deal of lip-service was paid to the seismic retrofit program, but very little has actually been done. In the introduction of its 1990 report, the governor's board of inquire observed that the Loma Prieta earthquake should serve as a powerful warning to the state's leadership that California's transportation system has been neglected for too long. The state currently estimates that for every 1.1 miles of freeway, there is a bridge (or an overpass) that needs to be retrofitted, rebuilt, or completely replaced. Yet, retrofit is just one program that we know has been delayed for political reasons. What other programs have been stopped or stalled for the same reasons? Are the freeways safe? As California's transportation system grows older, future events similar to the collapse of the Cypress may provide answers to these questions.

The New Urban Environmentalism

Mike Davis

> It is Nature's contours versus Man's ever stronger bulldozers, the historic past versus the politically expedient, the private vale versus the public highway, the orchard versus the subdivision, . . . the person versus the populace. . . . [Richard Lillard, 1966]

From Mike Davis, *City of Quartz* (London: Verso, 1990), pp. 169–73, 176–77, 179–81, 212–13. Reprinted by kind permission of the publisher.

The history of homeowner activism in Southern California divides into two epochs. In the [first] period—roughly the forty years between 1920 and 1960—homeowners' associations were overwhelmingly concerned with the establishment of what Robert Fishman has called "bourgeois utopia": that is, with the creation of racially and economically homogeneous residential enclaves glorifying the single-family home. In the subsequent period—roughly since the beginning of the Kennedy-Johnson boom—homeowner politics have focused on defense of this suburban dream against unwanted development (industry, apartments and offices) as well as against unwanted persons. The first epoch saw only episodic conflicts between developers and homeowners; indeed the former were frequently the mobilizers of the latter in the common cause of exclusionism. Homeowners had little material interest in opposing homevalue-raising "growth," except in occasional cases where it threatened to dump noxious uses on their doorsteps.

After 1965 the structural context of homeowner interests dramatically changed. On one hand, the open space amenities that supported the lifestyles and home values of wealthy hillside and beach dwellers were threatened by rampant, large-scale development; on the other, traditional single-family tracts were suddenly inundated by waves of apartment construction. New development was perceived as a categorical threat to the detached culture of low-density residential life. However reluctantly, in the face of entrenched conservative stereotypes and prejudices, elements of the environmental critique advanced by the Sierra Club and California Tomorrow gained currency amongst homeowner activists. . . .

This "new urban environmentalism" is usually recalled as a Bay Area invention associated with the 1960s movements to save the Bay and preserve hillside open space, which subsequently spilled over into statewide efforts at coastal conservation in the Jerry Brown era. In fact, identical concerns about deteriorating amenities produced parallel backlashes against growth in a number of wealthy Southern California communities. By the early 1970s, for example, environmental regulation of land use had become a potent, sometimes explosive, issue in the archipelago of "red-tile" communities from Coronado and Point Loma (in San Diego), to San Clemente, San Juan Capistrano, Newport Beach, Riverside, Redlands and Santa Barbara. . . .

But the best southern analogue to Bay Area patrician environmentalism was the broad-based homeowners' movement that emerged in the 1960s to "save" the Santa Monica Mountains. This famous range, from the movie colony at Malibu to the Griffith Observatory (including the Hollywood Hills), contains one of the largest concentrations of affluence on the planet: a unique ecology which Reyner Banham memorably described as the "fat life of the delectable mountains." Thousands of rambling split-levels, mansard-roofed mansions and mock Greek temples shelter in the artificial lushness of dozens of arroyos and canyons with world-famous names. But, as Banham pointed out, it is an ecology imperiled by its own desirability: on the one hand, by overdevelopment and "hill cropping"; on the other, by manmade disasters like slides and fires. With lifestyles and property values so dependent upon the preservation of a delicate balance, it is not surprising that wealthy homeowners emerged from their "thickets of privacy" to organize the earliest and most powerful coalition of homeowners' associations in the country.

Already in the early 1950s, the pioneer Federation of Hillside and Canyon Homeowners, founded in the gated movie colony of Bel-Air, was crusading against hotrodding on Mulholland Drive and lobbying for slope-density down-zoning and the establishment of minimum lot sizes to control new hillside development. With a dozen affiliated associations by the mid 1960s (grown to fifty in 1990), and armed with volunteer expertise in landuse law and planning, the Federation was an evolutionary leap beyond any homeowners' group in existence.

Moreover, at a time when academic opinion still visualized the typical Southern California homeowner as a yahoo with a power mower and a Goldwater bumpersticker, the Federation's world view was being represented by Richard Lillard's acclaimed *Eden in Jeopardy (Man's Prodigal Meddling With His Environment: The Southern California Experience)* (1966). A founder of the Federation and first president of the Residents of Beverly Glen, Inc., Lillard polemicized passionately (and at times almost radically) against a mechanized capitalism that seemed determined to turn nature into "one big parking lot" and to erase the past with a "quickened destruction more exact than wartime bombing":

> Allied on the one side have been love for unspoiled nature and adjustment to it, respect for the past, conservationism and conservatism, single-dwelling home life, agriculture, utopianism, the status quo, individual character, established wealth, traditional legality, privacy and private property, and nostalgia. . . . On the other side are concentration on development and alteration, immediate use and exploitation of nature or improvement on nature, emphasis on repetitive recreation for masses of people, the inalienable rights of all to the pursuit of happiness, adulation of novelty and the doings of the newly prominent, and a faith in force, machinery and progress.

For Lillard and the Federation, Eden's last-ditch defense was in the Santa Monicas, where a handful of large landowners—including Hilton Hotels, the Lantain Corporation, Castle and Cooke, Gulf-America and the Tucker Land Company—were threatening to "despoil" the hillsides west of Sepulveda Pass. They had capitalized the Las Virgenes and Triunfo water districts with the aim of bringing as many as 450,000 new residents into mountain tracts (including Lantain's proposed "20-square-mile Trousdale Estates" nightmare). Development, however, hinged on a plan by the State Division of Highways to cut four new freeways through wild canyons and to convert scenic Mulholland Drive, on the crestline of the Santa Monicas, into a four-lane, 120-foot-wide expressway. The Federation, in alliance with the Sierra Club and Friends of the Santa Monica Mountains, mobilized ten thousand homeowners to oppose this "lunatic" mountain freeway scheme. Their petition counterposed the creation of a regional park to permanently conserve open space.

The first chairperson of the resulting Santa Monica Mountain Regional Park Association was a wealthy electronics entrepreneur, Marvin Braude, who presided over the Crestwood Hills Homeowners Association, a Federation affiliate in Brentwood. Braude—who today relishes his reputation as "the sage of the slow-growth movement"—was the first standard-bearer for homeowners on the Los Angeles city council. With the ardent support of the Federation and the Santa Monicas movement, he ousted the bribe-tainted Incumbent in the ritzy eleventh councilmanic district (which included the Reagans' ould sod of Pacific Palisades) in 1967 and began his long, unbroken representation of the interests of Westside canyon and hill dwellers. . . .

The "greening" of the Santa Monicas, like growth-control initiatives in red-tile beach towns or Marin County villages, was widely seen as a hypocritical attempt by the rich to use ecology to detour Vietnam-era growth around their luxury enclaves. By 1972, however, this first wave of preservationist protest was reinforced by populist outbursts in dozens of flatland white-collar communities. Suddenly "slow growth" no longer seemed so socially precious or, for that matter, politically containable. . . .

Accumulated resentments against apartment construction and suburban "deruralization" vented themselves in the April and June 1972 local elections. . . . In Tustin, Brea, Yorba Linda, Orange, and Fullerton simmering grievances over apartment infill boiled over into bitter council contests. In unincorporated Laguna Niguel, residents appealed to county supervisors to prevent further increases in density, while the homeowners' associations of the Saddleback Valley banded together to seek federal funding for a study of how to restrict density and preserve open space. . . . Finally, in red-tiled Newport Beach and San Juan Capistrano, irate homeowners tossed out pro-growth mayors and their supporters. . . .

Far out on the subdivision frontier of eastern Los Angeles County, homeowners' associations from Hacienda Heights, Diamond Bar and Rowland Heights united. . . . Anti-density tremors continued eastward through red-tiled Riverside and Redlands before reaching a crescendo in Palm Springs where Desert People United confronted voters in the fall elections with the choice, "Carmel versus Las Vegas!" The first act of the newly elected slow-growth council majority was to impose a tough 120-day moratorium on multiple-unit construction.

The significance of these Southland skirmishes was amplified statewide by the parallel progress of local growth control in Northern California and, especially, by the passage in November of Proposition 20, which provided for coastal commissions to control beachfront development. . . . Faced with a spreading homeowners' revolt, the development industry suddenly sang the praises of regional government and housing the poor. . . .

The 1973 elections had mixed, and not entirely happy, results for the emerging politics of slow growth. In the mayoral race, Yorty's silly, McCarthyite fulminations against commie-environmentalists estranged many of the same Westside and Valley homeowners who had once warmed to his "just plain white folks" style. But the Unruh and Wachs forces underestimated the centrality of the Black Southcentral electorate to any realignment in city politics. Unruh's massive paper strength wilted into a poor third-place showing in the primary, while Wachs registered a barely discernible blip. Bradley, regrouping trade-union support and harvesting Westside homeowners, handily overcame Yorty's desperate, last-minute barrage of racist innuendo. But in any inventory of Bradley's political debts, homeowners and environmentalists were far down on the list. . . .

The paradoxical result of the 1972–3 density revolt, in short, was to reinforce pro-growth coalitions at both the city and county levels. The first wave of slow-growth protest galvanized land developers as effectively as homeowners, and their redoubled campaign contributions usually overwhelmed growth control initiatives. . . .

[But] the folk maxim that gaunt men rebel while fat men sleep was neatly reversed by the historic suburban protests of 1976–9. In face of a massive inflationary redistribution of wealth, it was the haves, not the have-nots, who raised their pikes in the great tax revolt and its kindred school and growth protests. Many of the actors in this drama were the direct beneficiaries of one of the largest mass windfalls of wealth in history. . . .

[The] Southern California land inflation of 1975–9 . . . enriched many tens of thousands of middle-class families beyond their wildest expectations. Yet the second inflation ultimately produced almost as much anxiety and political turmoil as the first. Homeowners experienced property inflation as a roller-coaster ride that unsettled traditional household accounting, raising unreasonable hopes and fears at the same time. Moreover their windfalls of wealth appeared precarious, while their bloated tax bills seemed all too real—especially for income-strapped retirees. Anxieties were particularly high in the San Fernando Valley where homeowners believing themselves to be little more than a tax colony of Downtown LA yearned for the kind of local control that their counterparts in the Lakewood Plan cities seemed to possess. To make matters worse, the escalated tax assessments arrived on their doorsteps in the same seasons as court-ordered school busing and a host of new growth-related complaints. It was this fusion of grievances in an unstable economic climate, and not just the tax crisis alone, that explains the extraordinarily high emotional temperature in the Southern California suburbs during the summer of 1978. . . .

Like all ideology, "slow growth" and its "pro-growth" antipode must be understood as much from the standpoint of the questions absent, as those posed. The debate between affluent homeowners and mega-developers is, after all, waged in the language of *Alice in Wonderland*, with both camps conspiring to preserve false opposites, "growth" versus "neighborhood quality." It is symptomatic of the current distribution of power (favoring both capital and the residential upper-middle classes) that the appalling destruction and misery within Los Angeles's inner city areas . . . became the great non-issue during the 1980s, while the impact of growth upon affluent neighborhoods occupied center-stage. The silent majority of non-affluent homeowners and renters have remained mere pawns in the growth power struggles, their independent social interests (for instance, economic justice and environmental protection, jobs and clean air, and so on) suppressed in civic controversy.

If the slow-growth movement, in other words, has been explicitly a protest against the urbanization of suburbia, it is implicitly—in the long tradition of Los Angeles homeowner politics—a reassertion of social privilege. . . . Growth politics, in general, seem to militate against class politics.

Further Readings

Abbott, Carl. *The New Urban America: Growth and Politics in Sunbelt Cities*. Chapel Hill: University of North Carolina Press, 1981.

Banham, Rayner. *Los Angeles: The Architecture of the Four Ecologies*. New York: Harper & Row, 1971.

Barth, Gunther. *Instant Cities: Urbanization and the Rise of San Francisco*. New York: Oxford University Press, 1975.

Bernard, Richard, and Bradley Rice, eds. *Sunbelt Cities: Politics and Growth since World War II*. Austin: University of Texas Press, 1983.

Blackford, Mansel G. *The Lost Dream: Businessmen and City Planning on the Pacific Coast, 1890–1920*. Columbus: Ohio State University Press, 1993.

Bottles, Scott L. *Los Angeles and the Automobile: The Making of the Modern City*. Berkeley: University of California Press, 1987.

Cherny, Robert W., and William Issel. *San Francisco: Presidio, Port, and Pacific Metropolis*. Sparks, NV: Materials for Today's Learning, 1988.

Davis, Mike. *City of Quartz: Excavating the Future in Los Angeles*. New York: Vintage Books, 1990.

Dowall, David E. *The Suburban Squeeze: Land Conversion and Regulation in the San Francisco Bay Area*. Berkeley: University of California Press, 1984.

Dumke, Glenn S. *The Boom of the Eighties in Southern California*. San Marino, CA: Huntington Library, 1944.

Findlay, John M. *Magic Lands: Western Cityscapes and American Culture after 1940*. Berkeley: University of California Press, 1992.

Fisher, Irving D. *Frederick Law Olmsted and the City Planning Movement in the United States*. Ann Arbor, MI: UMI Research Press, 1986.

Garreau, Joel. *Edge City: Life on the New Frontier*. New York: Doubleday, 1991.

Hundley, Norris, Jr. *The Great Thirst: Californians and Water, 1770s–1990s*. Berkeley: University of California Press, 1992.

Hynding, Alan. *From Frontier to Suburb: The Story of the San Mateo Peninsula*. Belmont, CA: Star Books, 1981.

Issel, William, and Robert W. Cherny. *San Francisco 1865–1932: Politics, Power, and Urban Development*. Berkeley: University of California Press, 1986.

Johnson, Marilynn. *The Second Gold Rush: Oakland and the East Bay in World War II*. Berkeley: University of California Press, 1994.

Kling, Rob, Spencer Olin, and Mark Poster, eds. *Postsuburban California: The Transformation of Orange County since World War II*. Berkeley: University of California, 1991.

Lotchin, Roger W. *Fortress California, 1911–1960: From Warfare to Welfare*. New York: Oxford University Press, 1992.

Lotchin, Roger W. *San Francisco, 1846–1856: From Hamlet to City, The Urban Life in America*. New York: Oxford University Press, 1974.
Ma, L. Eve Armentrout. *The Chinese of Oakland: Unsung Builders*. Oakland, CA: Oakland Chinese History Research Committee, 1982.
McWilliams, Carey. *Southern California: An Island on the Land*. Santa Barbara: Peregrine Smith, 1973.
McWilliams, Carey. *Southern California Country*. New York: Duell, Sloan & Pearce, 1946.
Nash, Gerald D. *The American West in the Twentieth Century: A Short History of an Urban Oasis*. Englewood Cliffs, NJ: Prentice-Hall, 1973.
Nash, Gerald D. *The American West Transformed: The Impact of the Second World War*. Bloomington: Indiana University Press, 1985.
Olmsted, Nancy. *Vanished Waters: A History of San Francisco's Mission Bay*. San Francisco: Mission Creek Conservancy, 1986.
Pomeroy, Earl S. *The Pacific Slope: A History of California, Oregon, Washington, Idaho, Utah, and Nevada*. New York: Knopf, 1965.
Reps, John W. *Cities of the American West: A History of Frontier Urban Planning*. Princeton, NJ: Princeton University Press, 1979.
Rice, Richard B., et al. *The Elusive Eden: A New History of California*. New York: McGraw-Hill, 1996.
Rolle, Andrew F. *Los Angeles: From Pueblo to City of the Future*. San Francisco: Boyd & Fraser, 1981.
Scott, Allen J., and Edward W. Soja, eds. *The City: Los Angeles and Urban Theory at the End of the Twentieth Century*. Berkeley: University of California Press, 1997.
Scott, Mel. *The San Francisco Bay Area: A Metropolis in Perspective*. Berkeley: University of California Press, 1959.
Shumate, Albert. *Rincon Hill and South Park: San Francisco's Early Fashionable Neighborhood*. Sausalito, CA: Windgate Press, 1988.
Starr, Kevin. *Material Dreams: Southern California through the 1920s*. New York: Oxford University Press, 1990.
Vance, Jr., James E. *Geography and Urban Evolution in the San Francisco Bay Area*. Berkeley: Institute of Government, University of California, 1964.
Wade, Richard C. *The Urban Frontier*. Cambridge, MA: Harvard University Press, 1986.
Wiley, Peter, and Robert Gottlieb. *Empires in the Sun: The Rise of the New American West*. New York: Putnam, 1982.
Wollenberg, Charles. *Golden Gate Metropolis: Perspectives on Bay Area History*. Berkeley: Institute of Governmental Studies, University of California, 1985.

Chapter 12
The Rise of Environmental Science

Documents

Eugene Hilgard Describes the Agricultural Experiment Station, 1890

The idea of providing for the work of an agricultural experiment station prevailed at the organization of the College of Agriculture of the University of California. In 1870, [Geology] Professor E[zra]. S. Carr, in an address at the State [Fair], made the following specific allusion: "The University proposes to furnish the facilities for all needful experiments; to be the station where tests can be made of whatever claims attention."

Ex-President [Daniel C.] Gilman, in his report dated December 1, 1873, alludes to progress in this work, as follows: "The University domain is being developed with a view to illustrate the capability of the State for special cultures, whether of forests, fruits, or field crops, and the most economical methods of production. It will be the station where new plants and processes will be tested and the results made known to the public. A fine estate has been provided, well adapted to the establish-

From Eugene Hilgard, *Report on the Agricultural Experiment Stations of the University of California . . .* (Sacramento: J. D. Young, 1890), pp. 19, 20, 24, 37, 38, 39, 40, 41, 48, 53.

ment of an experiment station in agriculture, a botanic garden, an arboretum, etc." . . .

In 1874 E. W. Hilgard was chosen Professor of Agriculture. . . .

As the increase of means afforded by the United States Experiment Station Fund inaugurates a new era in the work of the station, it is proper to give, at this time, a summary of the kind and amount of work done by the Berkeley station since its establishment.

The fact that the experimental work of greatest practical utility must differ in each State or climatic region, and with the greater or less progress of settlement or population, is too obvious to need extended comment and illustration. In the newer States and Territories the question is not how to maintain the fertility of the soil, but rather which soils are most likely to afford the settler a comfortable living, and what cultures are best adapted to the prevailing conditions of soil, climate, and market? Later in time, and with the incoming of a stable population, comes the question of the maintenance of profitable production by the cheapest and most effective means; and still later, the more minute discriminations as to the exact composition of fertilizers, feeding stuffs, and their most advantageous use, leads to such work as now chiefly occupies the attention of the stations of Europe, and of the older States in this country. California is still in the first and second stages; the questions to be determined are broad and fundamenta', and upon the answer frequently depends the weal or woe of whole communities, or even of extended regions. . . .

In view of the great diversity of soils and climates within the limits of the State of California, it is obviously of the first importance that the experiment station should be in possession of all the data needed to make intelligent recommendations in response to questions and inquiries of all kinds that are addressed to it, especially in regard to the probable value and the adaptation of lands to certain cultures. . . .

Among the more important general results obtained in the course of this work with regard to the character of the soils of the State are the following:

In nearly all cases they are calcareous, i.e., contain enough of carbonate of lime to impart to them the distinctive character of such soils, and to render a farther addition of that substance as a fertilizer superfluous and ineffective.

The great majority contain amounts of potash largely in excess of those found in the soils of the region east of the Mississippi; very often exceeding one per cent.

Potash salts are often found circulating in the soil water; the conclusion being that the use of potash as a fertilizer will likewise be uncalled for for a long time to come. On the other hand, it has been found that phosphoric acid exists in the soils of California in relatively small supply as compared with those of the East, and of Oregon, Washington, and Montana. . . .

The investigation of alkali lands has received much attention and study, with respect to their composition, mode of reclamation, and culture plants adapted to their peculiar conditions.

This subject is of growing importance, not only because of the great intrinsic fertility of the alkali lands, many of which have actually been shown to suffer from a surfeit of valuable plant food, but also because the practice of irrigation without a proper provision for drainage, and in some cases natural conditions in the subsoil under cultivation, cause alkali to rise where none was ever known before. . . .

Parallel with the investigation of the soils, the examination and analysis of the waters of the State have been pursued with a view to ascertaining their qualities for irrigation, as well as for domestic, manufacturing, and medicinal use. . . .

The examination of rocks, marls, gypsum, and other materials connected with agriculture, have also formed an important portion of this work; while, on the other hand, the examination of artificial or commercial fertilizers is but little called for, since as yet these fertilizers are but sparingly used, and so long as no fertilizer control is exercised by the State, the publication of such analysis is inexpedient. . . .

The examination (including analysis, when necessary) of agricultural products of various kinds, such as sorghums, sugar canes, sugar beets, and the by-products of sugar making therefrom; watermelons, oranges, lemons, etc., from different varieties and localities, has been more or less constantly part of the station work, as occasion demanded. . . .

A very large and exacting portion of the station's work is an extended correspondence in the domain of all the subjects mentioned . . . or at times with almost every branch of technical science . . . and thus serves for the information of the public at large.

An Insect Biologist Tells How Lady-Beetles Saved the Orange Groves, 1958

In 1887 the infant citrus industry in California was threatened with destruction because a massive infestation of cottony-cushion scale was forcing the farmers to abandon citrus growing as a commercial venture. The Convention of Fruit Growers meeting in Riverside, California, in April 1887, invited as their principal speaker Charles Valentine Riley, Chief of the Division of Entomology of the federal government, to whom the citrus growers looked to provide a remedy. . . .

When Riley addressed the Fruit Growers . . . he had two field agents working in California. One was Albert Koebele, a naturalized German immigrant whom he had first met at a meeting of the Brooklyn Entomological Society. . . . The other agent in California was D. W. Coquillett, a native of Illinois. . . . His publications in entomology attracted the attention of Riley who appointed him as field agent in 1885. . . .

Riley's address to the Fruit Growers' Convention contained the following major points:

1. The original home of *Icerya purchasi* [the cottony cushion scale] was stated as Australasia, but he was not sure whether it was Australia or New Zealand.

2. He believed that the scale had been accidentally introduced to California at Menlo Park on Acacia in 1868, but he was not certain of this as he had learned that all Acacias had been brought in as seed. . . .

Riley said, "It had doubtless occurred to many of you that it would be very desirable to introduce from Australia such parasites as serve to keep this fluted scale in check in its native land. . . . This State—yes, even Los Angeles County—could well afford to appropriate a couple of thousand dollars for no other purpose than the sending of an expert to Australia to devote some months to the study of these parasites there and to their artificial introduction here. . . .

From Richard L. Doutt, "Vice, Virtue, and the Vedalia," *Bulletin of the Entomological Society of America*, 4 (1958): 119–23. Reprinted by permission.

As a result of Riley's address and persuasion the Convention adopted a resolution favoring the idea of sending someone to Australia to seek out natural foes of *Icerya* and to bring them to California if any were found. . . . Riley selected Koebele to accompany the [U.S.] Commission [to the Melbourne Exposition], although the California people had suggested Coquillett. . . . Koebele sailed for Australia on August 25, 1888. . . .

In October 1888, he wrote Riley as follows, "So far my work has been much more successful than I expected. I not only found the dipterous [fly] parasite [*Cryptochaetum iceryae*] within *Icerya* in large numbers, but also three predaceous larvae feeding upon the eggs of *Icerya*. One of these is a *Chrysopa* [lacewing] larva;—the others are larvae of a small coccinella [the vedalia, a lady bettle]." . . .

Through Koebele's efforts a total of approximately 12,000 individuals of *Cryptochaetum* were sent to California, and relatively little attention was at first paid to his discovery of the vedalia feeding upon the *Icerya* in a North Adelaide garden on October 15, 1888. The actual introduction of the vedalia is best described in these words of Coquillett:

> The first consignment of these lady-birds reached me on the 30th of November (1888), and numbered 28 specimens; the second consignment of 44 specimens arrived December 29; and the third consignment of 57 specimens reached me on January 24, making 129 specimens in all. These, as received, were placed under a tent on an Icerya-infested orange tree, kindly placed at my disposal by J. W. Wolfskill, of this city (Los Angeles). Here they were allowed to breed unmolested, and early in April it was found that nearly all of the *Iceryas* on the enclosed tree had been destroyed by these voracious lady-birds. Accordingly, on the 12th of April, one side of the tent was removed, and the lady-birds were permitted to spread to the adjoining trees. At this date I began sending out colonies to various parts of the state, and in this work have been greatly aided by Wolfskill and his foreman, Alexander Craw, both of whom were well acquainted with the condition of the orchards in this part of the state. By the 12th of June we had thus sent out 10,556 of these lady-birds, distributing them to 228 different orchardists and in nearly every instance the colonizing of these lady-birds on Icerya infested trees in the open air proved successful. . . .

On July 2, 1889 citrus grower J.R. Dobbins had much to say about the work of the vedalia on his property:

> The Vedalia has multiplied in numbers and spread so rapidly that every one of my 3,200 orchard trees is literally swarming with them. All of my ornamental trees, shrubs, and vines which were infested with white scale, are practically cleansed by this wonderful parasite. . . . Over 50,000 have been taken away to other orchards during the past week, and there are millions still remaining, and I have distributed a total of 63,000 since June 1. I have a list of 130 names of persons who have taken colonies, and as they have been placed in orchards extending from South Pasadena to Azusa, over a belt of country 10 miles long and six or seven in width, I feel positive from my own experience, that the entire valley will be practically free from *Icerya* before the advent of the New Year.

On July 31, 1889 Dobbins' orchard was so completely clean of *Icerya* that he posted a notice saying that he had no more vedalias for distribution. The Los Angeles County Board of Horticultural Commissioners propagated the vedalia for distribution by caging five large orange trees heavily infested with *Icerya* and permitting the vedalia to increase within the enclosures. It is said that scores of people came each day, singly and in groups, with pill-boxes, spool-cotton boxes, or some sort of receptacle in which to place the beetles so that they could be carried home and placed in trees and on vines infested with the scale. . . .

Since the suppression of *Icerya* in California projects in biological control have been conducted all over the world, and many of these have been equally successful and effective. None, however, has equalled its drama nor its appeal to the public's fancy. It has remained *the* project in biological control and an important milestone in applied entomology.

A Conservationist Explains How State Forestry Matured, 1985

California was one of the first states to recognize a need for state policies on forests and forestry. The state Board of Forestry, established in 1885, accomplished much for those times in the way of investigations, surveys, nursery operations, and publications. However, California lost its leadership position when the legislature decommissioned that body in 1893. Although the board was reestablished in 1903, the Golden State did not become one of the forerunners in forestry affairs again until four decades later. During these interim years, efforts were mainly directed toward the development and maintenance of a fire protection system. Lesser attention was given to nursery production, pest control, and the operation of forestry work camps. These were years of slow but solid growth, especially after 1927 when the Division of Forestry was formed within the newly created California Department of Natural Resources. This reorganization enabled the governor to supervise forestry matters more closely. . . .

As in other states, there was a long-held dream in California to acquire some state forests. The first forestry board had expressed hopes for such a program; hopes that would reappear intermittently thereafter, but there was little progress. From 1930 to 1944, four tracts were donated to the Division of Forestry, but the aggregate area was only 1,188 acres, and the administration made no serious attempt to obtain funds for acquisition. . . .

One person who became increasingly involved at this stage . . . was Professor Emanuel Fritz of the forestry school at Berkeley. Fritz had been a teacher and researcher in forestry in California since 1920, a part-time consulting forester for forty years, a designated official consultant to the California Redwood Association since 1934, and a participant and adviser with the Save-the-Redwoods League. While pursuing his favorite interests in the coastal redwoods, Fritz built strong associations with the principals of the industry and an understanding of their operations. At the same time, he was also well acquainted with the damage to trees and sites caused by logging of very large timber on steep terrain in an area of heavy rainfall.

From T. F. Arvola, "The Maturing of California State Forestry, 1943–1947," *Journal of Forest History*, 29 (January 1985): 22–31. Reprinted by permission.

Conditions worsened as the country prepared for war. Timber cutting increased sharply, and more operators migrated to northwestern California. This situation, along with the question of the uncertain future of cutover timberland, bothered the professor. He observed that acreage of cutover lands was increasing faster than that being reforested, and large ownerships were being fragmented. Fritz was convinced that these lands had excellent potential for growing timber to sustain a permanent industry. . . . Fritz suggested that the state government should acquire and reforest cutover lands as demonstration sites throughout the state. . . . The Board of Forestry . . . recommended that Director Fulton initiate studies on establishing some state forests. . . . [A] bill passed on 5 May 1943 . . . created a Forestry Study Committee. . . .

The committee observed and received views about old- and young-growth forests, cutover lands, brush fields, forest fire burn, pest depredations, reforestation problems, timber harvesting and processing operations, and many other conditions and factors relating to forest production. . . . The problems in need of immediate attention were determined to be: (1) cutting old growth so as to maintain productivity of forestland and to conserve the supply of old-growth timber; (2) reforesting as much as possible of the cutover land that could regenerate; (3) reducing the amount of standing timber lost to forest fires, insects, and disease, both in old growth and second growth, and also protecting the growing capacities of non-reforested cutover lands from further fire damage; and (4) providing for the continuity of state forest policy. . . .

Fritz strongly criticized shortsighted methods of forest exploitation in California during World War II, but he traced the origins of this problem back to federal land policies in the nineteenth century:

> The Homestead Act [in the 1860s] made it possible for a citizen to obtain title to 160 acres of valuable timber. . . . One hundred and sixty acres might make a good farm, but it can't support a sawmill. . . . What Uncle Sam had fragmented, the timber investors reassembled. . . .
>
> [Then] came World War II with its tremendous lumber requirements. . . . Many of the small loggers of Oregon and Washington, finding themselves out of timber and hearing about the large area of "inaccessible" Douglas fir in northern California, looked it over and liked it. Much of it was owned by ranchers who had tried for years to get rid of it by burning. . . . Some sold their stumpage for as little as one dollar per

> [thousand] board feet, at which price even a small logger could afford to build roads into it. . . .

[Most] local officials and business people . . . had only the most meager concept of the possibilities of forest management for permanence. . . . I'll never forget what a rancher in Mendocino County said: "You're all wrong; cutover land should be converted into grazing land." . . . Someone asked a pine county tax assessor. . . . "Aren't you interested in this land being kept productive?" His answer was, "It'll take about a hundred years before you can get a crop, and I'm not going to live that long, so why should I worry about it?" . . .

Despite his criticisms of loggers and landowners, Fritz did not see public ownership of the forests as a panacea. He discussed the politics of public and private forestry as the background to the California state legislation on forestry enacted in the mid-1940s. . . .

> There were . . . those of us in forestry who believed in the private enterprise system. The U.S. Forest Service in those days was very socialistic, at least for forestry. Some were real socialists . . . [but others of us] couldn't see that the Forest Service should own and direct everything. If the Forest Service could dictate how a lumberman is going to cut his lands, when and where and how, then the government could also dictate to a farmer what crop he's going to plant and how he's going to do it and when he's going to harvest it, and so on. . . . If any kind of cutting laws are needed, they should be state laws . . . .

The 1945–7 study helped educate legislators, the state administration, and citizens in general about California forestry and its needs. . . . The progressive developments of the 1943–1947 period . . . built a solid foundation for later major improvements in California forestry. Professor Fritz [and others] all contributed to a brighter future for forestry in the state. Of course, they had a supportive governor and fine cooperation from the legislature and the entire Board of Forestry. This teamwork widened and intensified the charge to the state forestry agency in the years ahead so that it could again become a leader in the country just as it had been in infancy.

Starker Leopold on Preserving California's Wildlife, 1985

[The California] landscape today would scarcely be recognizable to those who saw it in its pristine form. Valley lands are almost entirely converted to agriculture. Most of the fresh water has been dammed, diverted, and ditched to produce power or irrigate crops. Forests have been cut and recut and now in great areas are being replaced with even-aged conifer plantations. Tremendous areas of formerly productive land have been converted to cities or paved over as highways, airports, or supermarket parking lots. Nor is the rate of change decreasing; it is, in fact, still accelerating as the human population continues to rise.

Fortunately, there are still recognizable patches of the old California left. Some of the original ecosystems are well represented in parks, national forests, and public domain lands. Mountain tops and some forest types are generously preserved. Big blocks of desert are held in reserved status, although much of the desert is being battered by motorcycles and vacationers. But other important landscapes have been reduced to tattered remnants. Riparian woodlands, for example, which played so important a role in supporting wildlife and the aboriginal peoples in the Central Valley, are nearly gone. So also are the perennial grasslands, the vernal pools, the interior tule marshes, and the valley woodlands. If samples of old California are to be saved, they will have to be identified and set aside quickly lest they disappear while we are preparing to act.

There are several good reasons for preserving adequate samples of all ecosystems, and not just those that are scenic or spectacular. The first and most obvious justification of natural area preservation is the aesthetic and educational value—we should know and be able to show future generations the landscapes from which modern California was carved. But of equal significance is the scientific, "baseline" value of retaining combinations of soils, plants, and animals that formed the fer-

From A. Starker Leopold and Tupper Ansel Blake, *Wild California: Vanishing Lands, Vanishing Wildlife* (Berkeley: University of California Press, 1985), pp. 2–3, 122–26.

tile ground we now exploit with such abandon. . . . In this sense, samples of arable land may prove to be the most important of natural areas. Similarly, native forest types of mixed conifers and hardwoods may retain fertility better than pure conifer plantations. Livestock ranges of shrubs and native grasses may in the end outproduce pure stands of crested wheatgrass. Finally, fully 50 percent of all pharmaceuticals have a natural component as the active ingredient; yet only 2 percent of the world's plants have been analyzed and tested. Countless potential benefits for mankind certainly lie locked within many of these untested species. Simply stated, society depends on the complete array of plants and animals found in whole, undisturbed ecosystems for food, medicine, clothing, and shelter.

There is growing recognition that natural areas should be preserved, although the diverse reasons may not always be understood. But who is responsible for the total preservation program? Various government agencies contribute in a number of ways, but some ecosystems, such as riparian woodlands, do not fit neatly into any government program. There remains an important role for private initiative, both in the purchase or leasing of lands for conservation purposes and in the management of privately owned lands to provide for long-term conservation values. . . .

The social values of wildlife and the desirability of retaining our heritage of native animals are recognized in a myriad of laws, both state and federal, that protect all but a few pest species from wanton destruction. The animals themselves receive adequate protection from the law; not so the habitat in which they live. It is illegal to kill a woodpecker but perfectly legal to cut down the dead tree harboring its nest, which in the end of course eliminates the woodpecker permanently. . . .

Forest management practices affect the diversity and abundance of wild animals by altering the successional stages of the forest. Some of the native American birds and mammals that evolved to fit the wide spectrum of niches the continent offers prefer or require a rather specialized habitat. Some species live only in tall mature forests. Others thrive in forest openings or in early brushland stages of forest regeneration. The point is evident that to maintain the full spectrum of native vertebrates, it is necessary to preserve or create areas representing all stages of forest succession, particularly the mature forest. Intensive forest practice tends to shorten rotations and to truncate succession, reducing or eliminating animals adapted to live in mature stands.

Perhaps the most widespread forest practice—and one of least benefit to wildlife—is the culture of conifers in pure, even-aged stands. . . . Among the most intensively managed forest areas in the country is the

Douglas-fir zone . . . where a great deal of forest land is devoted to the culture of fast-growing, even-aged stands of Douglas fir. The redwood region of northern California is maintained in dense, fully stocked stands with little attractiveness to wildlife.

These monocultures are virtually devoid of wildlife after the canopy closes. Shrub undergrowth is shaded out, or its elimination is hastened with herbicide spraying. Snags and mature trees are removed with the first cutting and are not allowed to regenerate. Hole-nesting birds and mammals are eliminated for lack of shelter, and most other kinds of wildlife find little or no food. I have driven through such plantations in northern California where a bluejay would have to pack a lunch to get across. . . .

Species dependent upon mature forest will inevitably disappear; the process will produce the "rare and endangered species" of tomorrow. The spotted owl . . . appears to live only in mature conifer stands. When the last of these are logged, the spotted owl will be a rare bird indeed. . . . Besides deer, some granivorous birds and various small mammals also respond favorably to the forb-shrub stage of succession. The balance is not entirely negative.

But in contrast to the clearcut, even-aged trend in silvicultural practice, selective logging of individual mature trees has far less adverse impact on wildlife. A selectively cut forest is actually a mosaic of large and small trees, openings, and thickets and sometimes includes a mixture of hardwoods and conifers. . . . It is relatively easy to adjust a selective cut to provide varied wildlife habitat while at the same time yielding a reasonable crop of logs. . . .

Many . . . climax bird species have decreased in range and abundance as the virgin forests have shrunk. In western conifer forests, the widespread pileated woodpecker is now declining, as are a number of insectivorous forest birds, including various parids [chickadees], flycatchers, thrushes, warblers, and woodpeckers.

If forest managers are willing, there are many ways in which timber practice can be modified on the ground to provide wildlife habitat, with moderate—but not excessive—sacrifice in board-feet production and economic yield. The problem is not beyond solution biologically, but it requires compromise, one component of which is reduced profit. . . . The management procedures that enhance wildlife habitat are nearly all of a sort that cut profits to the timber operator: leaving snags and potential snags in the forest, leaving strips or corners of mature trees uncut, keeping clearcut blocks small, desisting from excessive use of herbicides and pesticides, maintaining some uneven-aged stands when even-aged stands are simpler to manage mechanically. On public forests these silvicultural adjustments are possible and fully justified. . . .

In a frontier community, animal life is cheap and held in low esteem. Thus it was that a frontiersman would shoot a bison for its tongue or an eagle for amusement. In America we inherited a particularly prejudiced and unsympathetic view of animals that may at times be dangerous or troublesome. From the days of the mountain men through the period of conquest and settlement of the West, incessant war was waged against the wolf, grizzly, cougar, and the lowly coyote, and even today in the remaining backwoods the maxim persists that the only good varmint is a dead one.

But times and social values change. As our culture becomes more sophisticated and more urbanized, wild animals begin to assume recreational significance at which the pioneer would have scoffed. Americans by the millions swarm out of the cities on vacation seeking a refreshing taste of the wilderness, of which animal life is the living manifestation. Some come to hunt; others to look, or to photograph. Recognition of this reappraisal of animal value is manifest in the myriad restrictive laws and regulations that now protect nearly all kinds of animals from capricious destruction.

Only some of the predators and troublesome rodents and birds remain unprotected by law or public conscience. In many localities bounties are paid for their scalps, and government hunters are employed for their control. In point of fact, there are numerous situations where control of predators, rodents, and even some birds is essential to protect important agricultural and pastoral interests or human health and safety. The problem is to differentiate these local situations where control is justified from the numerous cases where the same species of animals have social values far in excess of the negligible damage they cause. The large carnivores in particular are objects of fascination to most Americans, and for every person whose sheep may be molested by a coyote there are perhaps a thousand others who would thrill to hear a coyote chorus in the night.

Three Scientists Tell How to Conserve Biodiversity, 1993

Habitats, ecosystems, and landscapes are rapidly gaining favor as appropriate scales for conservation actions. Attention to habitat conservation is simultaneously an outgrowth of species protection efforts, a recognition of the need to conserve biodiversity on the ecosystem scale, and a response to changing perceptions of acceptable losses. The question of the correct scale for conservation action is presently under discussion, with proposals to work only at the level of habitat conservation sometimes in conflict with efforts to work on both habitat and species scales.

Recent research in conservation biology suggests incorporating larger scales into our conservation actions, both to ensure the protection of individual species from habitat fragmentation and to ensure the perpetuation of ecosystem processes. Piecemeal development leads to incremental loss of habitat, which can only be avoided by planning over larger geographic areas and incorporating biodiversity considerations into land use planning. Conflicts are growing between conservation and land development as development spreads through the state and increasing numbers of species are endangered and then listed under the state or federal Endangered Species Act. For these reasons, we included in our recommendations a Habitat Protection Act to encourage county land use planning that incorporates ecosystem conservation.

In 1991, attempting to resolve conflicts between conservation and development while preventing further species endangerment, the California legislature established a new program for Natural Community Conservation Planning (NCCP) administered by the California Department of Fish and Game [CDFG]. The goal of a Natural Community Conservation Plan "is to conserve long-term viable populations of the State's native animal and plant species, and their habitats, in landscape units large enough to ensure their continued existence." To achieve this goal the CDFG will enter into agreements with other public agencies and pri-

From Debra Jensen, Margaret Torn, and John Harte, *In Our Own Hands: A Strategy for Conserving California's Biological Diversity* (Berkeley: University of California Press, 1993), pp. 240–43, 258–60.

vate interests to develop and implement a plan to protect certain biological communities adversely impacted by growth and development. A pilot project has been started for the coastal sage scrub community in southern California.

Some perceive this new approach as the best possible means of resolving conflicts between conservation and development, whereas others insist the NCCP process is simply a means of getting around the Endangered Species Act. These viewpoints are both accurate assessments of the potential directions the NCCP process may take. The consequences for California's biodiversity are quite different if state policy encourages choosing between species and habitats rather than choosing to protect both species and habitats.

In San Diego, Orange, Los Angeles, and Riverside counties (all areas undergoing rapid urban development) numerous species are threatened by habitat loss and fragmentation. Several of these species, in particular two birds, the California gnatcatcher and the coastal cactus wren, depend on coastal sage scrub habitats. Proposals to list these species under the federal and state ESAs have created controversy—if the birds were listed, the pace and pattern of development in these counties would be greatly restricted. . . .

The NCCP process establishes a Scientific Review Panel to identify the data collection needs and the guidelines for conserving the target natural community. Land owners and local governments voluntarily enter into agreements to be part of the planning area and to cooperate in surveys of their lands to determine the habitats and species present. . . . By participating in the NCCP process landowners expect to enjoy a streamlined environmental review process because concerns about endangered species and habitat loss will be addressed in the conservation plan. The NCCP is anticipated to be considered equivalent to completing appropriate sections of the environmental review requirements of CEQA and will provide state and federal permits for impacts to listed species.

Although modeled after habitat conservation efforts under the federal ESA, the NCCP process is quite different from what would occur if the gnatcatcher or cactus wren were federally listed. . . . Actions that eliminate either the habitat of the listed species or an individual animal are usually prohibited. To obtain permits for development, landowners, in conjunction with government agencies, are required to complete a Habitat Conservation Plan (HCP). Typically, HCPs allow a portion of the listed species' habitat to be lost, in exchange for ensuring long-term protection of the remainder of the habitat through acquisition and dedication of the area as a park or reserve. . . . Once a listed species is involved, HCPs are considered preferable to the uncertainty and confusion that would result from each individual landowner potentially violating the ESA and trying to resolve the problem alone.

NCCPs are intended to be an improvement on the HCP process. . . . HCPs are usually long and complex negotiations between several parties—conducting more than one in the same geographic area is time consuming and wasteful of resources. In comparison, NCCPs will also require significant data collection efforts, but will accommodate several species simultaneously, are expected to be limited to an eighteen-month planning period, and are voluntary. . . .

Conservation policy in California is changing quickly, a consequence of the state's search for a more sensible future for its human and wild inhabitants. Some current trends in conservation policy in California are promising, whereas others threaten to hinder progress toward a long-term biodiversity conservation strategy for the state. New endeavors such as the Biodiversity MOU and Natural Community Conservation Plans contain visions of long-term institutional changes that can improve the management of California's biodiversity. They show a willingness to anticipate conservation crises and to begin to establish proactive rather than reactive responses. Many people and projects are recognizing the need to work at several scales for successful conservation—both species and habitats must be included, and both local and bioregional approaches must be taken. In addition, disparate groups are trying to overcome their polarities and discuss means of reaching solutions acceptable to many parties.

Despite this evidence of progress and increased commitment to conservation, many other activities are antagonistic to new conservation efforts. . . .

One reason for the slow progress toward a conservation strategy is that the options to stem species' losses are politically much more difficult than in the past. For example, the plight of the Delta smelt, a declining species that dwells in the Sacramento-San Joaquin river delta, has been inaccurately compared to the snail darter. Although both are small endangered fish, the political circumstances and the policy choices in these two cases are very different. The snail darter case was a decision about whether to take a future action (building the Tellico dam) which would likely have an adverse effect on a species. The Delta smelt case, like many issues debated now, questions whether to stop present actions that are already having a known detrimental effect on a species. These are much harder decisions to make. There are many established interests at stake, including those whose current livelihoods are at risk. To protect the Delta smelt may require diverting less of the natural stream flows of the Sacramento and San Joaquin rivers for human use. Furthermore, the Delta smelt is only one of a suite of species adversely affected by the altered streamflows in the Bay-Delta. . . . The stakes are very high on both sides—no wonder the debate is loud.

These are fundamental questions: can we decide as a society to make changes in practices that we know cause environmental harm? how are we going to make these decisions? where on the conservation-development continuum will that decision fall? and who will decide? Our choices appear to be either reducing the damage per unit benefit, be that benefit water, timber, or agriculture, or reducing the benefit we receive now, or else accepting the loss of genes, species, and ecosystems, as well as reduced benefits for future generations. . . .

Changes will affect private property owners and will require planning and tough decisions. Most people are willing to support conservation if it affects land uses far from their daily lives. Saving endangered species is also easily supported unless it requires changing our plans for urban growth or changing the operation of our water delivery systems. We have already used up many of our easy choices. Now we are being tested by the difficult decisions.

Accomplishing our conservation goals and accommodating both human and ecosystem needs in California will take time, ideas, and information. Solutions will not be arrived at overnight, but must instead be achieved by a long-term commitment to California's future. The challenges California faces are large, yet many steps have been taken toward creating a strategy that will truly protect the biological heritage of California. Success will not be measured by rhetoric from the governor's office, the legislature, industry, or environmentalists. . . . The true test will be the census figures for the northern spotted owl, the winter-run chinook salmon, the California gnatcatcher, and other endangered species twenty years from now. We will know if we have moved forward on a course for conserving biodiversity by the acres of healthy forests, wetlands, and deserts left to our grandchildren.

ESSAYS

California Scientists and the Environment

Michael L. Smith

The conquest of California coincided with the golden age of scientific exploration. Belatedly following the lead of navies and merchants, naturalists in the late eighteenth century stepped out of their gardens and into the jungles and deserts of the New World. In the course of the nineteenth century they were joined by a growing number of pioneers in the geophysical sciences. Together they proposed to complete an inventory—and, they hoped, construct a design—for the entire planet and everything living on it. Their itinerary and their techniques, if not altogether their purpose, overlapped nicely with the needs of expansionist nations to catalogue nature's wealth and chart routes to reach it. . . . By the late 1840s dozens of naturalists and a handful of geographers had visited California, their numbers increasing as overland and seaborne routes became less imposing. . . .

Gold rush California's swift and single-minded growth created an instant practical need for reliable survey work—especially along the coast, where the incessant clamor for passage had converted anything vaguely afloat into a ship, some with nothing better than schoolbook maps to guide them. Luckily, the ablest scientific agency in the federal government was the United States Coast Survey (later the Coast and Geodetic Survey), another brainchild of Jefferson. . . .

With no roads or settlements disturbing the horizon, the first coast surveyors had to row ashore countless times, carrying two tons of instruments across nearly impassible terrain, where "wild animals and wilder

From Michael L. Smith, *Pacific Visions: California Scientists and the Environment, 1850–1915* (New Haven: Yale University Press, 1987), pp. 12–13, 18–19, 24–46, 114–17, 144–50, 191–92.

men" provided their only company. . . . For [geological surveyor] George Davidson, their assignment to chart the coast of California marked the beginning of a fifty-year career as a scientist in this "strangest of all countries." . . . After three months in the field, he had visited the pine forests of the Coast Range, gazed across the valley floor to the High Sierra, and observed the central coastline's varied microclimates of cypress-studded cliffs, coastal dunes and grasslands, and steep escarpments notched by lush, fogbound inlets. . . .

The Coast Survey's budget was linked to its ability to serve commercial development; more capital growth meant more Survey support. What distressed George Davidson was the corresponding absence of a cultural context for science. In the East, science occupied a more featured position in the cultural orchard of genteel Victorian society, where it could attract public recognition while filling the baskets of commerce. . . .

On November 6, 1860, . . . two ships passed each other along the Pacific Coast of Mexico early that November, perhaps on Election Day itself. The southbound vessel carried George Davidson and his family, en route from San Francisco to Philadelphia. On board the northbound steamer were Josiah Dwight Whitney, director of California's new State Geological Survey, and three of his assistants. Whitney approached California on the crest of a gradual wave of regional interest in scientific exploration. In addition to the meticulous shoreline reconnaissance by Davidson and the Coast Survey, tentative webs of surveyors' triangulation began to lace the state's interior in the 1850s. . . .

Perhaps most impressive to Californians was his 1854 publication, *The Metallic Wealth of the United States*, for many years the definitive work on the subject. . . . [The] survey undertook to provide a blueprint for rational development of California. Exhaustive, precise, unhindered by the tunnel vision of single-purpose industries, Whitney's overview actually served the state's economic interests on a greater scale than the legislators imagined. . . . By 1864 the Whitney Survey had completed its most important work. . . . Their efforts "to grapple with the geological structure of an unknown region of unlimited extent," as Whitney described it, had produced one of the most spectacular of all the state surveys. . . .

[In 1868] "fate and flowers carried [John Muir] to California." Muir's image as an unconventional inventor-genius provided a basis for many of the strengths and limitations of his later career. . . . His most influential teachers were geology professor Ezra Carr, who had studied under [Louis] Agassiz, and his wife, Jeanne Carr, who informally encouraged Muir's botanizing. . . .

A Yosemite journal entry in January 1870 suggested the blend of scientific inquiry and transcendental flight that comprised such a revelation for Muir. If he could only escape from the "so-called prison" of his body, he wrote, he would "follow [the earth's] magnetic streams to their source," "go to the very center of our globe and read the whole splendid page from the beginning." There he could "study Nature's laws in all their crossings and unions."

It was not by coincidence that Muir chose to combine the imagery of religious ecstasy with the aims of the earth scientist. His determination to "read the whole splendid page from the beginning" swept aside his father's denunciation of venturing "beyond what was written." It also described the field research that led him to his new occupation. Glacial scorings and moraines now replaced cogs and pulleys as the objects of his late-night flashes of inspiration. "Waking or sleeping, I have no rest," he wrote to Mrs. Carr in September 1871. "In dreams I read blurred sheets of glacial writing, or follow lines of cleavage, or struggle with the difficulties of some extraordinary rock-form." His discovery of living glaciers in Yosemite won Muir recognition from the nation's scientists. California's mountains had opened a new career for Muir. . . .

A key figure among Muir's "natural Elect" was Joseph Le Conte. The University of California's first professor of geology, botany, and natural history, Le Conte arrived in California when the school opened in September 1869. The following summer he accompanied eight students on a six-week excursion through Yosemite and the upper Sierra region. In the course of these ramblings, which he described as "almost an era in my life," Le Conte encountered two major influences on his future scientific career: the "mountain structure and mountain sculpture" of Yosemite, and John Muir. . . . In the days to follow, [Muir] and Le Conte exchanged theories of the valley's origin, enlivening Le Conte's geologic interests and bringing Muir a step closer to public recognition. Both men described in nearly identical terms their delight in the other's enthusiasm. "Mr. Muir gazes and gazes, and cannot get his fill," Le Conte wrote in his journal. "He is a most passionate lover of nature. Plants and flowers and forests, and sky and clouds and mountains seem actually to haunt his imagination." Muir, in turn, recalled that the professor "studied the grand show, forgetting all else, riding with loose, dangling rein, allowing his horse to go as it liked. He had a fine poetic appreciation of nature, and never tired of gazing at the noble forests and gardens, lakes and meadows, mountains and streams" of Yosemite. . . .

As a defender of science, the university, and the importance of preserving California's "magnificent field" of natural resources, Le Conte served as a bellwether for scientists in this "wonderful new country." Like George Davidson twenty years earlier and Stanford's David Starr Jordan twenty years later, he found in the unfinished quality of California's social structure, and in the great promise of its terrain and climate, a correlative to his own aspirations. The tasks before them were to construct, from this succession of arrivals, a coherent vision of the role of science in California and to convince Californians, as John Muir exhorted them, "to come and see." . . .

Like the earliest of the Eastern societies, the California Academy of Sciences was established in part because of its founders' desire to associate their names and their relatively undeveloped social community with science and advanced learning. Just as Whitney, King, and Davidson bestowed the names of scientists on California's mountains and trees, so did the academy's founders undertake to add the name of science to the social landscape of their new home.

None of the seven men who founded the academy in 1853 were professional scientists. . . . When they praised California's natural environment as "a field of richer promise in the department of natural history in all its variety than has previously been discovered," most of them spoke from personal observation. . . . [But] together they promoted the cause of science in California, calling for "a thorough survey of the State and the collection of a cabinet of her rare and rich productions." . . . In the 1880s, when its resources and collections warranted a staff of curators, the academy was the first scientific organization in America (and possibly the first in the world) to include women among its paid curators.

The most prominent woman among the academy's early curators was Alice Eastwood. She began to conduct botanical outings in the Bay Area and traveled the length and breadth of the state, often unaccompanied, to observe vegetation and collect flora until the academy's herbarium was the largest in the West. All of the academy's botanical collections were destroyed in the fire that followed the 1906 San Francisco earthquake—except for those Eastwood rescued by climbing the bannisters to the sixth floor herbarium (the stairs had collapsed in the quake) and pitching the most valuable specimens out the window. . . .

Of the 182 charter members [of the Sierra Club founded in 1892], 73 (or 40 percent) also belonged to the California Academy of Sciences. Among them were several talented amateur botanists, ornithologists, and lepidopterists. And as the club's publications and educational outings attest, the general membership demonstrated an active fascination for the natural history of the region, all of which contributed to their esteem for natural scientists. . . . Of the ten honorary members chosen at a general meeting in November 1892, five were scientists: J. D. Whitney and Clarence King for their years with the California Geological Survey (William Brewer later joined them on the list); John Wesley Powell, director of the U.S. Geological Survey; B. E. Fernow, director of the National Forestry Bureau; and Irish glaciologist John Tyndall. Of the remaining five honorary members, three occupied positions crucial to the political enactment of environmental reform: Secretary of the Interior John Noble; U.S. Senator A. Sidney Paddock of Nebraska, who had just introduced a bill calling for national forest reserves; and Robert Underwood Johnson, an editor of *Century* magazine who had first proposed to Muir the creation of a national park for Yosemite. Scientists and their relation to the politics of conservation figured prominently in the image that the Sierra Club members wished to project. . . . Muir remained president, and the club's most visible member, for twenty-two years until his death in 1914. . . .

Scientists . . . were featured in every volume of the [Sierra Club] *Bulletin* between 1893 and 1914. A few of them, like geologists Grove Karl Gilbert and Francois Matthes, offered detailed geological accounts of the Sierra—sometimes with titles suitable for scientific journals ("Domes and Dome Structures of the High Sierra," "Systematic Asymmetry of Crest-Lines in the High Sierra of California"). Others, like botanists John Lemmon, Willis Linn Jepson, and William Setchell, described the trees and plants of California in terms accessible to amateurs as well as scientists. Zoologists Joseph Grinnell and Vernon Kellogg wrote similar articles on the birds, insects, and butterflies of the Sierra.

More frequently, scientists popularized their knowledge of the region by writing descriptions of excursions that included information on the natural features of the area. At a time when almost no mountaineering information or equipment was available, the scientists' trail descriptions, maps, and advice on supplies were welcomed by a growing number of wilderness enthusiasts who otherwise might not have ventured into the High Sierra. For the scientists, anything that encouraged people "to come and see," in Muir's words, contributed to the environmental literacy of the region. Muir, Joseph LeConte, and Mark Kerr interspersed accounts of their mountain travels with geological or botanical details, hoping to lure the general reader into a fuller understanding of the terrain. Alice Eastwood solved the problem of combining descriptive narrative with technical information by appending to her account of the Trinity wilderness a list, with botanical classification, of all the trees and shrubs of the area. . . .

[1915] marked the founding of the Ecological Society of America, the first national organization devoted to the promotion of scientific ecology. Early and mid-twentieth century ecologists developed far more sophisticated techniques than those employed by the first California scientists: bionomics, Malthusian demography and population dynamics, and especially the quadrat field analysis developed by Frederick Clemens' "grassland school" of ecology at the University of Nebraska. The focus of these techniques and their relation to public policy differed considerably from the aims envisioned by California's first scientists. As a new scientific profession, ecology developed its theoretical framework in relative isolation both from the rest of the scientific community and from the general public. Like forestry, the ecological sciences affected society primarily as a managerial instrument, applying problem-solving techniques to enhance production in forestry, agriculture, and fish and range management. . . .

Pacific scientists had contributed comparatively little to theoretical science. Their social aims for California remained unrealized. . . . Yet their careers collectively pointed toward both the importance and the difficulty of developing an ecologically informed vision of science and society.

Eugene Hilgard and the Birth of Soil Science

Hans Jenny

When California awoke from the dazzling Gold Rush of the fifties and began to cast its eyes on the agricultural empire, the farmer immigrants were confronted by a host of unfamiliar and seemingly unsolvable soil problems. Paramount among these was the salt menace: square mile after square mile of salt-incrusted lands where crops would not germinate, or would wither and die before they fulfilled the life's cycle. Farmers desperately tried to cope with this enigma. As judged from the accounts in the press, the situation was chaotic and despondent.

[Eugene] Hilgard was ideally equipped to conquer the alkali plague. As

From Hans Jenny, *E. W. Hilgard and the Birth of Modern Soil Science* (Pisa, Italy: Collanda della Revista "Agrochimica," 1961), pp. 42–43, 45, 47–48, 80–81, 83, 85, 87–88, 123–24.

a geologist he had insight into soil origin and into important groundwater relations; as a botanist he could delineate alkali's connection with plant life; as a chemist he possessed the sharp tools of analysis that would probe into the salt mystery. . . .

In the [1880s] complaints of the rise of alkali on irrigated lands in the upper part of the San Joaquin Valley became more and more frequent. Many valuable tracts became unfit for production. The disaster was associated with the rapidly expanding and excessive irrigation with water from the mountains which brought the water table close to the soil surface.

Hilgard personally examined the region and amplified:

> It is emphatically true that the alkali salts rise from below, through the agency of the water evaporating upon the surface. . . . In some cases the leakage from the [irrigation] ditches has now so filled up the entire country around, that before the introduction of drainage ditches, the earlier, low lying properties were literally swamped, the water standing in puddles and ponds over the country. . . . But this water was not as pure as that running in the ditches; it had taken up the alkali salts developed by the weathering of forty feet of soil material for thousands of years, and had brought it to the surface by an easily intelligible process of upward leaching. As this water evaporated from the surface, its hoard of alkali salts was left behind and . . . wide stretches of land, where alkali had never been seen before, became spotted with alkali areas of greater or less extent, and in depressions alkali-water formed evil-smelling ponds of undrinkable water.

Through soil analysis Hilgard soon discovered that alkali soils very frequently contained copious amounts of soluble plant nutrients, such as nitrates, phosphates, and potash salts. Optimistically he announced that reclaimed alkali soils may be expected to possess extraordinary and lasting fertility, an attractive prospect which kindled the farmers' interests in practical sides of reclamation.

These Hilgard pursued at once with the aid of progressive, cooperative farmers who were willing to test his reclamation hypotheses. . . . Hilgard, blending field observations with laboratory analyses, made shrewd guesses by invoking principles of chemistry and the soil knowledge he himself had accumulated. These guesses his friends tried out.

His suggestion in the [1870s] that gypsum would cure black alkali, the most toxic and harmful of all, proved to be an immediate success. . . . As early as 1890 Hilgard could state that the successful use of gypsum had given rise "to so lively a search for a supply of that material that it can now be obtained almost anywhere in the State at rates which the farmer can afford." Up to this day the use of gypsum on California lands has remained popular, over a half-million tons being applied every year. . . .

Hilgard dispelled the fear of alkali soils. He opened ways to redeem and to control them. He proved that the sterile sand of the desert was a poetical myth, that the western sandy soils had no affinity with their impotent relatives of the forest region. He presented a rational account of the favorable fertility of arid soils in general—arid countries are rich counties—thus narrowing the problem of their utilization to that of handling the water regime. He devised irrigation practices that would ensure maximum benefit to the plant as well as to the soil. Unfortunately, he had little opportunity to explore the water economy of dry-land farming, so crucial to the nonirrigable areas, such as the Great Plains. . . .

In letters, newspaper articles, and pamphlets, Hilgard tried incessantly to change the provincial ideology of the Atlantic Coast farmer, and of his representatives in Congress and in the Department of Agriculture, who shaped the agricultural policies for the entire nation. He pleaded against blind adoption of findings and practices of European and eastern agricultural experiment stations. The arid farmer could profit little from the textbooks of the humid farmer. . . . Hilgard contended . . . that it was the great fertility of arid soils which drew prehistoric man away from the forest rim. . . .

Hilgard's second quarter-century of vocation, 1880–1905, was rich, but hectic. His activities and emotions were consumed by the struggles of the College of Agriculture, and the tribulations of the Agricultural Experiment Station. . . . While the Golden State's agriculture was still dominated by wheat, the fruit and vegetable industries were steadily expanding. Hilgard foresaw that fruit eating would increase very greatly, "to the benefit of the national health; the more fruit, the less whisky and doctor bills." His staff, especially G. E. Colby, analyzed fruits, rind, pulp, juice, for organic matter, ash, sugar, acid, and protein content to combat the widespread belief that fruits, especially oranges, were luxuries. . . . In conjunction with analyses of fruits, leaves, twigs, branches, and of soil, Hilgard forecast that in most citrus orchards lack of nitrogen would be among the first things to be apprehended, a prediction abundantly verified in subsequent years. . . .

Until 1871 . . . California orchards were remarkably free from maladies and depredations by insects. But soon diseases ravaged the crops and Hilgard encouraged active programs in economic entomology and plant pathology, and appointed [entomologist] C. W. Woodworth to be in charge. The orchard and agricultural grounds on the western portion of the Berkeley campus served as a testing site. Hilgard [also] initiated research and promotion in cotton growing and in sugar industry, and both expanded eventually. . . .

By 1900 [his] experiment of an Experiment Station had become a success, and had established solid partnership with California agriculture. The staff had risen from one to thirty, correspondence reached 15,000 letters annually, and research reports had to be printed in editions of 10,000 and more. . . . Funds for investigations and services flowed in from all sides. The Station blossomed forth with such profusion that it outgrew its founder. . . .

We regret that Hilgard could not pierce the future and see his Station staff swell to hundreds, with a budget of millions; and learn that his book was bought by seven thousand people—a great, silent tribute; and become aware that his name is honored as *Hilgardia*, the University of California's journal of agricultural research; and notice that the Berkeley Campus has a Hilgard Hall, and that California cities have Hilgard Avenues. To top it all, in the majestic Sierra Nevada of the Pacific Coast a lofty granite mountain, 13,357 feet, is named Mt. Hilgard. It is flanked by other giants, Mt. Darwin, Mt. Lyell, and Agassiz Needle to the south.

California Scientists and the Insecticide Crisis

John Perkins

Agricultural insect control practices after 1945 were heavily dominated by the use of insecticides. . . . The immediate source of the heavy interest in insecticides came from dramatic successes of DDT and other new insecticides invented during and following World War II. . . . [But] euphoria with insecticides was short lived. . . . [By] 1955, signs of tech-

From John Perkins, *Insects, Experts, and the Insecticide Crisis: The Quest for New Pest Management Strategies* (New York: Plenum, 1982), pp. 1, 57–58, 74, 79–81, 61, 84, 85–87, 201. Reprinted by permission of the author and publisher.

nological failure were clearly evident. Resistance, resurgence, and the outbreak of secondary pests created problems for the farming industries. Entomologists based largely in the land grant universities of California and the mid-South created [a new] strategy, Integrated Pest Management (IPM).

California research leading to integrated [pest] control [had begun] in the late nineteenth century with the arrival of Charles William Woodworth to the Berkeley campus. . . . Woodworth was the first scientifically trained entomologist to be hired in a permanent position within the University. . . .

Woodworth was in no way shy about testing and advocating the use of insecticides when he thought them warranted. In fact, he began work with arsenical sprays for codling moth during his first year in California. He was nonetheless a severe critic of the use of chemicals when he believed they were used improperly. Half of the costs of treating crops were wasted, he argued, owing to inappropriate application of materials. Woodworth argued that expert entomologists should help in making decisions in the field, and he was an early proponent of legislation to regulate insecticides.

Woodworth's complaints about the ineffective or useless application of insecticides were not prompted so much by the occurrence of resistance, outbreaks of secondary pests, or health and environmental hazards. That potpourri of problems resulting from insecticides was not yet recognized. Instead, he was probably more concerned with costs of treatment and use of ineffective insecticides. . . . Any expenditures made by farmers had to be economical if California growers were to remain competitive with their eastern counterparts.

Regardless of the reason for Woodworth's stand against indiscriminate spraying and dusting, he exercised a tremendous influence in the shaping of professional entomology within the state of California for many subsequent decades. . . . He [and his] colleagues, moved the critical stance toward insecticides from a mere attitude to a paradigm for guiding research and later to a foundation for public policy. . . .

Woodworth . . . helped to create a tradition within the University of California in which chemicals were studied critically for both their positive and negative impacts. [Riverside entomologist] Harry Scott Smith and his close associates created a center of expertise on biological control and kept alive the belief after 1945 that the importation and release of natural enemies was by no means obsolete in the new age of chemicals. . . . A body of scientific workers dedicated to integrated control [emerged in the 1950s].

[By] 1959, integrated control theory contained a number of essential components:

> 1. Recognition of the ecosystem. Crop plants, pest organisms (insects, weeds, plant pathogens, and others), and man, together with the physical environment, make up the unified ecosystem. Pest control had to be recognized as an activity that could alter the entire ecosystem with both beneficial and deleterious results. The control of one pest species could not be considered unrelated to the activities of other organisms. Furthermore, human economic activities and interests were integrally tied to the manipulation of the ecosystem. Insect control had to consider especially the fluctuations of the pest species and those of its parasites and predators.
>
> 2. Economic injury levels and economic thresholds. Mere presence of a pest organism in a farmer's field did not create cause for alarm. Only a population of the pest sufficiently high to cause crop injury equal to or greater than the cost of controlling the pest was deleterious; such a level was the economic injury level. Control measures in integrated control were to be taken at the economic threshold, a population density somewhat less than the economic injury level; thus action was indicated somewhat before the level at which actual damage was anticipated.
>
> 3. Augmentation of natural enemies. Classical biological control, in which a new parasite or predator was imported from another area, was the primary means of augmentation envisioned in . . . 1959. . . . Also important were (a) periodic release of natural enemies, (b) artificial release of the pest species when it was normally low to provide food for natural enemies, (c) selective breeding and release of natural enemies so that they could withstand adverse climates or insecticides, and (d) modification of the environment (such as dust control) to make it more suitable for the activities of natural enemies.
>
> 4. Selective insecticides. Chemicals were important but had to be used with care in order to preserve natural enemies. Selective killing of the pest could be accomplished with (a) materials of selective toxicity, (b) treatment of selected areas to leave reservoirs of natural enemies, (c) proper timing of application to spare natural enemies, and (d) use of short-lived compounds that break down in the environment before killing too many natural enemies. . . .
>
> 5. Supervised control. Integrated control presupposed substantial knowledge of the population dynamics of an insect pest and its nat-

ural enemies. . . . University of California scientists recommended that professional entomologists evaluate a grower's insect problems and recommend the proper course of action. . . .

[Ray] Smith and Hal T. Reynolds of the Riverside campus in 1965 refined the integrated control concept to include all methods of control: chemical, biological, cultural, genetic, attractants, and repellents. This enlargement began to change it from merely a philosophy and strategy of insect control into a scheme for production management on the farm. Cultural control was the key to this shift because it included selection of crop plant variety, fertilization, irrigation, weed control, planting dates, harvesting dates, and any other farm practice that had an impact on the size of a pest population. Significantly . . . Smith and Reynolds excluded eradication of a pest:

> The philosophy of pest control based on eradication of the pest species is the antithesis of integrated pest control. Nevertheless, eradication may be a legitimate goal under special circumstances. However, these circumstances do not prevail for most agricultural and forest pests. In most situations our goal should be to manage pest populations so as to eliminate them as pests but not to eradicate them.

The contemporary [Integrated Pest Management] IPM concept became a political and intellectual entity through a major research program during the 1970s, "The Huffaker Project," named after its director, [Berkeley entomologist] Carl Barton Huffaker. . . . The research community that developed around the IPM paradigm [employed] a set of assumptions about the natural world and man's role within it that was different from that implicitly held [by those who employed a] chemical control paradigm. . . .

The most important assumptions made in the IPM school were that (1) humans are a biological species firmly embedded in a complex ecosystem, (2) anything they do to control insects competing with them for resources must be based on the presupposition of man as an ecological entity, (3) man changes the environment with technology to meet his needs, and (4) those technologies are subject to limitations due to human ignorance about the complexity of the environment.

The IPM paradigm was . . . firmly based upon fundamental assumptions about the natural world and man's role in it. As the paradigm matured in the 1970s, some important additions were made. First, an explicit sense that man would achieve sound and safe pest control measures by mimicking nature was articulated. . . . Second, "We cannot afford any longer to disregard the considerable capabilities of pest organisms for countering control efforts." [This] suggests that man's technological powers may be limited by intrinsic biological factors.

Scientists in general and applied scientists in particular are ever reluctant to concede the existence of intrinsic limits to man's knowledge and power. Nevertheless, Robert van den Bosch, one of the foremost theoreticians of the IPM paradigm, moved to such a concession in 1978. He may not have represented a majority opinion within the IPM school, but his criticisms of chemical control were explicitly based on his sense of man's inability to dominate Nature:

> Our problem is that we are too smart for our own good, and for that matter, the good of the biosphere. The basic problem is that our brain enables us to evaluate, plan, and execute. Thus, while all other creatures are programmed by Nature and subject to her whims, we have our own gray computer to motivate, for good or evil, our chemical engine. Indeed, matters have progressed to the point where we attempt to operate independently of Nature, challenging her dominance of the biosphere. This is a game we simply cannot win, and in trying we have set in train a series of events that have brought increasing chaos to the planet.

It is important to note that those entomologists such as van den Bosch who had doubts about the ability of man to dominate the natural world based their pessimism on the ills associated with insecticides: resistance, resurgence, secondary-pest outbreaks, environmental damage, and health hazards. To these observations they added their convictions about the complexity of ecosystems and the evolutionary successes of the arthropods over the past 300 million years. In their own literature, they seldom resorted to explicit philosophical considerations about the nature of the man–environment relationship. Rather, they presented their conclusion that man was subject to domination as one derived from an objective consideration of empirical facts. . . .

IPM emphasized the interactions between the pest, its natural enemies, the plant, potential pests, and their natural enemies. IPM was clearly related to the notion of synecology, in which emphasis is given to the processes governing communities of organisms. . . .

Smith and van den Bosch clearly believed that "successful integrated pest control must be sensitive to . . . changes that occur both on the short-term dynamic basis and the long-term evolutionary basis." Their point was that entomologists needed to recognize the full complexity of ecosystems, and a historical-evolutionary perspective helped develop the needed sensitivities.

Restoring California's Vegetation

Michael Barbour, Bruce Pavlik, Frank Drysdale, and Susan Lindstrom

American beliefs, attitudes, and policies toward the environment are continually changing. As more of us have positive experiences in the natural environment, and as that natural environment shrinks, our society no longer fears it nor endeavors to subdue it. The closer we come to comprehending that we are an integral part of nature, the better we understand the consequences of our actions in nature. We have evolved from a society that instinctively cleared natural cover—often impoverishing the best land—to one that is beginning to conserve and enhance that cover. . . .

Until the past two decades disruption was minimally repaired at best. According to the 1970 California Environmental Quality Act (CEQA), it is the state's goal "to develop and maintain a high-quality environment," "to protect, rehabilitate, and enhance the environmental quality of the state," and "to preserve for future generations representations of all plant and animal communities." . . .

Much of the earth's surface today requires rehabilitation and enhancement, and California is an important place to begin. California's 100 million acres are only 1/360th of the world's thirty-six billion terrestrial acres, but we are entrusted with a landscape and a collection of organisms whose unique importance represents much more than 0.3 percent of the world's land area. Within its unique ecosystems, California has ninety-nine percent of all the redwoods (*Sequoia sempervirens*) on earth—the

From Michael Barbour, Bruce Pavlik, Frank Drysdale, and Susan Lindstrom, *California's Changing Landscapes* (Sacramento: California Native Plant Society, 1993), pp. 185–87, 203–5. Reprinted by permission.

tallest life form; all of the giant sequoias (*Sequoiadendron giganteum*)—the most massive life form; and many of the bristlecone pines (*Pinus longaeva*)—the oldest tree life form. Hundreds of less dramatic plant and animal species are found in California and nowhere else. . . .

There are global reasons for restoring California's plant cover. The concentration of carbon dioxide in the atmosphere has increased by about ten percent in this century alone, with a consequent rise in world temperature of 1 degree Fahrenheit. Natural vegetation serves as a correcting balance to carbon dioxide released in the burning of fuel, but when natural vegetation is degraded the balance is lost and the consequences of pollution are more immediately felt. In global context, Californians' treatment of vegetation is the same as the Brazilian harvest of tropical forests and the clearing of woodlands and savannas by Asians and Africans.

There are sound cultural and evolutionary reasons to restore the environment. For hundreds of thousands of years the human species and its predecessors evolved in the context of wilderness; our agricultural, pastoral, and urban history is, in comparison, only a moment of time. . . .

California is rich. Economically, the annual Gross State Product places the state among the top dozen nations in the world. This is a short-term, temporary kind of richness. Ecologically, the state is also wealthy, but this wealth has been declining for two centuries. The GNP increases yearly at the expense of ecological resources. . . .

Economic prosperity requires ecological richness. Long-term benefits, as well as short-term employment opportunities, will come to human populations from restoration activities. One of the primary lessons of ecology is that everything is connected to everything else. There is no free lunch for any of us. We may be sure that there is a cost for every alteration to the environment. Alterations must be sensibly and thoughtfully weighed.

There are also ethical considerations. We have to question our present relationship with the environment. Is it right to destroy vegetation, degrading that upon which life depends? Is it right that long-term ecosystem stability be sacrificed for short-term benefits? Is it right that we exploit other species? Is it right for one generation to extinguish forever habitats that future generations will never experience?

People interested in protecting and enhancing habitat quality must scrutinize land use plans. Management proposals for public lands must be monitored for conflict with natural vegetation. Many public agencies permit logging, grazing, and mining in direct conflict with protection of the environment. The activities are not economically justified; they are heavily subsidized by taxes, and they provide limited benefits to the public. Agency plans should fairly state the economic and environmental benefits of alternative, non-destructive land uses. Often, these non-destructive uses are undervalued, poorly quantified, or ignored. It is difficult to place a price on undisturbed watershed, high biomass, photosynthesis, non-game animal habitat, rich species diversity, and ecological complexity. We have never before in human existence gone without them. We do not know the cost of their destruction, but it may be a very high one. . . . If the past one hundred years of conservation efforts had not occurred in California, we would not have today's state and national parks, wilderness areas, and other protected lands.

The public has to make it known to policy makers that protection and enhancement of the environment is politically acceptable. . . . In September 1991 ten state and federal agencies signed a memorandum of understanding with several important features. The signers agreed to place "maintenance and enhancement of biological diversity a preeminent goal," and to join with others as members of a Biodiversity Council chaired by the secretary of the Resources Agency, . . . Douglas Wheeler. The council will recommend goals, standards, and guidelines for conserving biodiversity, and it will divide the state into ten to fifteen bioregions, each with its own regional council and agenda. We have progressed, writes Marc Hoshovsky, from protecting only game animals early in this century, to rare species by the 1970s, and now to biotic regions in the 1990s; from conservation initially to preservation now.

But the work is only just started. We must continue the efforts to protect remaining undisturbed areas, certainly; but we must also begin a process of restoration and enhancement of much larger areas in California, those beyond the pristine preserves. As all of California's cover regains its diversity, integrity, complexity, and stability in the coming century, so shall we.

Further Readings

Bailey, Gilbert E., and Paul S. Thayer. *California's Disappearing Coast: A Legislative Challenge*. Berkeley: Institute of Governmental Studies, University of California, 1971.

Barbour, Michael G. *Coastal Ecology: Bodega Head*. Berkeley: University of California Press, 1973.

Barbour, Michael, et al. *California's Changing Landscapes*. Sacramento: California Native Plant Society, 1993.

Carson, Rachel. *Silent Spring*. Boston: Houghton Mifflin, 1962.

Commoner, Barry. *The Closing Circle: Nature, Man, and Technology*. New York: Knopf, 1971.

Conrad, Les. *Desperate Remedies: The Tragedy of Santa Maria, California*. Santa Barbara: Atlas Signs, 1987.

DeBach, Paul. *Biological Control of Natural Enemies*. Cambridge: Cambridge University Press, 1974.

Dunlap, Thomas R. *DDT: Scientists, Citizens, and Public Policy*. Princeton, NJ: Princeton University Press, 1981.

Essig, Edward O. *A History of Entomology*. New York: Macmillan, 1931.

Goetzmann, William. *Exploration and Empire: The Explorer and Scientist in the Winning of the American West*. New York: Knopf, 1966.

Hart, John. *Storm over Mono: The Mono Lake Battle and the California Water Future*. Berkeley: University of California Press, 1996.

Hays, Samuel P. *Beauty, Health, and Permanence: Environmental Politics in the United States, 1955–1985*. New York: Cambridge University Press, 1987.

Helms, Douglas, and Susan L. Flader, eds. *The History of Soil and Water Conservation*. Washington, DC: Agricultural History Society, 1985.

Huffaker, Carl B., ed. *New Technology of Pest Control, Sponsored by the International Center for Integrated and Biological Control*. New York: John Wiley, 1980.

Jenny, Hans. *E. W. Hilgard and the Birth of Modern Soil Science*. Pisa, Italy: Collanda della Rivista "Agrochimica," 1961.

Jensen, Deborah B., Margaret S. Torn, and John Harte. *In Our Own Hands: A Strategy for Conserving California's Biological Diversity*. Berkeley: University of California Press, 1993.

Knox, Joseph B., ed. *Global Climate Change and California*. Berkeley: University of California Press, 1991.

Leopold, Starker, and Tupper Ansel Blake. *Wild California: Vanishing Lands, Vanishing Wildlife*. Berkeley: University of California Press, 1985.

Lewis, Richard S. *The Nuclear-Power Rebellion: Citizens vs. the Atomic Industrial Establishment*. New York: Viking, 1972.

MacLeod, R., and P. F. Rehbock. *Nature in Its Greatest Extent: Western Science in the Pacific*. Honolulu: University of Hawaii Press, 1988.

Marshall, Robert. *The People's Forests*. New York: H. Smith and R. Haas, 1933.

McEvoy, Arthur. *The Fisherman's Problem: Ecology and Law in the California Fisheries, 1850–1980*. New York: Cambridge University Press, 1986.

Melosi, Martin V. *Garbage in the Cities: Refuse, Reform, and the Environment, 1880–1980*. College Station, TX: Texas A&M University Press, 1981.

Mitchell, Lee Clark. *Witnesses to a Vanishing America: The Nineteenth Century Response*. Princeton, NJ: Princeton University Press, 1981.

Nash, Roderick. *Wilderness and the American Mind*. New Haven: Yale University Press, 1967.

Perkins, John H. *Insects, Experts, and the Insecticide Crisis: The Quest for New Pest Management Strategies*. New York: Plenum, 1982.

Smith, Michael L. *Pacific Visions: California Scientists and the Environment, 1850–1915*. New Haven: Yale University Press, 1987.

van den Bosch, Robert. *The Pesticide Conspiracy*. Garden City, NY: Doubleday, 1978.

Vernon, Raymond. *The Oil Crisis*. New York: Norton, 1976.

Wells, George Stevens. *Garden in the West; A Dramatic Account of Science in Agriculture*. New York: Dodd, Mead, 1969.

Whorton, James. *Before Silent Spring: Pesticides and Public Health in Pre-DDT America*. Princeton, NJ: Princeton University Press, 1974.

Worster, Donald. *Nature's Economy: A History of Ecological Ideas*. Cambridge: Cambridge University Press, 1977.

Worster, Donald, ed. *American Environmentalism: The Formative Period, 1860–1915*. New York: Wiley, 1973.

Chapter 13

Revisioning California: Contemporary Environmental Movements

Documents

Ernest Callenbach on Population in Ecotopia, 1975

The *Times-Post* is at last able to announce that William Weston, our top international affairs reporter, will spend six weeks in Ecotopia, beginning next week. This unprecedented journalistic development has been made possible through arrangements at the highest diplomatic level. It will mark the first officially arranged visit by an American to Ecotopia since the secession [of northern California, Oregon, and Washington] cut off normal travel and communications in 1980.

San Francisco, May 20, [1999]. Ecotopia's population is slowly declining, and has done so for almost 15 years. This startling fact—which by itself would set Ecotopia apart from the U.S. and all other nations except Japan—has led to speculation that rampant abortion and perhaps even infanticide may be practiced here. However, I have now investigated suf-

From Ernest Callenbach, *Ecotopia: The Notebooks and Reports of William Weston* (Berkeley: Banyan Tree Books, 1975), pp. 61–64. Reprinted by permission of the author.

ficiently to report that Ecotopia's decline in population has been achieved through humane measures.

We tend to forget that even before Independence the rate of population growth in the area that became Ecotopia had slowed, as it had in most of the rest of the U.S. This was due, according to American demographers, partly to the persistent inflation-recession, partly to the relaxation of abortion laws, and perhaps most of all to increased recognition that additional children, in a highly advanced industrial society, could be more of a burden than an advantage to a family—the reverse of the situation in agricultural or less advanced societies. In addition, the horrible "Green Revolution" famines, in which tens of millions perished in Pakistan, India, Bangladesh, and Egypt, had provided new and grim lessons in the dangers of overpopulation.

After secession, the Ecotopians adopted a formal national goal of a declining population—though only after long and bitter debate. It was widely agreed that some decline was needed, to lessen pressure on resources and other species and to improve the comfort and amenity of life. But opinions differed widely on exactly how a decline could be achieved, and how far it should go. Deep fears of national extinction gave heavy ammunition against advocates of population decline, and economists warned of fiscal dislocations.

Finally a three-stage program was adopted. The first stage, to last through 1982, was a massive educational and medical campaign aimed at providing absolutely all women with knowledge of the various birth-control devices. Abortion upon demand was legalized; its cost soon became very low, and it was practiced in local clinics as well in hospitals. . . .

The second stage, 1983 through 1984, was linked to the radical decentralization of the country's economic life, and was thus was more political in nature. During this period the Ecotopians largely dismantled their national tax and spending system, and local communities regained control over all basic life systems. This enabled people to deliberately think about how they now wished to arrange their collective lives, and what this meant in population levels and distribution. With better conditions in the countryside, the great concentrations of people in San Francisco, Oakland, Portland, Seattle, and even the smaller metropolitan areas began to disperse somewhat. New minicities grew up in favorable locations, with their own linkage necklaces of transit lines: Napa, on its winding, Seine-like River, at last pollution free; Carquinez-Martinez, stretching out along rolling hills dropping down to the Strait; and others throughout the country. Some old city residential areas were abandoned and razed, and the land turned into parks or reforested. Some rural towns,

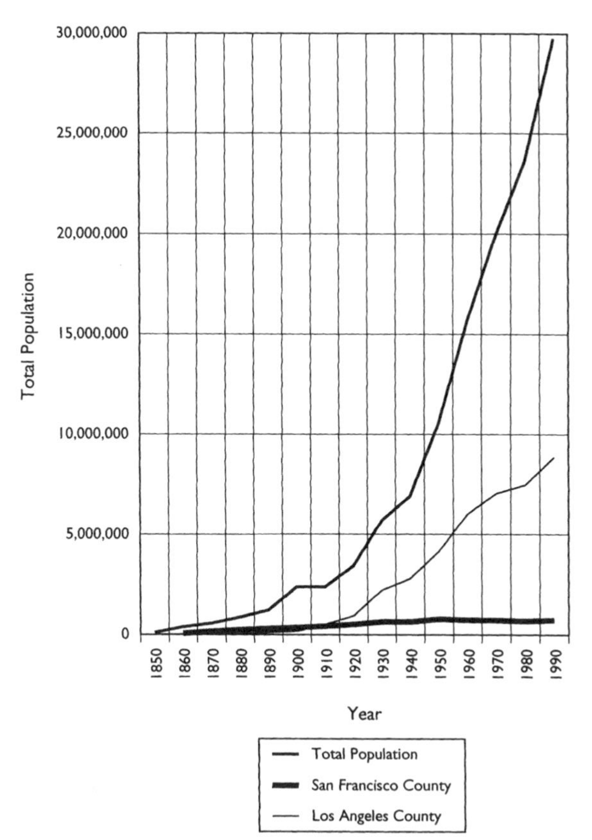

FIGURE 10. CALIFORNIA'S CHANGING POPULATION, 1850–1990.
California's population grew exponentially between 1850 and 1990, when it was close to 30 million. Los Angeles County grew at a similarly rapid rate while San Francisco County's population increased more slowly. Graph by Jessica Teisch; data from David Hornbeck, California Patterns: A Geographical and Historical Atlas. *Mountain View, CA: Mayfield Publishing Company, 1983, pp. 94–95, and 1990 U.S. Bureau of the Census.*

like Placerville, which had been in the 10–20,000 people range, gained satellite minicities that would in a decade bring them to a total of 40–50,000—which was felt to be about ideal for an urban constellation.

Decentralization affected every aspect of life. Medical services were dispersed. . . . Schools were broken up and organized on a novel teacher-controlled basis. Agricultural, fishery, and forestry enterprises were also reorganized and decentralized. Large factory-farms were broken up through a strict enforcement of irrigation acreage regulations which had been ignored before Independence, and commune and extended-family farms were encouraged. . . .

Thus, the pressures for further population control measures waned during 1984. When final statistics were in, however, the population had indeed taken its first actual drop—by about 17,000 people for Ecotopia as a whole. This fact was not greeted by the hysteria that had been widely predicted, and people probably took grim satisfaction from the news that American society, with its widely publicized overpopulation, had grown by another three million during the same period.

The third stage, if we can call it that, was one of watchful waiting, which has continued to the present. Abortion costs have fallen further, and the number per year has stabilized. The use of contraceptive devices now seems universal. (They are all, incidentally, female-controlled; there is no "male pill" here.) Population has tended to drop gently at a rate of around 65,000 per year, so that the original Ecotopian population of some 15 million has now declined to about 14 million. It is argued by some extremists that the declining population provides a substantial annual surplus per capita and helps account for the vitality of Ecotopian economic life. . . .

What will happen to Ecotopian population levels in the future? Most people here foresee a continued slow decline. They consider that a more rapid drop might endanger the nation, making it more vulnerable to attack by the United States—which is still widely feared to be desirous of recapturing its "lost territories." On the other hand, some people hope that American population will itself soon begin to decline—and if that happens, many Ecotopians are prepared to accept an indefinite drop in their own numbers. In fact, some radical Survivalist Party thinkers believe that a proper population size would be the number of Indians who inhabited the territory before the Spaniards and Americans came—something less than a million for the whole country, living entirely in thinly scattered bands! Most Ecotopians, however, contend that the problem is no longer numbers as such. They place their faith for improvement of living conditions in the further reorganization of their cities into constellations of minicities, and in a continued dispersion into the countryside.

Raymond Dasmann and Peter Berg Advocate Bioregionalism, 1980

Living-in-place means following the necessities and pleasures of life as they are uniquely presented by a particular site, and evolving ways to ensure long-term occupancy of that site. A society which practices living-in-place keeps a balance with its region of support through links between human lives, other living things, and the processes of the planet—seasons, weather, water cycles—as revealed by the place itself. It is the opposite of a society which "makes a living" through short-term destructive exploitation of land and life. Living-in-place is an age-old way of existence, disrupted in some parts of the world a few millennia ago by the rise of exploitative civilization, and more generally during the past two centuries by the spread of industrial civilization. It is not, however, to be thought of as antagonistic to civilization in the more humane sense of that word, but may be the only way in which a truly civilized existence can be maintained. . . .

Once all California was inhabited by people who used the land lightly and seldom did lasting harm to its life-sustaining capacity. Most of them have gone. But if the life-destructive path of technological society is to be diverted into life-sustaining directions, the land must be reinhabited. Reinhabitation means learning to live-in-place in an area that has been disrupted and injured through past exploitation. It involves becoming native to a place through becoming aware of the particular ecological relationships that operate within and around it. It means undertaking activities and evolving social behaviour that will enrich the life of that place, restore its life-supporting systems, and establish an ecologically and socially sustainable pattern of existence within it. Simply stated it involves becoming fully alive in and with a place. It involves applying for membership in a biotic community and ceasing to be its exploiter. . . .

Reinhabitation involves developing a bioregional identity, something most North Americans have lost, or have never possessed. . . . The term [bioregion] refers both to geographical terrain and a terrain of conscious-

From Raymond Dasmann and Peter Berg, "Reinhabiting California," *The Ecologist*, 7, no. 10 (1980): 399–410. Reprinted by permission of the authors.

ness—to a place and the ideas that have developed about how to live in that place. . . .

A bioregion can be determined initially by use of climatology, physiography, animal and plant geography, natural history and other descriptive natural sciences. The final boundaries of a bioregion are best described by the people who have long lived within it, through human recognition of the realities of living-in-place. All life on the planet is interconnected in a few obvious ways, and in many more that remain barely explored. But there is a distinct resonance among living things and factors which influence them that occurs specifically within each separate place on the planet. Discovering and describing that resonance is a way to describe a bioregion.

The realities of a bioregion are obvious in a gross sense. Nobody would confuse the Mojave desert with the fertile valley of Central California, nor the Great Basin semi-arid land with the California coast. Between the major bioregions the differences are sufficiently marked that people do not usually attempt to practice the Sonoran desert way of life in the Oregonian coastal area. But there are many intergradations. The chaparral-covered foothills of southern California are not markedly distinct from those of the coast ranges of northern California. But the attitudes of people and the centers to which they relate (San Francisco vs. Los Angeles) are different and these can lead to different approaches to living on the land.

The northern California bioregion is ringed by mountains on the north, east, and south and extends some distance into the Pacific Ocean on the west. Since the boundaries depend in part on human attitudes they cannot be clearly mapped. These attitudes, however, have been persistent since prehistoric times. The region is separated from Southern California by the barrier of the Tehachapi Mountains and their extension through the Transverse Ranges to Point Conception on the seaward side. Flora and fauna change to some extent on either side of this boundary, but human attitudes are more important in the separation. Eastward, the region is enclosed by the Sierra Nevada which stops the rain and defines the dry Nevadan bioregion. Northward the volcanic Cascade Range and the geologically ancient Klamath Mountains separate the Oregonian bioregion. Along the coast the boundaries are fuzzy, but one could draw a line at the northern limit of the coastal redwood forests, at Oregon's Chetco River.

Within the bioregion is one major watershed, that of the Sacramento-San Joaquin river system which drains from all of the Sierra-Nevada, Cascade, and interior Coast Ranges and flows through the broad plain of the Central Valley. Coastally, smaller watersheds are significant, those of the Salinas, Russian, Eel, Mad, Klamath and Smith rivers. The Klamath River is anomalous in that it drains from an area that belongs to a different bioregion. So too does the Pit River which joins the Sacramento. Otherwise the drainage systems help to define and tie together the life of the bioregion, and the characteristics of watersheds point out the necessities which those who would live-in-place must recognize.

Biologically the California biotic province, which forms the heart of the bioregion, is not only unique but somewhat incredible—a west coast refuge for obscure species full of endemic forms of plants and animals. It is a Mediterranean climatic region unlike any other in North America. It is a place of survival for once widespread species as well as a place where other distinct forms evolved. Anthropologically it is also unique, a refuge for a great variety of non-agricultural peoples on a continent where agriculture had become dominant. . . .

Our real "period of discovery" has just begun. The bioregion is only barely recognized in terms of how life systems relate to each other within it. It is still an anxious mystery whether we will be able to continue living here. How many people can the bioregion carry without destroying it further? . . .

Natural watersheds could receive prominent recognition as the frameworks within which communities are organized. The network of springs, creeks, and rivers flowing together in a specific area exerts a dominant influence on all non-human life there; it is the basic designer of local life. Floods and droughts in northern California remind us that watersheds affect human lives as well, but their full importance is more subtle and pervasive. Native communities were developed expressly around local water supplies and tribal boundaries were often set by the limits of watersheds. Pioneer settlements followed the same pattern, often displacing native groups with the intention of securing their water.

Defining the local watershed, restricting growth and development to fit the limits of water supplies, planning to maintain these and restore the free flowing condition of tributaries that are blocked or the purity of any which have been polluted, and exploring relationships with the larger water systems connecting to it could become primary directions for reinhabitory communities. They could view themselves as centered on and responsible for the watershed.

The Central Valley has become one of the planet's food centers. The current scale of agriculture there is huge. . . . It's a naturally productive place. Northern California has a temperate climate, a steady supply of water, and the topsoil is some of the richest in North America. But the current scale of agriculture is untenable in the long term. Fossil fuel and chemical fertilizer can only become more expensive, and the soil is simultaneously being ruined and blown away.

There needs to be massive redistribution of land to create smaller farms. They would concentrate on growing a wider range of food species (including native food plants, increasing the nutritional value of crops, maintaining the soil, employing alternatives to fossil fuels, and developing small-scale marketing systems). More people would be involved, thereby creating jobs and lightening the population load on the cities.

Forests have to be allowed to rebuild themselves. Clearcutting ruins their capability to provide a long-term renewable resource. Watershed-based reforestation and stream restoration projects are necessary everywhere that logging has been done. Cut trees are currently being processed wastefully; tops, stumps and branches are left behind, and whole logs are shipped away to be processed elsewhere and sold back in the region. Crafts that use every part of the tree should be employed to make maximum use of the materials while employing a greater number of regional people. Fisheries have to be carefully protected. They provide a long-term life-support of rich protein, if used correctly, or a quickly emptied biological niche, if mishandled. Catching fish and maintaining the fisheries have to be seen as parts of the same concern.

Reinhabitory consciousness can multiply the opportunities for employment within the bioregion. New reinhabitory livelihoods based on exchanging information, cooperative planning, administering exchanges of labor and tools, intra- and inter-regional networking, and watershed media emphasizing bioregional rather than city-consumer information could replace a few centralized positions with many decentralized ones. The goals of restoring and maintaining watersheds, topsoil, and native species invite the creation of many jobs to simply un-do the bioregional damage that invader society has already done.

Beginning with the Spanish Occupation, the distinctiveness of northern California's ongoing bioregional life has been obscured by a succession of alien super-identities. The place to fit into simply wasn't recognized. First, it was part of "New Spain," a designation that tells nothing of this specific place and lumps it with a dozen barely related bioregions radiating out from the Caribbean. "California" was a fictional island created by a 16th century Spanish novelist and it became the next rough label pasted over the bioregion when it was adopted for the Pacific side of New Spain. "Alta California" actually approximated the bioregion by accident; its real use was simply to acknowledge further Spanish explorations above the "baja." Mexico held it (along with half the western

U.S.) in the early 19th century, but since the middle of last century almost the whole bioregion has been included in the annexed portion of Mexican territory that was sliced out as the State of California along with totally foreign pieces of the Great Basin desert and, similarly dry, stretches below the Tehachapi Mountains.

The bioregion that exists largely in what is now called northern California has now become visible as a separate whole, and, for purposes of reinhabiting the place, it should have a political identity of its own. It is predictable that as long as it belongs to a larger state it will be subject to southern California's demands on its watersheds. . . . From a reinhabitory point of view, both are bioregional death threats. Elections over the last decade have shown a distinct difference in voting sentiments between northern and southern California. It is likely that this difference will continue and increase on vital bioregional issues on which the population weight of southern California will prevail.

The bioregion cannot be treated with regard for its own life continuities while it is part of and administered by a larger state government. It should be a separate state. As a separate state, the bioregion could redistrict its counties to create watershed governments appropriate to maintaining local life-places. City–country divisions could be resolved on bioregional grounds. Perhaps the greatest advantage of separate statehood would be the opportunity to declare a space for addressing each other as members of a species sharing the planet together and with all the other species.

Bill Devall and George Sessions Explain Deep Ecology, 1982

Deep ecology is emerging as a way of developing a new balance and harmony between individuals, communities and all of Nature. It can potentially satisfy our deepest yearnings: faith and trust in our most basic intuitions; courage to take direct action; joyous confidence to dance with the sensuous harmonies discovered through spontaneous, playful intercourse

From Bill Devall and George Sessions, *Deep Ecology: Living as if Nature Mattered* (Salt Lake City: Peregrine Smith books, 1982), pp. 7–10. Reprinted by permission.

with the rhythms of our bodies, the rhythms of flowing water, changes in the weather and seasons, and the overall processes of life on Earth. We invite you to explore the vision that deep ecology offers.

The deep ecology movement involves working on ourselves, what [California] poet Gary Snyder calls "the real work," the work of really looking at ourselves, of becoming more real.

This is the work we call cultivating ecological consciousness. This process involves becoming more aware of the actuality of rocks, wolves, trees, and rivers—the cultivation of the insight that everything is connected. Cultivating ecological consciousness is a process of learning to appreciate silence and solitude and rediscovering how to listen. It is learning how to be more receptive, trusting, holistic in perception, and is grounded in a vision of nonexploitive science and technology.

This process involves being honest with ourselves and seeking clarity in our intuitions, then acting from clear principles. It results in taking charge of our actions, taking responsibility, practicing self-discipline and working honestly within our community. It is simple but not easy work. Henry David Thoreau, nineteenth-century naturalist and writer, admonishes us, "Let your life be a friction against the machine."

Cultivating ecological consciousness is correlated with the cultivation of conscience. [California] historian Theodore Roszak suggests in *Person/Planet* (1978), "Conscience and consciousness, how instructive the overlapping similarity of those two words is. From the new consciousness we are gaining of ourselves as persons perhaps we will yet create a new conscience, one whose ethical sensitivity is at least tuned to a significant good, a significant evil."

We believe that humans have a vital need to cultivate ecological consciousness and that this need is related to the needs of the planet. At the same time, humans need direct contact with untrammeled wilderness, places undomesticated for narrow human purposes.

Many people sense the needs of the planet and the need for wilderness preservation. But they often feel depressed or angry, impotent and under stress. They feel they must rely on "the other guy," the "experts." Even in the environmental movement, many people feel that only the professional staff of these organizations can make decisions because they are experts on some technical scientific matters or experts on the complex, convoluted political process. But we need not be technical experts in order to cultivate ecological consciousness. Cultivating ecological consciousness, as Thoreau said, requires that "we front up to the facts and determine to live our lives deliberately, or not at all." We believe that people can clarify their own intuitions, and act from deep principles.

Deep ecology is a process of ever-deeper questioning of ourselves, the assumptions of the dominant worldview in our culture, and the meaning and truth of our reality. We cannot change consciousness by only listening to others, we must involve ourselves. We must take direct action. . . .

One hopeful political movement with deep ecology as a base is the West German Green political party. They have as their slogan, "We are neither left nor right, we are in front." . . . The Greens present a promising political strategy because they encourage the cultivation of personal ecological consciousness as well as address issues of public policy. If the Greens propagate the biocentric perspective—the inherent worth of other species besides humans—then they can help change the current view which says we should use Nature only to serve narrow human interests.

[California philosopher] Alan Watts, who worked diligently to bring Eastern traditions to Western minds, used a very ancient image for this process, "invitation to the dance," and suggests that "the ways of liberation make it very clear that life is not going anywhere, because it is already *there*. In other words, it is playing, and those who do not play with it, have simply missed the point." . . .

The trick is to trick ourselves into reenchantment. As Watts says, "In the life of spontaneity, human consciousness shifts from the attitude of strained, willful attention to *koan*, the attitude of open attention or contemplation." This is a key element in developing ecological consciousness. This attitude forms the basis of a more "feminine" and receptive approach to love, an attitude which for that very reason is more considerate of women.

In some Eastern traditions, the student is presented with a *koan*, a simple story or statement which may sound paradoxical or nonsensical on the surface, but as the student turns and turns it in his or her mind, authentic understanding emerges. This direct action of turning and turning, seeing from different perspectives and from different depths, is required for the cultivation of consciousness. The *koan*-like phrase for deep ecology, suggested by prominent Norwegian philosopher Arne Naess, is: "simple in means, rich in ends."

Cultivating ecological consciousness based on this phrase requires the interior work of which we have been speaking, but also a radically different tempo of external actions, at least radically different from that experienced by millions and millions of people living "life in the fast lane" in contemporary metropolises. As Theodore Roszak concludes, "Things move slower; they stabilize at a simpler level. But none of this is experienced as a loss or a sacrifice. Instead, it is seen as a liberation from waste and busywork, from excessive appetite and anxious competition that allows one to get on with the essential business of life, which is to work out one's salvation with diligence."

Irene Diamond and Gloria Orenstein on the Emergence of Ecofeminism, 1990

Today, more than twenty-five years after Rachel Carson's *Silent Spring* first raised a passionate voice of conscience in protest against the pollution and degradation of nature, an ecofeminist movement is emerging globally as a major catalyst of ethical, political, social, and creative change. Although Carson was not an avowed feminist, many would argue that it was not coincidental that a woman was the first to respond both emotionally and scientifically to the wanton human domination of the natural world. Carson's 1962 text prefigured a powerful environmental movement that culminated in the nationwide Earth Day of 1970, but the notion that the collective voices of women should be central to the greening of the Earth did not blossom until the mid to late 1970s.

Ecofeminism is a term that some use to describe both the diverse range of women's efforts to save the Earth and the transformations of feminism in the West that have resulted from the new view of women and nature. With the birth of the Women's Movement in the late 1960s, feminists dismantled the iron grip of biological determinism that had been used historically to justify men's control over women. Feminists argued that social arrangements deemed to be timeless and natural had actually been constructed to validate male superiority and privilege. They asserted that women had the right to be full and equal participants in the making of culture. In this process writers and scholars documented the historical association between women and nature, insisting that women would not be free until the connections between women and the natural world were severed.

But as the decade advanced and as women began to revalue women's cultures and practices, especially in the face of the twin threats of nuclear

From Irene Diamond and Gloria Orenstein, *Reweaving the World: The Emergence of Ecofeminism* (San Francisco: Sierra Club Books, 1990), pp. ix–xii. Reprinted by permission.

annihilation and ecocide, many women began to understand how the larger culture's devaluation of natural processes was a product of masculine consciousness. Writers as diverse as Mary Daly, Elizabeth Dodson Gray, Susan Griffin, Carolyn Merchant, Maria Mies, Vandana Shiva, Luisah Teish, and Alice Walker demonstrated that this masculine consciousness denigrated and manipulated everything defined as "other" whether nature, women, or Third World cultures. In the industrialized world, women were impelled to act, to speak out against the mindless spraying of chemicals, the careless disposal of toxic wastes, the unacknowledged radiation seepage from nuclear power plants and weapons testing, and the ultimate catastrophe—the extinction of all life on Earth. In the Third World, women had still more immediate concerns. For women who had to walk miles to collect the water, fuel, and fodder they needed for their households, the devastation wrought by patriarchal fantasies of technological development (for example, the Green revolution, commercial forest management, and mammoth dam projects) was already a daily reality.

In many ways, women's struggle in the rural Third World is of necessity also an ecological struggle. Because so many women's lives are intimately involved in trying to sustain and conserve water, land, and forests, they understand in an immediate way the costs of technologies that pillage the Earth's natural riches. By contrast, in the industrialized world, the connections between women's concerns and ecological concerns were not immediately apparent to many feminists. Community activists such as Rachel E. Bagby, Lois Gibbs, and Carol Von Strom, who were struggling to protect the health of their families and neighborhoods, were among the first to make the connections. . . . Through the social experience of caretaking and nurturing, women become attentive to the signs of distress in their communities that might threaten their households. When environmental "accidents" occur, it is these women who are typically the first to detect a problem. Moreover, because of women's unique role in the biological regeneration of the species, our bodies are important markers, the sites upon which local, regional, or even planetary stress is often played out. Miscarriage is frequently an early sign of the presence of lethal toxins in the biosphere.

Feminists who had been exploring alternatives to the traditional "woman is to nature as man is to culture" formulation, who were seeking a more fundamental shift in consciousness than the acceptance of women's participation in the marketplace of the public world, began to question the nature versus culture dichotomy itself. These activists, theorists, and artists sought to consciously create new cultures that would embrace and honor the values of caretaking and nurturing—cultures that would not perpetuate the dichotomy by raising nature over culture or by raising women over men. Rather, they affirmed and celebrated the embeddedness of all the Earth's peoples in the multiple webs and cycles of life.

In their hope for the creation of new cultures that would live with the Earth, many women in the West were inspired by the myths and symbols of ancient Goddess cultures in which creation was imaged as female and the Earth was revered as sacred. Others were inspired by the symbols and practices of Native-American cultures that consider the effects on future generations before making any community decision. The sources of inspiration were many and varied and led to a diverse array of innovative practices—from tree-planting communities, alternative healing communities, organic food coops, performance art happenings. . . . The languages they created reached across and beyond the boundaries of previously defined categories. These languages recognized the lived connections between reason and emotion, thought and experience. They embraced not only women and men of different races, but all forms of life—other animals, plants, and the living Earth itself.

Carl Anthony Explains Why African Americans Should Become Environmentalists, 1990

When Martin Luther King Jr. decided to raise his voice in opposition to the war in Vietnam, many of his friends, as well as his critics, told him he ought to stick to domestic issues. He should concentrate on securing civil rights of African Americans in the South and leave foreign policy to the professionals who knew best. But King decided to oppose the war because he knew it was morally wrong and because he understood the link between the brutal exploitation and destruction of the Vietnamese people and the struggle of African Americans and others for justice and freedom in our own land.

Today, African American leadership and the African American com-

From Carl Anthony, "Why African Americans Should Become Environmentalists," *Earth Island Journal* (Winter 1990): 43–44. Reprinted by permission.

munity face a similar situation. Every day the newspapers carry stories about the changing atmosphere and climate, threats to the world's water supply, threats to the biodiversity of the rainforest, and the population crisis in poor nations that are growing too fast to be supported by the carrying capacity of their lands. Can we afford to view the social and economic problems of African American communities in isolation from these global trends?

African Americans could benefit from expanding their vision to include greater environmental awareness. For example, a recent study of the deteriorating conditions within the African American community termed young African American males in America "an endangered species." "This description applies, in a metaphorical sense, to the current status of young African American males in contemporary society," writes Jewelle Taylor Gibbs. Her study . . . presents a comprehensive interdisciplinary perspective of the social and economic problems of these young people, providing valuable statistics on high school dropout rates, work skills and attitudes, unemployment, robbery, rape, homicide and aggravated assault, drug addiction as well as teenage parenthood. But Gibbs makes no mention of the utter alienation of these young people from the natural environment, which is, after all, the source of Earth's abundance and well-being. The loss of this contact with living and growing things, even rudimentary knowledge of where food and water comes from, must present serious consequences that we, as yet, have no way of measuring.

The study said nothing of the difficult days ahead as American society seeks to make the transition from its current levels of consumption of resources to the more sustainable patterns of the future. Developing an environmental perspective within the African American community could help smooth this transition in several ways:

- by promoting greater understanding of the productive assets of society, including land, water and natural resources,
- by strengthening collaboration with groups seeking to redirect public investment and economic development away from wasteful exploitation of nature toward urban restoration and meeting basic human needs,
- by gaining access to information and resources which enhance the potential of community survival,
- by developing new knowledge and skills to be shared by groups of people who live in the city,

- by strengthening social and political organization and creating new opportunities for leadership within the community.

Environmental organizations in the United States should also modify recruitment efforts in order to expand their constituency to include African Americans and members of other minority groups as participants in shaping and building public support for environmental policies. With the exception of limited collaboration between environmentalists and Native American groups, as well as anti-toxics campaigns, there has been little communication between environmentalists and non-European minority groups in the U.S. Critical issues—such as population control, limiting human intervention in the ecosystem, or rebuilding our cities in balance with nature—have been discussed almost entirely from a European and often elitist perspective. . . .

The principle of social justice, however, must be at the heart of any effort aimed at bringing African Americans into the mainstream of environmental organizations in the United States. . . . Environmental protection must be understood as intimately connected to efforts to eradicate injustice. Solutions must offer a practical guide for goals which can be accomplished in the short run as we seek a path toward a more sustainable future. Environmental organizations can no longer afford to take the view that they are unconcerned about who benefits and who loses from restrictions on economic growth. Shifting resources away from projects which are damaging to the ecosystem toward programs and projects which meet basic human needs must become the highest priority for the environmental movement. In the United States, organizations such as the National Association for the Advancement of Colored People and the Urban League have a real stake in these outcomes. They should be part of the environmental dialogue. New organizations dealing explicitly with urban habitat are needed. . . .

The American inner-city was once a wilderness. Today, islands, estuaries, forests, and riparian habitats that once existed in these locations have been replaced by asphalt, concrete, barbed wire fences, boarded-up stores, crack houses, abandoned factories, landfills and pollution. After generations of isolation and manipulation, the people who live in these places rarely remember what it once was—or speculate on what it might become.

Isolation of African Americans from stewardship of the environment has deep historic roots. It is hard to keep the faith. The African American population migrated to the cities to escape the four centuries of exploitation on the plantations, crop farms, and in the coal mines of the South. Displacement from rural countryside is parallel to similar experiences in the Third World. Understanding of these experiences, however painful, is an important resource as we seek a path towards sustainable development. . . .

In order to meet responsibilities for citizenship, African Americans must have opportunities and learn to play a greater role in formulating environmental policies which affect all members of the community. We must find new ways to bridge the gap between environmental advocates and African American communities.

Magdelena Avila Tells How Hispanic Farmworkers Stopped an Incinerator, 1992

[When Chemical Waste Management wanted to locate a toxic waste incinerator in Kettleman City, California, it] initially assessed [the community] as [one] of low resistance, one without political power. They felt they were dealing with a "sleepy lagoon": "No problem, we're going to pass right through." But this Goliath company, playing the role of corporate monster, completely miscalculated the community's resources and capacity. Kettleman City includes many individuals with tremendous experience in leadership roles, in grassroots organizing, much of it stemming from the farmworkers struggles. All these years of experience and skills have been drawn upon.

El Pueblo para el Aire y Agua Limpio—People for Clean Air and Water [founded in 1989]—pulled together all the different sections of the community: the campesinos, the growers, the residents who aren't involved with agriculture, the youth. The organization has been bilingual and bicultural from the beginning.

The two main people in organizing *El Pueblo* have been Esperanza Maya and Mary Lou Mares, two extremely strong *mujeres*. Espy and her husband Joe Maya have a grower background and brought in that

From Magdelena Avila, "David Versus Goliath," *Forward Motion/Crossroads* (April 1992): 13–15. Reprinted by permission.

strength and that perspective. Mary Lou and Ramon Mares are real grassroots, they knew the campesinos and were working with them. They brought in that dimension. . . . Then you have people like Jose and Esperanza Cuevas, and my father, representing the senior elders. One thing about comunidad, especially in this case, is that El Pueblo pulled in respected individuals whose words are listened to. The youth also were not ignored; Mary Lou's daughters and Espy's children have been involved.

El Pueblo has sustained the struggle for over three years, keeping the community organized and maintaining momentum. There is a core group of 10 to 15 that sometimes grows to 35. When you call huge membership meetings, the whole town shows up, 300 or more people. Of course there are lulls when there's no mass activity, but those times are just as crucial as the peak moments. During the lulls the core people continue working on coalition building. . . . During these periods, though, there is sometimes despair, the community feels a sense of desolation—where will we get the resources to continue? Are we by ourselves?

But landmark events follow the valleys. . . . A major landmark activity took place October 12, 1991, on the *Dia de la Raza*. It was a march for environmental justice with a twofold purpose: to focus on the struggle of Kettleman City, and at the same time to give a platform to all the different communities struggling in similar ways. Over 85 organizations came on board—American Indian organizations fighting the attempt by toxic waste companies to hit the reservations, Mothers of East L.A., . . . students, the Berkeley Coalition for Environmental Justice and many more. About 1,000 people from five states came to Kettleman. Jesse Jackson and Maxine Waters were main speakers.

The relationship between the community and outsiders has been crucial. [A] toxic waste dump . . . already exists in Kettleman—in 1978 it was literally snuck in. The community had no idea what was happening. But when Chem Waste tried to sneak in the incinerator three years ago, Greenpeace had access to information that the community did not have. Greenpeace came to Kettleman City and said that Chem Waste is here, this is what they're trying to do. Overnight, the people said it's not going to happen and the organizing began. . . .

I've been a critic of the major environmental organizations for their failure to recognize and address issues of concern to people of color beyond mere tokenistic mention and participation. My earlier perception of [Greenpeace] was that this was an organization about saving the whales and the dolphins. When Greenpeace came to Kettleman, there was some sense of—I won't say distrust, but not immediate *confianza*, not immediately accepting somebody into your house. But I never saw Greenpeace try to take the agenda away from the community. I remember planning meetings where Greenpeace was there to provide the resources. They would say let's put out a flyer. Espy or Mary Lou would say, we don't want the Greenpeace name on it because it's important that this be conveyed in a certain way. And Greenpeace would without question accept their decision. Bradley Angel of Greenpeace has been a tremendous source of resources and support.

California Rural Legal Assistance (CRLA), which took on the legal aspect of the case, has also played a key role. Attorney Luke Cole, who has dealt with poverty law and has worked with the communities at risk for toxic waste dumping, came in and offered to represent us. The community met and asked what it was going to cost. CRLA said it's not going to cost anything. Between Greenpeace and CRLA, you had the involvement of two well-established organizations at the state and also the national level. They were key in feeding information to Kettleman City, in connecting us to other individuals and communities involved in similar struggles, in providing technical assistance.

Altogether, what you have in Kettleman City is a case of coalition building, not only among grassroots organizations but also with professional organizations and the academic community, tying in researchers and people who can provide information and expertise. It's the coalition building model that has made this battle so effective. . . . This is a coalition that's building into a major social movement.

A social movement doesn't take off just like that. It's like a volcano. It's there, it's happening, things are working underneath to build the foundation. In this case, the key factor is that people of color in the environmental movement recognize that environmental issues are not the same for us as they are for the so-called mainstream. We look at the civil rights movement, at struggles against discrimination, at all the struggles that have been key to obtaining justice. Around the dangers of environmental racism, people of color feel there's a common net pulling everything together. . . .

There are still not enough [people], but there is now a critical mass of people of color in positions that are strategic for assisting community empowerment. And that's what we're looking at, self-determination and empowerment, as opposed to the sense of dependency that has been created for so long. The fight against environmental racism is shaping up as one of the key social movements of the 1990s.

I've experienced all this first-hand—I've come full circle. I was involved with my family in the United Farmworkers movement in Kettleman City, fighting against pesticides and organizing labor unions. That's why I don't see environmental racism as a new or "trendy" phenomenon. It's been an issue for a long time, it just didn't have the name "environmental racism." And today it's a point of convergence where the African American community, the Native American and Latino and Asian communities are all coming together. . . .

That strength . . . won a big victory. In November [1991], Kettleman City's legal case went before the Superior Court of Sacramento County. The judge first had to decide if Chem Waste and the county were required to do another Environmental Impact Report (EIR) because the first one had been deficient in presenting the information to the public—because it was not translated into Spanish. On January 2 [1992], the judge ruled in favor of Kettleman City, requiring Chem Waste and the county to conduct another EIR. The ruling indicated that they needed to translate all documents and the materials—that the community had been denied due process by not being fairly informed. . . .

Kettleman City is a community in education, a community of self-determination, and this education has really helped to empower the residents. This is a landmark fight not only for Kettleman, but for the whole Central Valley and the Latino community, and the movement against environmental racism by people of color nationally and internationally in the [1990s].

Essays

Chaos and California

Mike Davis

Once or twice each decade, Hawaii sends Los Angeles a big, wet kiss. Sweeping far south of its usual path, the westerly jetstream hijacks water-laden tropical air from the Hawaiian archipelago and hurls it toward the Southern California coast. This "Kona" storm system—dubbed the "Pineapple Express" by television weather reporters—often carries several cubic kilometers of water, or the equivalent of half of Los Angeles's annual precipitation. And when the billowing, dark turbulence of the storm-front collides with the high mountain wall surrounding the Los Angeles Basin, it sometimes produces rainfall of a ferocity that is unrivaled anywhere on earth, even in the tropical monsoon belts.

The two-week-long Kona storm of January 1995 differed little from the classic pattern, except perhaps in the unusual intensity of rainfall in the South Bay area, forcing the evacuation of low-lying neighborhoods in Long Beach, Carson, Torrance and Hawaiian Gardens. Otherwise, the scenes were those of an ordinary, familiar disaster: Power was cut off to tens of thousands of homes. Sinkholes mysteriously appeared in front-yards. Pet animals and several children were sucked into the deadly vortices of the flood channels. Reckless motorists were drowned in flooded intersections. Lifeguards had to rescue shoppers in downtown Laguna Beach. Million-dollar homes tobogganed off their hill-slope perches.

What was exceptional was not the storm itself (a "20-year event" according to meteorologists), but the way in which it was instantly assimilated to other recent disasters as a malevolent omen. There is growing

From Mike Davis, "Los Angeles after the Storm," *Antipode*, 27, no. 3 (1995): 221–41. Reprinted by permission.

popular apprehension that the former Land of Sunshine is reinventing itself (to use a fashionable gerund) as a Book of the Apocalypse theme-park. First the natives rioted, then Nature. In less than three years, the megalopolis has endured three of the ten most costly domestic disasters since the Civil War. The fierce February 1992, January 1993 and January and March 1995 floods (approximately $400 million damage) were mere brackets around the April 1992 insurrection ($1 billion), the October-November 1993 firestorms ($1 billion), and the January 1994 earthquake ($20 billion). . . .

This virtually biblical conjugation of disaster is unique in American history, and it has purchased thousands of one-way tickets to Seattle, Portland, and Santa Fe. After a century of population influx, there is now a net exodus out of Southern California. Middle-class apprehensions about the angry, abandoned underclasses are only exceeded by anxieties about blind thrust faults and hundred-year floods. Meanwhile, Caltech seismologists warn that the Pacific Rim is only beginning its long overdue rock and roll: Kobe [Japan] may be a 3-D postcard of Los Angeles 2000. And waiting in the wings are the locusts and killer bees.

It is unclear, moreover, whether this vicious circle of disaster is coincidental or eschatological. Could this be merely what nonlinear statisticians wave away as the "Joseph effect" of fractal geometry: "the common clustering of catastrophe"? Or are these the Last Days, as prefigured so often in the genre of L.A. disaster fiction and film (from Day of the Locust to Earthquake and Wilshire Boulevard)? Either way—[Benoit] Mandelbrot or Nathanael West—millions of Angelenos have become genuinely terrified of their environment.

Paranoia about nature, of course, distracts attention from the obvious fact that Los Angeles has deliberately put itself in harm's way. For generations, market-driven urbanization has transgressed environmental common-sense. Historic wild-fire corridors have been turned into view-lot suburbs, wetland liquefaction zones into marinas, and flood plains into industrial districts and housing tracts. Monolithic public works have been substituted for regional planning and a responsible land ethic. As a result, Southern California has reaped flood, fire, and earthquake tragedies that were as avoidable, and unnatural, as the beating of Rodney King and the ensuing explosion in the streets.

But the social construction of "natural" disaster is largely hidden from view by a perverse ideology that simultaneously imposes false categories and expectations on the environment, and then explains the inevitable discrepancies as proof of a malign and hostile Nature. Pseudo-science, in the service of rampant greed, has militarized Anglo-American perceptions of the regional landscape. Southern California, in the most profound sense, suffers a crisis of identity.

No belief, for instance, is more deeply rooted than the self-serving conviction that Los Angeles would be Death Valley except for the great aqueducts that transfer the stolen snow-melt of the Sierras and Rockies to its lawns and pools. The city is advertised as the triumph of super-engineers like William Mulholland who built rivers in the desert. A corollary of this promethean claim, of course, is the idea that beneath the artificial landscape there is something sinister and barren, incapable of sustaining even a tiny fraction of the current multitudes.

Yet Los Angeles County, while semi-arid, is no more desert than Murcia or the Côte d'Azur (which have the same annual rainfall). Technology and concrete have alienated any clear view of its natural history. In contrast, the earliest written descriptions of the region, the eighteenth-century diaries of the Franciscan padres, eulogized its waterscapes and natural fertility.

"All the soil is black and loamy, and is capable of producing every kind of grain and fruit which may be planted," Father Juan Crespi wrote in 1769. "We went west, continually over good land well covered with grass. . . . All the land we saw this morning seemed admirable to us." The diaries of Fathers Francisco Palou and Pedro Font also extolled the "abundant springs [*cienegas*]," "beautiful rivers," and the valleys "green and flower-strewn." After crossing the true deserts of Sonora and Antigua California, these Mediterranean men (Majorcans) were delighted by the familiar oak savannahs and the "infinity of wild rosebushes in full bloom." From their cultural perspective, it was land "well-watered."

The Anglo-American conquistadors three quarters of a century later, however, were riven by confusion and ambivalence. Boosterism coexisted with an almost irrational fear of aridity, and there were great debates during the 1850s and 1860s about whether California as a whole was Eden or worthless desolation. . . .

In the most fundamental sense, language and cultural inheritance failed the newcomers. English terminology, specific to a humid climate, was incapable of describing the dialectic of water and drought that shaped Mediterranean environments. By no stretch of the imagination, for example, is an arroyo merely a "glen" or "hollow"—they are the results of radically different hydrological processes. . . . The Anglos were thus forced to preserve the more accurate Spanish terms without appreciating their larger environmental context.

It was not until the discovery of the great artesianal basins—millions of acre-feet of subterranean water—during the 1870s, and the subsequent growth of the citrus industry, that a luxuriant stereotype of Southern California became possible. Even then, there was the initial mistake of advertising it as "sub-tropical": a term that aroused in nervous Eastern imaginations the image of a malarial swamp with green tree snakes coiled in the branches. . . .

The extreme events that shape the Southern Californian environment tend to be organized in surprising and powerfully-coupled causal chains. Drought, for example, dries fuel for wildfires which, in turn, remove ground-cover and make soils impermeable to rain. (Earthquakes may have already created new erosion surfaces and increased stream-power by elevation change.) In such conditions, storms produce sheet flooding, landslides and debris flows that result in dramatic erosion and landform change. Vast volumes of sediment realign river channels, and, before the advent of 20th-century flood-control engineering, even switched river courses between alternate deltas. Sedimentation also creates sandbars that temporarily cut off tidal flow to coastal marshes; initiating, in turn, a 50–75 year cycle of ecological readjustment.

This is not random disorder, but a hugely complicated system of positive and negative feedback loops, geomorphic and biological, that channel powerful pulses of climatic or tectonic energy ("disasters") into environmental work. The outcomes, however, are radically unpredictable. The Southern California environment exemplifies the non-linear rules of "deterministic chaos," governed by both probabilistic and deterministic elements. In such systems, entropy (in the mathematical rather than thermodynamic sense) measures the relative degree of unpredictability or the range of possible states. Southern California, accordingly, might be described as having high "landscape entropy" vis-à-vis most temperate regions. . . .

This temporal/structural complexity of landforms, in turn, contributes to the relatively greater spatial hetereogeneity, and, thus, biotic diversity, of Mediterranean environments. "Their complex physical geography makes possible," explains [fire historian Steven] Pyne," . . . a complex geography of life, the mosaic of microclimates sustaining a mosaic of microbiotas." The biodiversity of Mediterranean regions is second only to tropical rainforests. California alone has more than 5,000 native plants.

Finally, landscape and biome have coevolved. Each of the ecosystems comprising the original biological mosaic of the Los Angeles region was evolutionarily adapted to conditions of high entropy and periodic convulsion. . . .

Some researchers, it should be added, are trying to use Complexity

Theory to model the behavior of high-energy Mediterranean environments as "nonlinear dynamical systems." But if Ilya Prigogine and the Santa Fe Institute provide a framework for thinking about chaotic dynamics, then so does the Prophet John of the *Revelation*. . . .

The need for a new environmental epistemology is most evident in any discussion of the massive, single-purpose infrastructures that supposedly leash nature to the will of Los Angeles. As John McPhee has pointed out so eloquently in *The Control of Nature* (1989), the struggle between human and natural wills is totally one-sided. Bureaucratic faith in the immortality of public works derives exclusively from attenuated, almost meaningless, official time-frames. In effect, we think ourselves gods upon the land, but are, at most, mere tourists. . . .

But metropolitan Los Angeles fatally lacks the emergency capacity—engineers would call it "redundancy"—to deal with earthquake clusters or inevitable 200- or 500-year droughts and floods. The cheapest and most sensible form of redundancy, of course, would have been hazard zoning (to exclude development from the most disaster-prone terrains) combined with a conservative water ethic. Time and again, prophetic voices urged the region's leaders to mitigate inevitable disaster and enhance quality-of-life through restricting urbanization in the wetlands and foothills. They were ignored, and, as a result, there is scarcely any open space left (except at the outer edges) to preserve as a buffer against temperamental Mediterranean nature.

In the meantime, it is unclear how Southern California will afford the escalating costs of ordinary disaster, much less the immense damage of probable mega-events. The region's pharaonic public works are rapidly aging. Indeed, in some cases their functional working lives are proving to be less than an average human lifetime. John McPhee tells the sobering story of Pasadena's Devil's Gate Dam, designed to last generations, which filled up with debris in less than fifty years. It is now a sand-and-gravel quarry. . . . As urban growth, moreover, ceaselessly replaces porous soil surface with concrete and asphalt, the flood-control system loses its capacity to contain another 100-year storm event.

Likewise, as Los Angeles' urban fabric continues to be extended into new floodplains, liquefaction zones and mountain fire ecologies, the social costs of protecting private development from natural disaster are exploding. To deal with the impact of built-out urbanization in the eastern San Gabriel Valley and the western Inland Empire, the Army Corps of Engineers is spending almost $1.5 billion on new flood-control works in the Los Angeles and Santa Ana River channels. In the wake of Northridge, the California insurance industry is conducting a ruthless, and thus far successful, campaign to shift the burden of homeowner earthquake insurance to the public sector. And the continuing growth of white-flight suburbs in the chapparal belt is raising the public costs of fire protection to unforeseen levels.

So far the Clinton administration—mesmerized by the electoral prize of California in 1996—has been willing to spend more than $12 billion to finance the Los Angeles area's recovery from riot, fire, flood and earthquake. At the time of writing, however, a full-fledged Congressional rebellion, led by senior Democrats, has broken out against providing further federal disaster relief to California or Florida. Their bill would replace FEMA subsidies with a voluntary state-level insurance program whose premiums would cost billions of dollars a year in Southern California. . . .

Without FEMA to provide Keynesian injections of aid, the continuing clustering of disaster—on ordinary or extraordinary scales—inevitably will erode many of the comparative advantages of the Southern California economy. However, despite the wishful thinking of some "deep ecologists," who think that Gaia would be happiest with a thin sprinkling of hunter-gatherers, mega-cities like Los Angeles will never simply collapse and disappear. Rather, we will simply stagger on, with higher body counts and greater distress, through a chain of more frequent and destructive disasters. . . . Afficionados of complexity theory will marvel at the "nonlinear resonances" of unnatural disaster and social breakdown as the Golden Age is superseded forever by strange new chaotic attractors.

Grassroots Environmentalism

Robert Gottlieb

On the face of it, the Golden Arches, a block from the ocean in Santa Monica, California, seemed an unlikely place for Penny Newman to start her week of organizing for what she and others called the Movement for

Excerpt granted with permission from *Forcing the Spring: The Transformation of the American Environmental Movement* by Robert Gottlieb. Copyright © 1993 by Robert Gottlieb, pp. 162–64, 168–69, 170. Published by Island Press, Washington, DC and Covelo, CA.

Environmental Justice. But McDonald's had become a key target for dozens of grassroots groups, several of them high school-based, that were involved in the McToxics Campaign. Launched in early 1987, the campaign was initially designed to pressure McDonald's to eliminate its trademark clamshell package, a polystyrene foam container produced with the use of chlorofluorocarbons (CFCs), an ozone-destroying compound. . . .

The McToxics Campaign was a colorful and effective affair. It included boycotts, guerrilla theater (protesters dressed in styrofoam suits), hit-and-run tactics (hundreds of used clamshells left on countertops or mailed to McDonald's owner Joan Kroc), and related efforts at education and mobilization. Flustered by the campaign's tactics, McDonald's officials first denied there were problems. They argued that styrofoam was just "basically air" and performed a valuable function by "[aerating] the soil." Sensitive to issues of image, the company eventually eliminated the use of CFCs and, in a November 1990 agreement with the Environmental Defense Fund, shifted entirely away from polystyrene. . . .

For the activists, the real value of the campaign had been its ability to demonstrate how ordinary people could get involved and "make things happen," as Newman put it. The fast food industry as well as other toxics producers and users were coming under scrutiny for the first time from "average people" who were relaying a potentially radical message, Newman declared in her interview. "What we're saying," Newman continued, looking up at the Golden Arches, "is that you don't have to be an elected official or an industry executive to have an impact on waste policy." . . .

After checking the mail and the continuous flow of messages from the answering machine in her storefront office, Newman was off to a meeting in Riverside concerning the Stringfellow Acid Pits, the high-profile Superfund hazardous waste site that had consumed more than ten years of Newman's life. This meeting would plan for the next day's gathering of the Stringfellow Advisory Committee, on which Newman served as a representative of the community along with industry and government officials.

The industrial waste dump site at Stringfellow had opened in 1956 and was less than two miles from Newman's home. The first major dumper at the site was the U.S. Air Force, which deposited chemicals used for refurbishing missiles at nearby Norton Air Force Base. For more than sixteen years Stringfellow had received huge quantities of hazardous materials, including more than 34 million gallons of liquid wastes. The wastes included heavy metals, solvents, pesticides (including DDT), and large amounts of sulfuric, nitric, and hydrochloric acid (thus its name, the Acid Pits). These wastes were deposited by chemical, aerospace, steel, and aluminum companies as well as by plating operations, agricultural concerns, and the U.S. Air Force. Fifteen of these companies would eventually be held liable for cleanup costs estimated then to exceed $600 million. The California Department of Health Services would also be held liable for negligently overseeing construction of the dump site and then delaying taking action to rehabilitate it. . . .

During the 1970s, the Acid Pits became a political football. Various government agencies entered the picture, hiring consultants, issuing reports that included numerous health risk disclaimers, and focusing on containing the wastes rather than pursuing a more expensive cleanup process. In 1979 and 1980, heavy rains and floods caused another 5.5 million gallons of liquid hazardous wastes to spill from the site, an event that had a dramatic effect on Newman's family. One of Newman's sons, both during and after the discharges from the site, began to suffer serious asthma-related breathing problems, dizzy spells, headaches, and blurred vision. Many residents had similar complaints. Livid at the inaction of the agencies and their reassurances about no risk, the residents intensified their own efforts. A new organization, Concerned Neighbors in Action, was formed, and Newman emerged as one of its leaders. . . .

As Newman settled into her organizer's role, she became a prominent figure in the organization and a charismatic speaker who could inspire her audiences of like-minded activists. One such moment had been a keynote speech at [a Citizen's Clearinghouse for Hazardous Waste] convention in 1989 that had brought delegates out of their seats. . . . "The lesson of the Movement is that we are the power," Newman had told the gathering. "We are the experts. We are the ones who have watched our community devastated. We're the ones who have watched our life's investment in our homes disappear. We are the ones who lie awake at night listening to our children struggling to breathe." . . .

Resonating with some of the language and spirit of the 1960s movements, these networks have helped bring about specific victories by mobilizing communities, especially to block waste sites. Despite the fact that several of the community groups have disappeared once proposed facilities have been stopped, the network itself has continued to grow. More directly, the Movement for Environmental Justice exists primarily as an idea, a possibility more than a fully defined organizational approach. Dwarfed by the resources and media recognition of the mainstream environmental organizations, these grassroots groups, part of a potential social justice movement, have had to struggle to keep afloat and maintain their vision of community defense and environmental transformation. Still, as Penny Newman prepared that chilly October day to undertake her long drive back to her home by the Acid Pits, she knew that her week of organizing was continuing the process of constructing a new kind of environmentalism, a movement involved in issues of everyday life made up of advocates like herself.

Since the 1970s, there has emerged, distinct from the mainstream groups, a powerful current in contemporary environmentalism focused on issues of empowerment, environmental justice, equity, and urban and industrial restructuring. . . . This alternative movement is predominantly local in nature, more participatory and focused on action, and critical of the roles of expertise and lobbying in defining environmental agendas. With a direct lineage to earlier urban and industrial movements, the alternative groups have sought to develop a new framework for environmental change, relying on constituencies often underrepresented or excluded from environmental decision making and drawing on such critical concepts as citizen empowerment and the prevention or reduction (rather than management or control) of pollution.

A Place in Space

Gary Snyder

Jets heading west on the Denver-to-Sacramento run start losing altitude east of Reno, and the engines cool as they cross the snowy Sierra crest. They glide low over the west-tending mountain slopes, passing above the canyon of the north fork of the American River. If you look north out the window you can see the Yuba River country, and if it's really clear you can see the old "diggings"—large areas of white gravel laid bare by nineteenth-century gold mining. On the edge of one of those is a little hill where my family and I live. It's on a forested stretch between the South Yuba Canyon and the two thousand treeless acres of old mining gravel, all on a forty-mile ridge that runs from the High Sierra to the valley floor near Marysville, California. You're looking out over the northern quarter

From Gary Snyder, "Kitkitdizze: A Node in the Net," in *A Place in Space* (Washington, DC: Counterpoint, 1995), pp. 252, 258–63. Reprinted by permission.

of the greater Sierra ecosystem: a vast summer-dry hardwood-conifer forest, with drought resistant shrubs and bushes in the canyons, clear-cuts, and burns. . . .

The whole Sierra is a mosaic of ownership—various national forests, Bureau of Land Management, Sierra Pacific Industries, state parks, and private holdings—but to the eye of a hawk it is one great sweep of rocks and woodlands. We, along with most of our neighbors, were involved in the forestry controversies of the last decade, particularly in regard to the long-range plans for the Tahoe National Forest. The county boosters still seem to take more pleasure in the romance of the gold era than in the subsequent processes of restoration. The Sierra foothills are still described as "Gold Country," the highway is called "49," there are businesses called "Nugget" and "Bonanza." I have nothing against gold—I wear it in my teeth and in my ear—but the real wealth here is the great Sierran forest. My neighbors and I have sat in on many hearings and had long and complicated discussions with silviculturalists, district rangers, and other experts from the Forest Service. All these public and private designations seem to come with various "rights." With just "rights" and no land ethic, our summer-dry forests could be irreversibly degraded into chaparral over the coming centuries. We were part of a nationwide campaign to reform forest practices. . . .

We next turned our focus to the nearby public lands managed by the BLM. It wasn't hard to see that these public lands were a key middle-elevation part of a passageway for deer and other wildlife from the high country to the valleys below. . . . [We] launched a biological inventory, first with older volunteers and then with our own wild teenagers jumping in. We studied close to three thousand forested acres. We bushwhacked up and down the canyons to find out just what was there, in what combinations, in what quantity, in what diversity. Some of it was tallied and mapped (my son Kai learned Geographical Information Systems techniques and put the data into a borrowed Sun Sparc workstation), and the rest of our observations were written up and put into bundles of notes on each small section. We had found some very large trees, located a California spotted owl pair, noted a little wetland with carnivorous sticky sundew, described a unique barren home with serpentine endemics (plants that grow only in this special chemistry), identified large stands of vivid growing forest, and were struck by the tremendous buildup of fuel. The well-intended but ecologically ignorant fire-exclusion policies of the government agencies over the last century have made the forests of California an incredible tinderbox. . . .

The biological inventory resulted in the formation of the Yuba Watershed Institute, a nonprofit organization made up of local people, sponsoring projects and research on forestry, biodiversity, and economic sustainability with an eye to the larger region. One of the conclusions of the joint management plan, unsurprisingly, was to try to reduce fuel load by every available means. We saw that a certain amount of smart selective logging would not be out of place, could help reduce fuel load, and might pay some of the cost of thinning and prescriptive burning. We named our lands, with the BLM's blessing, the Inimin Forest, from the Nisenan word for pine, in recognition of the first people here.

The work with fire, wildlife, and people extends through public and (willing) private parcels alike. Realizing that our area plays a critical biological role, we are trying to learn the ground rules by which humans might live together with animals in an "inhabited wildlife corridor." . . . Our cooperative efforts here can be seen as part of the rapidly changing outlook on land management in the West, which is talking public-private partnership in a big way. Joint-management agreements between local communities and other local and committee interests, and their neighboring blocks of public lands, are a new and potent possibility in the project of responsibly "recovering the commons" region by region. The need for ecological literacy, the sense of home watershed, and a better understanding of our stake in public lands are beginning to permeate the consciousness of the larger society.

Lessons learned in the landscape apply to our own lands, too. So this is when my family and I are borrowing from the watershed work as our own Three-Hundred-Year Kitkitdizze Plan: We'll do much more understory thinning and then a series of prescribed burns. Some patches will be untouched by fire, to provide a control. We'll plant a few sugar pines, and incense cedars where they fit (ponderosa pines will mostly take care of themselves), burn the ground under some of the oaks to see what it does for the acorn crop, burn some bunchgrass patches to see if they produce better basketry materials (an idea from the Native basket-weaving revival in California). We'll leave a percentage of dead oak in the forest rather than take it all for firewood. In the time of our seventh-generation granddaughter there will be a large area of fire-safe pine stands that will provide the possibility of the occasional sale of an incredibly valuable huge, clear, old-growth sawlog.

We assume something of the same will be true on surrounding land. The wildlife will still pass through. And visitors from the highly crowded lowlands will come to walk, study, and reflect. A few people will be resident on this land, getting some of their income from forestry work. The rest may come from the information economy of three centuries hence. There might even be a civilization with a culture of cultivating wildness.

You can say that this is outrageously optimistic. It truly is. But the possibility of saving, restoring, and wisely (yes!) using the bounty of wild nature is still with us in North America. My home base, Kitkitdizze, is but one tiny node in an evolving net of bioregional homesteads and camps.

Beyond all this studying and managing and calculating, there's another level to knowing nature. We can go about learning the names of things and doing inventories of trees, bushes, and flowers, but nature as it flits by is not usually seen in a clear light. . . . One must be tuned to hints and nuances.

After twenty years of walking right past it on my way to chores in the meadow, I actually paid attention to a certain gnarly canyon live oak one day. Or maybe it was ready to show itself to me. I felt its oldness, suchness, inwardness, oakness, as if it were my own. Such intimacy makes you totally at home in life and in yourself. But the years spent working around that oak in that meadow and not really noticing it were not wasted. Knowing names and habits, cutting some brush here, getting firewood there, watching for when the fall mushrooms bulge out are skills that are of themselves delightful and essential. And they also prepare one for suddenly meeting the oak.

Further Readings

Adams, Robert. *Los Angeles Spring*. New York: Aperture, 1986.

Athearn, Robert G. *The Mythic West in Twentieth-Century America*. Lawrence: University Press of Kansas, 1986.

Baldwin, Donald N. *The Quiet Revolution: Grass Roots of Today's Wilderness Preservation Movement*. Boulder, CO: Pruett Pub. Co., 1972.

Barrett, T., and P. Livermore. *The Conservation Easement in California*. Covelo, CA: Island Press, 1983.

Brown, Michael. *Laying Waste: The Poisoning of America by Toxic Chemicals*. New York: Pantheon Book, 1980.

Cawley, R. McGreggor. *Federal Land, Western Anger: The Sagebrush Rebellion and Environmental Politics*. Lawrence: University of Kansas Press, 1993.

Commoner, Barry. *The Closing Circle: Nature, Man, and Technology*. New York: Knopf, 1971.

Conomos, T.J., ed. *San Francisco Bay: The Urbanized Estuary*. San Francisco: The Pacific Division, 1979.

Cook, Martin L. *Saving the Earth: The History of a Middle-Class Millenarian Movement*. Berkeley: University of California Press, 1990.

Cornford, Daniel, ed. *Working People of California*. Berkeley: University of California Press, 1995.

Davis, Charles, ed. *Western Public Lands and Environmental Politics*. Boulder, CO: Westview Press, 1997.

Dowie, Mark. *American Environmentalism at the Close of the Twentieth Century*. Cambridge, MA: MIT Press, 1995.

Dunlap, Thomas. *Saving America's Wildlife*. Princeton, NJ: Princeton University Press, 1988.

Epstein, Barbara. *Political Protest and Cultural Revolution: Nonviolent Direct Action in the 1970s and 1980s*. Berkeley: University of California Press, 1991.

Etulain, Richard. *Re-Imagining the American West: A Century of Fiction, History and Art*. Tucson: University of Arizona Press, 1996.

Gottlieb, Robert. *Forcing the Spring: The Transformation of the American Environmental Movement*. Washington, DC: Island Press, 1993.

Hanson, Dirk. *The New Alchemists: Silicon Valley and the Microelectronics Revolution*. Boston: Little, Brown, 1982.

Hart, John, ed. *The New Book of California Tomorrow: Reflections and Projections from the Golden State*. Los Altos, CA: W. Kaufman, 1984.

Hays, Samuel P. *Beauty, Health, and Permanence: Environmental Politics in the United States, 1955–1985*. New York: Cambridge University Press, 1987.

Helvarg, David. *The War against the Greens: The Wise Use Movement, the New Right, and Anti-Environmental Violence*. San Francisco: Sierra Club Books, 1994.

Hine, Robert V. *California's Utopian Colonies*. Berkeley: University of California Press, 1983.

Hofrichter, Richard, ed. *Toxic Struggles: The Theory and Practice of Environmental Justice*. Philadelphia: New Society Publishers, 1993.

Hynes, H. Patricia. *The Recurring Silent Spring*. New York: Pergamon Press, 1989.

Jensen, Deborah B., Margaret S. Torn, and John Harte. *In Our Own Hands: A Strategy for Conserving California's Biological Diversity*. Berkeley: University of California Press, 1993.

Knox, Joseph B., and Ann Foley Scheuring, eds. *Global Climate Change and California: Potential Impacts and Responses*. Berkeley: University of California Press, 1991.

Kohm, Kathryn A. *Balancing on the Brink of Extinction: The Endangered Species Act and Lessons for the Future*. Washington, DC: Island Press, 1991.

Lillard, Richard G. *Eden in Jeopardy: Man's Prodigal Meddling with His Environment: The Southern California Experience*. New York: Knopf, 1966.
Manes, Christopher. *Green Rage: Radical Environmentalism and the Unmaking of Civilization*. Boston: Little, Brown, 1990.
Merchant, Carolyn. *Earthcare: Women and the Environment*. New York: Routledge, 1996.
Pendley, William Perry. *War on the West: Government Tyranny on America's Great Frontier*. Washington, DC: Regnery Publishing, Inc., 1995.
Scott, Mel. *The Future of San Francisco Bay*. Berkeley: Institute of Governmental Studies, University of California, 1963.
Scott, Stanley, ed. *Coastal Conservation: Essays on Experiments in Governance*. Berkeley: Institute of Governmental Studies, University of California, 1981.
Short, C. Brant. *Ronald Reagan and the Public Lands: America's Conservation Debate, 1979–1984*. College Station, TX: Texas A&M University Press, 1989.
Snyder, Gary. *A Place in Space*. Washington, DC: Counterpoint, 1995.
Trefethen, James B. *An American Crusade for Wildlife*. New York: Winchester Press, 1975.
Truettner, William, ed. *The West as America: Reinterpreting Images of the Frontier*. Washington, DC: Smithsonian Institution Press, 1991.
Walters, Dan. *The New California: Facing the 21st Century*. Sacramento: California Journal Press, 1986.
Wood, Samuel E., and Alfred E. Heller. *California, Going, Going, . . .* , Sacramento, CA: California Tomorrow, 1962.
Worster, Donald. *Nature's Economy: A History of Ecological Ideas*. Cambridge: Cambridge University Press, 1977.
Wyant, William K. *Westward in Eden: The Public Lands and the Conservation Movement*. Berkeley: University of California Press, 1982.
Wyatt, David. *The Fall into Eden: Landscape and Imagination in California*. New York: Cambridge University Press, 1986.

Conclusion: Environmental Ethics and California's Future

In his *Nichomachean Ethics*, Aristotle noted that "all knowledge and every pursuit aims at some good." But whether this is an individual, social, or environmental good lies at the basis of many real world ethical dilemmas. Three forms of ethics—egocentric, homocentric, and ecocentric—often underlie the positions of private individuals (and corporations), government agencies, and environmentalists engaged in struggles over land and natural resources in California. Here I exemplify these ethical dilemmas and propose a new partnership ethic for the future that may help to resolve them.

Egocentric Ethics

An egocentric ethic is grounded in the self. It is based on an individual thought focused on an individual good. In its applied form, it involves the claim that what is good for the individual will benefit society in general. The individual good is thus prior to the social good which follows from it as a necessary consequence. An egocentric ethic's orientation does not derive from selfishness or narcissism, but rather is based on a philosophy that treats individuals (or private corporations) as separate,

Revised from Carolyn Merchant, "Environmental Ethics and Political Conflict: A View From California," *Environmental Ethics*, 12, no. 1 (Spring 1990): 45–68.

but equal, social atoms. It is often expressed as trickle-down economics: what is good for the individual is good for society as a whole.

How have egocentric ethics been actuated with respect to the California environment? In *The Fisherman's Problem*, environmental historian Arthur McEvoy describes the management of the California fisheries in terms of the problem of the depletion of the commons. After the settlement of California by Euro-Americans in the eighteenth and nineteenth centuries, individual exploitation of river and ocean fish superseded the communal management of fishing by native American groups. Fish, like gold nuggets, were commodities to be extracted from the state of nature and turned into profits. As in Santa Barbara ecologist Garrett Hardin's "tragedy of the commons," McEvoy notes that "American authorities recognized . . . that pollution and overharvesting could degrade inland fisheries. But the problem was that those forces were so diffused over society, every individual contributing a negligible share, as to be legally uncontrollable." By the late nineteenth century, depletion of the rivers made it essential that fishing be regulated through laws and managed by government agencies—Hardin's "mutual coercion mutually agreed upon." The law as a form of rational human cognition regulated exploitation. Conflict of interest cases resulted in the curtailing of fishing by minority groups such as Indians and the Chinese. To manage the fisheries, the newly created federal fishing agency and the California state Board of Fish Commissioners studied the problem scientifically and stocked the rivers with exotic fish.

A second example involving the egocentric ethic is the Santa Barbara oil spill. Union Oil Company of California, part of the consortium that had leased rights from the Federal government to drill for oil in a tract off the Santa Barbara coast, experienced a blowout of one of its deep water wells on January 28, 1969. Union's development reflected an egocentric ethic of self-interest. A corporation founded in the Santa Barbara area with assets of $2.4 billion, its directors sought to maximize profits and to elevate it from the eleventh largest oil company in the United States to a place among the Big Ten. Its oil drilling, petrochemical, tanker, and manufacturing operations made it an industrial giant. The blowout caused a large oil slick to spread toward Santa Barbara, invading the commons of water, air, and public beaches. Ecological effects included the damaging of barnacles, surf grass, California Sea Lions, and thousands of birds including grebes, loons, murres, cormorants, brown pelicans, and sea gulls, as well as introducing aromatic hydrocarbons into the food chain. Garrett Hardin's analysis also applies to this "tragedy of the commons." The advantage to Union Oil in using the ocean commons to drill for oil was +1, while the environmental consequences to them of polluting the commons were much less than –1 because the costs were shared by the oil companies and the public. The oil spill resulted in stricter controls and fines on environmental pollution, the development of a growing body of environmental law, and a "Declaration of Environmental Rights" that included the statement, "We must extend ethics beyond social relations to govern man's contact with all life forms and with the environment itself." . . .

From an environmental point of view, the egocentric ethic that legitimates *laissez-faire* capitalism has a number of limitations. Because egocentric ethics is based on the assumption that the individual good is the highest good, the collective behavior of human groups or business corporations is not considered a legitimate subject of investigation. Moreover, because it includes the assumption that humans are "by nature" competitive and capitalism is the "natural" form of economics, ecological effects are external to human economics and cannot be adjudicated. In the nineteenth century, however, the first of these problems was dealt with through a new form of environmental ethics—the homocentric or utilitarian ethic. In the twentieth century, the problem of internalizing ecological externalities was addressed through the development of ecocentric ethics.

Homocentric Ethics

A homocentric (or anthropocentric) ethic is grounded in society. A homocentric ethic underlies the social interest model of politics and the approach of environmental regulatory agencies that protect human health. The utilitarian ethics of Jeremy Bentham (1789) and John Stuart Mill (1861), for example, advocate that a society ought to act in such a way as to ensure the greatest good for the greatest number of people. The social good should be maximized, social evil minimized. For both Bentham and Mill, the utilitarian ethic has its origins in human sentience. Feelings of pleasure are good, those of pain are evil and to be avoided. Because people have the capacity for suffering, society has an obligation to reduce suffering through policies that maximize social justice for all. In the late nineteenth century conservationists such as WJ McGee, Gifford Pinchot, and Theodore Roosevelt added the concept of time, stating the conservation ethic as "the greatest good for the greatest number for the longest time."

How have homocentric ethics been actuated in California? A particularly salient example is the building of dams for water and hydraulic power for cities and states. The controversy in the early twentieth century over whether to dam Hetch-Hetchy Valley in Yosemite Park as a source for water and power for the city of San Francisco is a case in point. Gifford Pinchot, arguing for San Francisco, pointed out that a water supply for the city was a greater good for a greater number of people than leaving the valley in the state of nature for a few hikers and nature lovers. John Muir on the other hand viewed the valley as one of God's cathedrals and the proponents of the dam as temple destroyers, an ethic based on the valley's intrinsic right to remain as created. Today water control agencies such as the San Francisco Water Department or the Metropolitan Water District of Los Angeles are quite explicit in their claim that they must consider the greatest good for the greatest number of people in distributing water to their customers in time of shortages.

The controversy over the damming of the Stanislaus River in the 1970s is another example. Federal officials wanted to provide flood control and water delivery for farmers, whereas environmentalists asserted that the river had a right to continue in its own state of nature as a wild river. New Melones dam was proposed as part of the Bureau of Reclamation's Central Valley Project in the 1930s to control flooding and to recharge ground water sources, and in 1962 the plans were expanded to include hydropower, irrigation, and recreation. Congressman John McFall, who fought for authorization to build the dam, adopted a utilitarian stance, arguing that a "larger project will bring more benefits for my people." After lengthy planning, review, and litigation involving public agencies such as the federal Bureau of Reclamation, the Army Corps of Engineers, and the State Water Resources Control Board and environmental groups such as the Environmental Defense Fund, the Friends of the River, and the Sierra Club, the dam was finally authorized and built, with high waters reaching and covering the white waters of the Stanislaus in the spring of 1983.

In his 1979 protest over the dam, environmentalist Mark Dubois chained himself to a rock to prevent the river, endangered wildlife, and the rocks from losing their rights to remain free. "All the life of this canyon, its wealth of archaeological and historical roots to our past, and its unique geological grandeur are enough reasons to protect this canyon just for itself," he wrote to the Army Corps of Engineers. "But in addition, all the spiritual values with which this canyon has filled tens of thousands of folks should prohibit us from committing the unconscionable act of wiping this place off the face of the earth." The controversy was a conflict among interest groups with different underlying ethics. Here farmers and corporate agribusiness ventures, whose egocentric ethics promoted the individual's good, along with federal water control agencies, whose homocentric ethics saw water development as the greatest good for the greatest number, conflicted with the ecocentric ethics of those environmentalists who supported the river's intrinsic right to remain wild.

The latter conflict points to one of the main problems of both egocentric and homocentric ethics—their failure to internalize ecological externalities. Ecological changes and their long-term effects are outside the human/society framework of these ethics. The effects of ecological changes, such as salinity buildup in farming soils that use the dam's water or the loss of indigenous species when a valley is flooded, are not part of the human-centered calculus of decision making. One approach offered by ethicists is to extend homocentric ethics to include other sentient species. An alternative, however, is to formulate a radically different form of environmental ethics—ecocentric ethics.

Ecocentric Ethics

An ecocentric ethic is grounded in the cosmos. The whole environment, including inanimate elements, rocks, and minerals along with animate plants and animals, is assigned intrinsic value. The ecoscientific form of this ethic draws its "ought" from the science of ecology. Recognizing that science can no longer be considered value-free, as the logical positivists of the early twentieth century had insisted, proponents of ecocentric ethics look to ecology for guidelines on how to resolve ethical dilemmas. Maintenance of the balance of nature and retention of the unity, stability, diversity, and harmony of the ecosystem are its overarching goals. Of primary importance is the survival of all living and nonliving things as components of healthy ecosystems. All things in the cosmos as well as humans have moral considerability.

The modern form of ecocentric ethics was first formulated by Aldo Leopold during the 1930s and 1940s and published as "The Land Ethic," the final chapter of his posthumous *A Sand County Almanac* (1949). Some of Leopold's inspiration for the land ethic seems to have derived from Mill's *Utilitarianism*. Like Mill, who wrote about the "influences of advancing civilization" and the utilitarian Golden Rule as superseding basic prohibitions against robbing and murdering, Leopold thought ethics developed in sequence: "The first ethics," he wrote, "dealt with the relation between individuals; the Mosaic Decalogue is an example. Later accretions dealt with the relation between the individual and society. The Golden Rule tries to integrate the individual to society." The land ethic, he argued, extends the sequence a step further. It enlarges the bounds of the community to include "soils, waters, plants, and animals, or collectively, the land." It "changes the role of *homo sapiens* from conqueror of the land-community to plain member and citizen of it. It implies respect for his fellow members and also respect for the community itself." Perhaps influenced by Mill's phraseology that "actions are right in proportion as they tend to promote happiness; wrong as they tend to produce the reverse of happiness," Leopold wrote: "A thing is right when it tends to preserve the integrity, beauty, and stability of the biotic community. It is wrong when it tends otherwise." . . .

At the University of California, Santa Barbara, environmental historian Roderick Nash elaborated Leopold's land ethic in a 1977 article "Do Rocks Have Rights?" Rocks, he stated, are part of the pyramid of animate and inanimate things governed by the laws of ecology. Even though rocks are not sentient like animals, rocks as well as plants can be assigned interests that can be represented and ajudicated. Yet such a concept might still be used to protect rocks in the interest of humans. We can "suppose that rocks, just like people, do have rights in and of themselves. It follows that it is the rock's interest, not the human interested in the rock, that is being protected." Other cultures such as Native Americans, Zen Buddhists, and Shintos, Nash pointed out, assume that rocks are alive—a mystical religious belief not usually held by Western philosophers and scientists.

Ecocentric ethics are rooted in a holistic, rather than mechanistic, metaphysics. The assumptions of holism are:

1. Everything is connected to everything else. The whole qualifies each part; conversely, a change in one of the parts will change the other parts and the whole. No part of an ecosystem can be removed without altering the dynamics of the cycle.

2. The whole is greater than the sum of the parts. Unlike the concept of identity, in which the whole equals the sum of the parts, ecological systems exhibit synergy: the combined action of separate parts may produce an effect greater than the sum of the individual efforts.

3. Meaning is context dependent. As opposed to the context independence assumptions of mechanistic science, in holism each part at any instant takes its meaning from the whole.

4. There is a primacy of process over parts. Living systems are open, steady-state systems in which matter and energy are constantly being exchanged with the surroundings. Living things are dissipative struc-

tures, resulting from a continual flow of energy, just as a vortex in a stream is a structure arising from the continually changing water molecules swirling through it.

5. There is a unity of humans and nonhuman nature. As opposed to nature–culture dualism, in holism humans and nature are part of the same organic, cosmological system.

In California, the philosophical change from the dominant mechanistic worldview to an ecological worldview, or "deep ecology" (a term coined by Norwegian philosopher Arne Naess), is a subject investigated by sociologist Bill Devall of Humboldt State University in Arcata and philosopher George Sessions of Sierra College. Devall and Sessions put forward eight basic principles of deep ecology, including the idea that "the well-being and flourishing of human and nonhuman Life on Earth have value in themselves (synonyms: intrinsic value, inherent value). These values are independent of the usefulness of the non-human world for human purposes." They argue that policies should be implemented that both maintain the richness and diversity of life and allow for the fulfillment of basic human needs.

In the 1960s and 1970s holistic ideas blossomed with small-scale, back-to-the-land communes and households in which decision making was vested in the consensus of the whole group. Drawing on holistic assumptions, the bioregional movement in California likewise emphasizes living within the resources of the local watershed and developing them to sustain the human and nonhuman community as an ecological whole. Recently the emergence of green politics has given rise to a California political movement dedicated to the establishment of an ecologically viable society.

Examples of scientific applications of the ecocentric ethic in California include: (1) restoration ecology, (2) the biological control of insect pests, and (3) sustainable agriculture.

Restoration is the process of restoring human-disturbed ecosystems to earlier pristine forms. Over time synergistic relationships are reestablished among soils, plants, insect pollinators, and animals to recreate ecosystems. An ecocentric ethic guides the restoration of forests, marshes, prairies, and rivers. An example of restoration in California is the replanting of the redwoods in Big Basin Redwoods State Park, in the Santa Cruz mountains. Set aside in 1902 after it had been scarred by lumber operations, the park had seen heavy use, soil compaction, and erosion. As the old trees died new ones did not regenerate. In 1968, the Santa Cruz Lumber Company, which had held off cutting a stand of old-growth redwoods in the park's interior core, went out of business and threatened to cut the timber if the state did not immediately exercise its option to purchase the land. Successful efforts to purchase and protect the threatened areas were followed by restoration. Guided by an implicit ecocentric ethic of management, restorers planted young trees, ferns, huckleberries, and ground cover, enriched the soil with redwood chips, and removed old parking lots and remnants of lumber operations. Restoring the native plant species helped to establish the ecological conditions under which insect, mammal, and bird communities could also regenerate themselves. A new whole was created, helping to recreate the major elements of the presettlement ecosystem.

Biological control is a second example of an ecocentric ethic of management. Using ecological guidelines, natural insect enemies are introduced into the ecosystem to control population levels of pests. In one of the first successful uses of biological control in California, the vedalia, a lady beetle from Australia, was introduced to feed on the cottony-cushion scale that was destroying the state's citrus crop. One thousand beetles soon cleared acres of orange groves, saving the industry. This ecological strategy was vindicated in the 1940s when DDT killed so many of the vedalia that a resurgence of the scale occurred.

Sustainable agriculture, a third example of an ecocentric ethic, is an ecologically based form of farm management. This strategy is posited in opposition to the industrial approach to agriculture rooted in optimizing purchased inputs to produce outputs at the least cost. Sustainable agriculture is instead based on an ecocentric ethic of management in which the land is considered as a whole, its human components being only one element. Policy decisions must be based on considerations of what is best for the soil, vegetation, and animals (including humans) on the farm as well as outside sources of water, air, and energy. As a result, humans and the land will be sustained together.

Whereas the ecoscientific form of the ethic is rooted in the science of ecology, the ecoreligious form is based on the faith that all living and nonliving things have value. In California, one such religious formulation is process theology developed by John Cobb, Jr., David Ray Griffin, and others of the Center for Process Studies at Claremont Graduate School in southern California. Process theology owes its origins to British philosopher Alfred North Whitehead, who taught at Harvard University, and to philosopher Charles Hartshorne, a teacher of Cobb at the University of Chicago. According to Cobb and Griffin, process philosophy asserts that "process is fundamental. It does not assert that everything is in process . . . but to be *actual* is to be a process." Process philosophy implies an ecological ethic and a policy of social justice. "The whole of nature participates in us and we in it. We are diminished not only by the misery of the Indian peasant but also by the slaughter of whales and porpoises, and even by the "harvesting" of the giant redwoods. We are diminished still more when the imposition of temperate-zone technology on to tropical agriculture turns grasslands into deserts that will support neither human nor animal life."

These three dominant forms of environmental ethics, however, all have conceptual and practical shortcomings. Egocentric ethics are criticized for privileging the few at the expense of the many (narcissistic, cut-throat individualism), homocentric ethics for privileging majorities at the expense of minorities (tyranny of the majority, environmental racism), and ecocentric ethics for privileging the whole at the expense of the individual (holistic fascism). Egocentric and homocentric ethics often are lumped together as anthropocentrism, masking the role of economics and placing the onus on human hubris, rather than the capitalist appropriation of both nature and labor. Ecocentrism fails to recognize the positive contribution of the social-justice approach to homocentric ethics. On the other hand, the ecocentric approach of many environmentalists suggests the possibility of incorporating the intrinsic value of nature into an emancipatory green politics.

Partnership Ethics

As an alternative that transcends many of these problems, I propose a partnership ethic. A partnership ethic considers the human community *and* the biotic community to be in a mutual relationship with each other. It states that "the greatest good for the human and the nonhuman community in to be found in their mutual, living interdependence."

Just as egocentric ethics is grounded in the principle of self-interest, homocentric ethics in the concept of utility, and ecocentric ethics in intrinsic value, so partnership ethics is grounded in the concept of relation. A relation is a mode of connection. This connection may be between people or kin in the same family or community, between men and women, between people, other organisms, and inorganic entities, or between specific places and the rest of the earth. A relation also is a narrative; to relate is to narrate. A narrative connects people to a place, to its history, and to its multileveled meanings. It is a story that is recounted and told, in which connections are made, alliances and associations established. A partnership ethic is an ethic of connections between a human and a nonhuman community. The relationship is situational and contextual within the local community, but the community also is embedded in and connected to the wider earth, especially national and global economies. In a relationship there is give and take, offering and restraint, acknowledgment of the needs of other entities.

A partnership ethic has five precepts:

1. Equity between the human and nonhuman communities
2. Moral consideration for humans and nonhuman nature
3. Respect for cultural diversity and biodiversity
4. Inclusion of women, minorities, and nonhuman nature in the code of ethical accountability
5. An ecologically sound management consistent with the continued health of both the human and nonhuman communities

A partnership ethic draws on a homocentric ethic of partnership among human groups and an ecocentric ethic of partnership with non-human nature. But it also entails a new consciousness about nature as an actor and equal subject. Chaos theory suggests that the human ability to predict the outcome of natural processes is limited. Disorderly order, the world represented by chaos theory, is the nonhuman component of the partnership ethic.

The disorderly, ordered world of nonhuman nature must be acknowledged as a free autonomous actor, just as humans are free autonomous agents. Nature limits human freedom to totally dominate and control it, just as human power limits Nature's and other humans' freedom. Science and technology can tell us that an event such as a hurricane, earthquake, flood, or fire is likely to happen in a certain locale, but not when it will happen. Because nature is fundamentally chaotic, it must be respected and related to as an active partner through a partnership ethic.

If we know that an earthquake in Los Angeles is likely in the next 75 years, a utilitarian, homocentric ethic would state that the government ought not to license the construction of a nuclear reactor on the fault line. But a partnership ethic would say that, we, the human community, ought to respect nature's autonomy as an actor by limiting building and leaving open space. If we know there is a possibility of a 100-year flood on the San Joaquin River, we respect human needs for navigation and power, but we also respect nature's autonomy by limiting our capacity to dam every tributary that feeds the river and to build homes on every flood plain. We leave some rivers wild and free and leave some flood plains as wetlands and use others to fulfill human needs. If we know that forest fires are likely in the Sierras, we do not build cities along forest edges. We limit the extent of development, leave open spaces, plant fire resistant vegetation, and use tile rather than shake roofs. If cutting old-growth forests creates problems for both the global environment and local communities, but we cannot adequately predict the outcome or effects of those changes, we need to conduct partnership negotiations in which nonhuman nature and the people involved are equally represented.

Each of these difficult, time-consuming ethical and policy decisions will be negotiated by a human community in a particular place, but the outcome will depend on the history of people and nature in the area, the narratives they tell themselves about the land, vital human needs, past and present land-use patterns, the larger global context, and the ability or lack of it to predict nature's events. Each human community is in a changing, evolving relationship with a nonhuman community that is local, but also connected to global environmental and human patterns. Each ethical instance is historical, contextual, and situational, but located within a larger environmental and economic system. Implementing a partnership ethic may help people and nature to survive together in a green and golden California.

Further Readings

Bailes, Kendall E. *Environmental History: Critical Issues in Comparative Perspective.* Lanham, MD: University Press of America, 1985.

Berg, Peter. *Reinhabiting a Separate Country: A Bioregional Anthology of Northern California.* San Francisco: Planet Drum Foundation, 1978.

Capra, Fritjof, and Charlene Spretnak. *Green Politics: The Global Promise.* New York: Dutton, 1984.

Daly, Herman E., ed. *Economics, Ecology, Ethics: Essays toward a Steady-State Economy.* San Francisco: W. H. Freeman, 1973.

Douglass, Gordon K., ed. *Agricultural Sustainability in a Changing World Order.* Boulder, CO: Westview Press, 1984.

Easton, Robert. *Black Tide: The Santa Barbara Oil Spill and Its Consequences.* New York: Delacorte, 1972.

Griffin, David, ed. *The Reenchantment of Science: Postmodern Proposals.* Albany: State University of New York Press, 1988.

Hardin, Garrett. *Promethean Ethics: Living with Death, Competition, and Triage.* Seattle: University of Washington Press, 1980.

Hays, Samuel P. *Conservation and the Gospel of Efficiency: The Progressive Conservation Movement, 1890–1920.* New York: Atheneum, 1975.

Leopold, Aldo. *A Sand County Almanac*. London: Oxford University Press, 1949.

McEvoy, Arthur. *The Fisherman's Problem: Ecology and Law in the California Fisheries: 1850–1980*. New York: Cambridge University Press, 1980.

Merchant, Carolyn. *The Death of Nature: Women, Ecology, and the Scientific Revolution*. San Francisco: Harper & Row, 1980.

Nash, Roderick. "The Santa Barbara Oil Spill." In Roderick Nash, ed., *The American Environment: Readings in the History of Conservation*. Reading, MA: Addison-Wesley, 1976.

Pinchot, Gifford. *Breaking New Ground*. New York: Harcourt Brace Jovanovich, 1947.

Prigogine, Ilya, and Isabelle Stengers. *Order out of Chaos: Man's New Dialogue with Nature*. New York: Bantam, 1984.

Regan, Tom. *All That Dwell Therein—Essays on Animal Rights and Environmental Ethics*. Berkeley: University of California Press, 1982.

Singer, Peter. *Animal Liberation: A New Ethics for Our Treatment of Animals*. New York: Avon, 1975.

Todd, John. *The Sunset Land or the Great Pacific Slope*. Boston: Lee and Shepard, 1870.

Author Index

Subject Index